T0181661

Introduction to Surface and Thin Film Processes

This book covers the experimental and theoretical understanding of surface and thin film processes. It presents a unique description of surface processes in adsorption and crystal growth, including bonding in metals and semiconductors. Emphasis is placed on the strong link between science and technology in the description of, and research for, new devices based on thin film and surface science. Practical experimental design, sample preparation and analytical techniques are covered, including detailed discussions of Auger electron spectroscopy and microscopy. Thermodynamic and kinetic models of electronic, atomic and vibrational structure are emphasized throughout. The book provides extensive leads into practical and research literature, as well as to resources on the World Wide Web. Each chapter contains problems which aim to develop awareness of the subject and the methods used.

Aimed as a graduate textbook, this book will also be useful as a sourcebook for graduate students, researchers and practioners in physics, chemistry, materials science and engineering.

JOHN A. VENABLES obtained his undergraduate and graduate degrees in Physics from Cambridge. He spent much of his professional life at the University of Sussex, where he is currently an Honorary Professor, specialising in electron microscopy and the topics discussed in this book. He has taught and researched in laboratories around the world, and has been Professor of Physics at Arizona State University since 1986. He is currently involved in web-based (and web-assisted) graduate teaching, in Arizona, Sussex and elsewhere. He has served on several advisory and editorial boards, and has done his fair share of reviewing. He has published numerous journal articles and edited three books, contributing chapters to these and others; this is his first book as sole author.

Introduction to
Surface and Thin Film Processes

JOHN A. VENABLES

*Arizona State University
and
University of Sussex*

CAMBRIDGE UNIVERSITY PRESS
Cambridge, New York, Melbourne, Madrid, Cape Town, Singapore, São Paulo

Cambridge University Press
The Edinburgh Building, Cambridge CB2 2RU, UK

Published in the United States of America by Cambridge University Press, New York

www.cambridge.org
Information on this title: www.cambridge.org/9780521624602

First published 2000
Reprinted 2001

A catalogue record for this publication is available from the British Library

Library of Congress Cataloguing in Publication data

Venables, John, 1936–
Introduction to surface and thin film processes / John Venables.
p. cm.
Includes bibliographical references.
ISBN 0 521 62460 6 (hc.) – ISBN 0 521 78500 6 (pbk).
1. Thin films. 2. Surface chemistry. I. Title.
QC176.83.V46 2000
530.4′275–dc21   99–087614

ISBN-13 978-0-521-62460-2 hardback
ISBN-10 0-521-62460-6 hardback

ISBN-13 978-0-521-78500-6 paperback
ISBN-10 0-521-78500-6 paperback

Transferred to digital printing 2006

Contents

Preface

This book is about processes that occur at surfaces and in thin films; it is based on teaching and research over a number of years. Many of the experimental techniques used to produce clean surfaces, and to study the structure and composition of solid surfaces, have been around for about a generation. Over the same period, we have also seen unprecedented advances in our ability to study materials in general, and on a microscopic scale in particular, largely due to the development and availability of many new types of powerful microscope.

The combination of these two fields, studying and manipulating clean surfaces on a microscopic scale, has become important more recently. This combination allows us to study what happens in the production and operation of an increasing number of technologically important devices and processes, at all length scales down to the atomic level. Device structures used in computers are now so small that they can be seen only with high resolution scanning and transmission electron microscopes. Device preparation techniques must be performed reproducibly, on clean surfaces under clean room conditions. Ever more elegant schemes are proposed for using catalytic chemical reactions at surfaces, to refine our raw products, for chemical sensors, to protect surfaces against the weather and to dispose of environmental waste. Spectacular advances in experimental technique now allow us to observe atoms, and the motion of individual atoms on surfaces, with amazing clarity. Under special circumstances, we can move them around to create artificial atomic-level assemblies, and study their properties. At the same time, enormous advances in computer power and in our understanding of materials have enabled theorists and computer specialists to model the behavior of these small structures and processes down to the level of individual atoms and (collections of) electrons.

The major industries which relate to surface and thin film science are the micro-electronics, opto-electronics and magnetics industries, and the chemistry-based industries, especially those involving catalysis and the emerging field of sensors. These industries form society's immediate need for investment and progress in this area, but longer term goals include basic understanding, and new techniques based on this understanding: there are few areas in which the interaction of science and technology is more clearly expressed.

Surfaces and thin films are two, interdependent, and now fairly mature disciplines. In his influential book, *Physics at Surfaces*, Zangwill (1988) referred to his subject as an interesting adolescent; so as the twenty-first century gets underway it is thirty-something. I make no judgment as to whether growing up is really a maturing process, or whether the most productive scientists remain adolescent all their lives. But the various stages of a subject's evolution have different character. Initially, a few academics and industrial researchers are in the field, and each new investigation or experiment opens many new possibilities. These people take on students, who find employment in closely related areas. Surface and thin film science can trace its history back to Davisson and Germer, who in effect invented low energy electron diffraction (LEED) in 1927, setting the scene for the study of surface structure. Much of the science of electron emission dates from Irving Langmuir's pioneering work in the 1920s and 1930s, aimed largely at improving the performance of vacuum tubes; these scientists won the Nobel prize in 1937 and 1932 respectively.

The examination of surface chemistry by Auger and photoelectron spectroscopy can trace its roots back to cloud chambers in the 1920s and even to Einstein's 1905 paper on the photo-electric effect. But the real credit arguably belongs to the many scientists in the 1950s and 1960s who harnessed the new ultra-high vacuum (UHV) technologies for the study of clean surfaces and surface reactions with adsorbates, and the production of thin films under well-controlled conditions. In the past 30 years, the field has expanded, and the 'scientific generation' has been quite short; different sub-fields have developed, often based on the expertise of groups who started literally a generation ago. As an example, the compilation by Duke (1994) was entitled *Surface Science: the First Thirty Years*. The Surface Science in question is the journal, not the field itself, but the two are almost the same. That one can mount a retrospective exhibition indicates that the field has achieved a certain age.

Over the past ten years there has been a period of consolidation, where the main growth has been in employment in industry. Scientists in industry have pressing needs to solve surface and thin film processing problems as they arise, on a relatively short timescale. It must be difficult to keep abreast of new science and technology, and the tendency to react short term is very great. Despite all the progress in recent years, I feel it is important not to accept the latest technical development at the gee-whizz level, but to have a framework for understanding developments in terms of well-founded science. In this situation, we should not reinvent the wheel, and should maintain a reasonably reflective approach. There are so many forces in society encouraging us to communicate orally and visually, to have our industrial and international collaborations in place, to do our research primarily on contract, that it is tempting to conclude that science and frenetic activity are practically synonymous. Yet lifelong learning is also increasingly recognized as a necessity; for academics, this is itself a growth industry in which I am pleased to play my part.

This book is my attempt to distill, from the burgeoning field of *Surface and Thin Film Processes*, those elements which are scientifically interesting, which will stand the test of time, and which can be used by the reader to relate the latest advances back to his or her underlying knowledge. It builds on previous books and articles that perhaps emphasize the description of surfaces and thin films in a more static, less process-oriented sense. This previous material has not been duplicated more than is necessary; indeed, one of the aims is to provide a route into the literature of the past 30 years, and to relate current interests back to the underlying science. Problems and further textbook reading are given at the end of each chapter. These influential textbooks and monographs are collected in Appendix A, with a complete reference list at the end of the book, indicating in which section they are cited. The reader does not, of course, have to rush to do these problems or to read the references; but they can be used for further study and detailed information. A list of acronyms used is given in Appendix B.

The book can be used as the primary book for a graduate course, but this is not an exclusive use. Many books have already been produced in this general area, and on specialized parts of it: on vacuum techniques, on surface science, and on various aspects of microscopy. This material is not all repeated here, but extensive leads are given into the existing literature, highlighting areas of strength in work stretching back over the last generation. The present book links all these fields and applies the results selectively to a range of materials. It also discusses science and technology and their inter-relationship, in a way that makes sense to those working in inter-disciplinary environments. It will be useful to graduate students, researchers and practitioners educated in physical, chemical, materials or engineering science.

The early chapters 1–3 underline the importance of thermodynamic and kinetic reasoning, provide an introduction to the terms used, and describe the use of ultra-high vacuum, surface science and microscopy techniques in studying surface processes. These chapters are supplemented with extensive references and problems, aimed at furthering the students' practical and analytical abilities across these fields. If used for a course, these problems can be employed to test students' analytical competence, and familiarity with practical aspects of laboratory designs and procedures. I have never required that students do problems unaided, but encouraged them to ask questions which help towards a solution, that they then write up when understanding has been achieved. This allows more time in class for discussion, and for everyone to explore the material at their own pace. A key point is that each student has a different background, and therefore finds different aspects unfamiliar or difficult.

The following chapters 4–8 are each self-contained, and can be read or worked through in any order, though the order presented has a certain logic. Chapter 4 treats adsorption on surfaces, and the role of adsorption in testing interatomic potentials and lattice dynamical models, and in following chemical reactions. Chapter 5 describes the modeling of epitaxial crystal growth, and the experiments performed to test these ideas; this chapter contains original material that has been featured in recent multi-author compilations. Further progress in understanding cannot be made without some understanding of bonding, and how it applies to specific materials systems. Chapter 6 treats bonding in metals and at metallic surfaces, electron emission and the operation of electron sources, and electrical and magnetic properties at surfaces and in thin films. Chapter 7 takes a similar approach to semiconductor surfaces, describing their reconstructions and the importance of growth processes in producing semiconductor-based thin film device structures. Chapter 8 concentrates on the science needed to understand electronic, magnetic and optical effects in devices. The short final chapter 9 describes briefly what has been left out of the book, and discusses the roles played by scientists and technologists from different educational backgrounds, and gives some pointers to further sources of information. Chapters 4–7 give suggestions for projects based on the material presented and cited. Appendices C–K give data and further explanations that have been found useful in practice.

In graduate courses, I have typically not given all this material each time, and certainly not in this 4–8 order, but have tailored the choice of topics to the interests of the students who attended in a given term or semester. Recently, I have taught the material of chapters 1 and 2 first, and then interleaved chapter 3 with the most pressing topics in chapters 4–8, filling in to round out topics later. Towards the end of the course, several students have given talks about other surface and/or microscopic techniques to the class, and yet others did a 'mini-project' of 2000 words or so, based on references supplied and suggested leads into the literature.

With this case-study approach, one can take students to the forefront of current research, while also relating the underlying science back to the early chapters. I am personally very interested in models of electronic, atomic and vibrational structure, though I am not expert in all these areas. As a physicist by training, heavily influenced by materials science, and with some feeling for engineering and for physical/analytical chemistry, I am drawn towards nominally simple (elemental) systems, and I do not go far in the direction of complex chemistry, which is usually implicated in real-life processes such as chemical vapor deposition or catalytic schemes. With so much literature available one can easily be overwhelmed; yet if conflicts and discrepancies in the original literature are never mentioned, it is too easy for students, and indeed the general public, to believe that science is cut and dried, a scarcely human endeavor. In the workplace, employees with graduate degrees in physics, chemistry, materials science or engineering are treated as more or less interchangeable. Understanding obtained via the book is a contribution to this interdisciplinary background that we all need to function effectively in teams.

Having extolled the virtues of a scholarly approach to graduate education in book form, I also think that graduate courses should embrace the relevant possibilities opened up by recent technology. I have been using the World Wide Web to publish course notes, and to teach students off-campus, using e-mail primarily for interactions, in addition to taking other opportunities, such as meeting at conferences, to interact more personally. Writing notes for the web and interacting via e-mail is enjoyable and informal. Qualitative judgments trip off the fingers, which one would be hard put to justify in a book; if they are shown to be wrong or inappropriate they can easily be changed. Perhaps more importantly, one can access other sites for information which one lacks, or which colleagues elsewhere have put in a great deal of time perfecting; my web-based resources page can be accessed via Appendix D. One can be interested in a topic, and refer students to it, without having to reinvent the wheel in a futile attempt to become the world's expert overnight. And, as I hope to show over the next few years, one may be able to reach students who do not have the advantages of working in large groups, and largely at times of their choosing.

It seems too early to say whether course notes on the web, or a book such as this will have the longer shelf life. In writing the book, after composing most but not all of the notes, I am to some extent hedging my bets. I have discovered that the work needed to produce them is rather different in kind, and I suspect that they will be used for rather different purposes. Most of the notes are on my home page http://venables.asu.edu/ in the /grad directory, but I am also building up some related material for graduate courses at Sussex. Let me know what you think of this material: an e-mail is just a few clicks away.

I would like to thank students who have attended courses and worked on problems, given talks and worked on projects, and co-workers who have undertaken research projects with me over the last several years. I owe an especial debt to several friends and close colleagues who have contributed to and discussed courses with me: Paul Calvert (now at University of Arizona), Roger Doherty (now at Drexel) and Michael Hardiman at Sussex; Ernst Bauer, Peter Bennett, Andrew Chizmeshya, David Ferry, Bill Glaunsinger, Gary Hembree, John Kouvetakis, Stuart Lindsay, Michael Scheinfein, David Smith, John Spence and others at ASU; Harald Brune, Robert Johnson and Per Stoltze in and around Europe. They and others have read through individual chapters and sections and made encouraging noises alongside practical suggestions for improvement. Any remaining mistakes are mine.

I am indebted, both professionally and personally, to the CRMC2-CNRS laboratory in Marseille, France. Directors of this laboratory (Raymond Kern, Michel Bienfait, and Jacques Derrien) and many laboratory members have been generous hosts and wonderful collaborators since my first visit in the early 1970s. I trust they will recognize their influence on this book, whether stated or not.

I am grateful to many colleagues for correspondence, for reprints, and for permission to use specific figures. In alphabetical order, I thank particularly C.R. Abernathy, A.P. Alivisatos, R.E. Allen, J.G. Amar, G.S. Bales, J.V. Barth, P.E. Batson, J. Bernholc, K. Besocke, M. Brack, R. Browning, L.W. Bruch, C.T. Campbell, D.J. Chadi, J.N. Chapman, G. Comsa, R.K. Crawford, H. Daimon, R. Del Sole, A.E. DePristo, P.W. Deutsch, R. Devonshire, F.W. DeWette, M.J. Drinkwine, J.S. Drucker, G. Duggan, C.B. Duke, G. Ehrlich, D.M. Eigler, T.L. Einstein, R.M. Feenstra, A.J. Freeman, E. Ganz, J.M. Gibson, R. Gomer, E.B. Graper, J.F. Gregg, J.D. Gunton, B. Heinrich, C.R. Henry, M. Henzler, K. Hermann, F.J. Himpsel, S. Holloway, P.B. Howes, J.B. Hudson, K.A. Jackson, K.W. Jacobsen, J. Janata, D.E. Jesson, M.D. Johnson, B.A. Joyce, H. von Känel, K. Kern, M. Klaua, L. Kleinman, M. Krishnamurthy, M.G. Lagally, N.D. Lang, J. Liu, H.H. Madden, P.A. Maksym, J.A.D. Matthew, J-J. Métois, T. Michely, V. Milman, K. Morgenstern, R. Monot, B. Müller, C.B. Murray, C.A. Norris, J.K. Nørskov, J.E. Northrup, A.D. Novaco, T. Ono, B.G. Orr, D.A. Papaconstantopoulos, J. Perdew, D.G. Pettifor, E.H. Poindexter, J. Pollmann, C.J. Powell, M. Prutton, C.F. Quate, C. Ratsch, R. Reifenburger, J. Robertson, J.L. Robins, L.D. Roelofs, C. Roland, H.H. Rotermund, J.R. Sambles, E.F. Schubert, M.P. Seah, D.A. Shirley, S.J. Sibener, H.L. Skriver, A. Sugawara, R.M. Suter, A.P. Sutton, J. Suzanne, B.S. Swartzentruber, S.M. Sze, K. Takayanagi, M. Terrones, J. Tersoff, A. Thomy, M.C. Tringides, R.L. Tromp, J. Unguris, D. Vanderbilt, C.G. Van de Walle, M.A. Van Hove, B. Voightländer, D.D. Vvedensky, L. Vescan, M.B. Webb, J.D. Weeks, P. Weightman, D. Williams, E.D. Williams, D.P. Woodruff, R. Wu, M. Zinke-Allmang and A. Zunger.

Producing the figures has allowed me to get to know my nephew Joe Whelan in a new way. Joe produced many of the drawings in draft, and some in final form; we had some good times, both in Sussex and in Arizona. Mark Foster in Sussex helped effectively with scanning original copies into the computer. Publishers responded quickly to my requests for permission to reproduce such figures. Finally I thank, but this is too weak a word, my wife Delia, whose opinion is both generously given and highly valued. In this case, once I had started, she encouraged me to finish as quickly as practicable: aim for a competent job done in a finite time. After all, that's what I tell my students.

John A. Venables (john.venables@asu.edu *or* john@venables.co.uk)
Arizona/Sussex, November/December 1999

References

Duke, C.B. (Ed.) (1994) *Surface Science: the First Thirty Years* (*Surface Sci.* **299/300** 1–1054).
Zangwill, A. (1988) *Physics at Surfaces* (Cambridge University Press, pp. 1–454).

Web-based comments, corrections and updates

This book is supplemented by a limited number of web-based comments, corrections and updates. The entry point for these web pages can be Appendix D, as mentioned on page 313 in the text at http://venables.asu.edu/grad/appweb1.html or perhaps more easily at http://venables.asu.edu/book/contents.html These pages are both accessible via my home page at http://venables.asu.edu/ Because of the flexibility of web-based corrections, only a small number of (mostly typographic) errors have been corrected prior to this reprinting.

1 Introduction to surface processes

In this opening chapter, section 1.1 introduces some of the thermodynamic ideas which are used to discuss small systems. In section 1.2 these ideas are developed in more detail for small crystals, both within the terrace–ledge–kink (TLK) model, and with examples taken from real materials. Section 1.3 explores important differences between thermodynamics and kinetics; the examples given are the vapor pressure (an equilibrium thermodynamic phenomenon) and ideas about crystal growth (a non-equilibrium phenomenon approachable via kinetic arguments); both discussions include the role of atomic vibrations.

Finally, in section 1.4 the ideas behind reconstruction of crystal surfaces are discussed, and section 1.5 introduces some concepts related to surface electronics. These sections provide groundwork for the chapters which follow. You may wish to come back to individual topics later; for example, although the thermodynamics of small crystals is studied here, we will not have covered many experimental examples, nor more than the simplest models. The reason is that not everyone will want to study this topic in detail. In addition to the material in the text, some topics which may be generally useful are covered in appendices.

1.1 Elementary thermodynamic ideas of surfaces

1.1.1 Thermodynamic potentials and the dividing surface

The idea that thermodynamic reasoning can be applied to surfaces was pioneered by the American scientist J.W. Gibbs in the 1870s and 1880s. This work has been assembled in his collected works (Gibbs 1928, 1961) and has been summarized in several books, listed in the further reading at the end of the chapter and in Appendix A. These references given are for further exploration, but I am not expecting you to charge off and look all of them up! However, if your thermodynamics is rusty you might read Appendix E.1 before proceeding.

Gibbs' central idea was that of the 'dividing surface'. At a boundary between phases 1 and 2, the concentration profile of any elemental or molecular species changes (continuously) from one level c_1 to another c_2, as sketched in figure 1.1. Then the extensive thermodynamic potentials (e.g. the internal energy U, the Helmholtz free energy F, or the Gibbs free energy G) can be written as a contribution from phases 1, 2 plus a surface

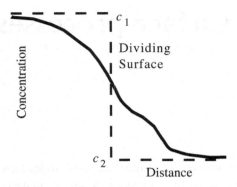

Figure 1.1. Schematic view of the 'dividing surface' in terms of macroscopic concentrations. See text for discussion.

term. In the thermodynamics of bulk matter, we have the bulk Helmholtz free energy $F_b = F(N_1, N_2)$ and we know that

$$dF_b = -SdT - pdV + \mu dN = 0,\tag{1.1}$$

at constant temperature T, volume V and particle number N. In this equation, S is the (bulk) entropy, p is the pressure and μ the chemical potential. Similar relationships exist for the other thermodynamic potentials; commonly used thermodynamic relations are given in Appendix E.1.

We are now interested in how the thermodynamic relations change when the system is characterized by a surface area A in addition to the volume. With the surface present the total free energy $F_t = F(N_1, N_2, A)$ and

$$dF_t = dF_b(N_1, N_2) + f_s dA.\tag{1.2}$$

This f_s is the extra Helmholtz free energy per unit area due to the presence of the surface, where we have implicitly assumed that the total number of atomic/molecular entities in the two phases, N_1 and N_2 remain constant. Gibbs' idea of the 'dividing surface' was the following. Although the concentrations may vary in the neighborhood of the surface, we consider the system as uniform up to this ideal interface: f_s is then the surface excess free energy.

To make matters concrete, we might think of a one-component solid–vapor interface, where c_1 is high, and c_2 is very low; the exact concentration profile in the vicinity of the interface is typically unknown. Indeed, as we shall discuss later, it depends on the forces between the constituent atoms or molecules, and the temperature, via the statistical mechanics of the system. But we can define an imaginary dividing surface, such that the system behaves as if it comprised a uniform solid and a uniform vapor up to this dividing surface, and that the surface itself has thermodynamic properties which scale with the surface area; this is the meaning of (1.2). In many cases described in this book, the concentration changes from one phase to another can be sharp at the atomic level. This does not invalidate thermodynamic reasoning, but it leads to an interesting

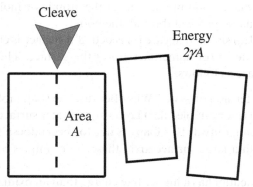

Figure 1.2. Schematic illustration of how to create new surface by cleavage. If this can be done reversibly, in the thermodynamic sense, then the work done is $2\gamma A$.

dialogue between macroscopic and atomistic views of surface processes, which will be discussed at many points in this book.

1.1.2 Surface tension and surface energy

The surface tension, γ, is defined as the reversible work done in creating unit area of new surface, i.e.

$$\gamma = \lim (dA \to 0) \, dW/dA = (dF_t/dA)_{T,V}. \tag{1.3}$$

In the simple illustration of figure 1.2, $\Delta F = F_1 - F_0 = 2\gamma A$; $dF_t = \gamma dA$. At const T and V,

$$dF_t = -SdT - pdV + \Sigma \mu_i dN_i + f_s dA = f_s dA + \Sigma \mu_i dN_i. \tag{1.4}$$

Therefore,

$$\gamma dA = f_s dA + \Sigma \mu_i dN_i. \tag{1.5}$$

In a one-component system, e.g. metal–vapor, we can choose the dividing surface such that $dN_i = 0$, and then γ and f_s are the same. This is the sense that most physics-oriented books and articles use the term. In more complex systems, the introduction of a surface can cause changes in N_i, i.e. we have $N_1 + N_2$ in the bulk, and $dN_i \to$ surface, so that dN_i, the change in the bulk number of atoms in phase i, is negative. We then write

$$dN = -\Gamma dA \text{ and } \gamma = f_s - \Sigma \Gamma_i \mu_i, \tag{1.6}$$

where the second term is the free energy contribution of atoms going from the bulk to the surface; γ is the surface density of $(F - G)$ (Blakely 1973, p. 5). An equivalent view is that γ is the surface excess density of Kramers' grand potential $\Omega = -p(V_1 + V_2) + \gamma A$, which is minimized at constant T, V and μ (Desjonquères & Spanjaard 1996, p. 5). You might think about this – it is related to statistical mechanics of open systems using the grand canonical ensemble . . .! Realistic models at $T > 0$ K need to map onto the relevant statistical distribution to make good predictions at the atomic or molecular level; such points will be explored as we proceed through the book.

The simple example leading to (1.6) shows that care is needed: if a surface is created, the atoms or molecules can migrate *to* (or sometimes *from*) the surface. The most common phenomena of this type are as follows.

(1) A soap film lowers the surface tension of water. Why? Because the soap molecules come out of solution and form (mono-molecular) layers at the water surface (with their 'hydrophobic' ends pointing outwards). Soapy water (or beer) doesn't mind being agitated into a foam with a large surface area; these are examples one can ponder every day!

(2) A clean surface in ultra-high vacuum has a higher free energy than an oxidized (or contaminated) surface. Why? Because if it didn't, there would be no 'driving force' for oxygen to adsorb, and the reaction wouldn't occur. It is not so clear whether there are exceptions to this rather cavalier statement, but it is generally true that the surface energy of metal oxides are much lower than the surface energy of the corresponding metal.

If you need more details of multi-component thermodynamics, see Blakely (1973, section 2.3) Adamson (1990, section 3.5) or Hudson (1992, chapter 5). For now, we don't, and thus $\gamma = f_s$ for one-component systems. We can therefore go on to define surface excess internal energy, e_s; entropy s_s, using the usual thermodynamic relationships:

$$e_s = f_s + Ts_s = \gamma - T(d\gamma/dT)_V; \; s_s = -(df_s/dT)_V. \tag{1.7}$$

The entropy s_s is typically positive, and has a value of a few Boltzmann's constant (k) per atom. One reason, not the only one, is that surface atoms are less strongly bound, and thus vibrate with lower frequency and larger amplitude that bulk atoms; another reason is that the positions of steps on the surface are not fixed. Hence $e_s > f_s$ at $T > 0$ K. The first reason is illustrated later in figure 1.17 and table 1.2.

1.1.3 Surface energy and surface stress

You may note that we have not taken the trouble to distinguish surface energy and surface stress at this stage, because of the complexity of the ideas behind surface stress. Both quantities have the same units, but surface stress is a second rank tensor, whereas surface energy is a scalar quantity. The two are the same for fluids, but can be substantially different for solids. We return to this topic in chapter 7; at this stage we should note that surface stresses, and stresses in thin films, are not identical, and may not have the same causes; thus it is reasonable to consider such effects later.

1.2 Surface energies and the Wulff theorem

In this section, the forms of small crystals are discussed in thermodynamic terms, and an over-simplified model of a crystal surface is worked through in some detail. When this model is confronted with experimental data, it shows us that real crystal surfaces have richer structures which depend upon the details of atomic bonding and temperature; in special cases, true thermodynamic information about surfaces has been obtained by observing the shape of small crystals at high temperatures.

1.2.1 General considerations

At equilibrium, a small crystal has a specific shape at a particular temperature T. Since $dF = 0$ at constant T and volume V, we obtain from the previous section that

$$\gamma dA = 0, \text{ or } \int \gamma dA \text{ is a minimum,} \tag{1.8a}$$

where the integral is over the entire surface area A. A typical *non-equilibrium* situation is a thin film with a very flat shape, or a series of small crystallites, perhaps distributed on a substrate. The *equilibrium* situation corresponds to *one* crystal with {hkl} faces exposed such that

$$\int \gamma(\text{hkl}) dA(\text{hkl}) \text{ is minimal,} \tag{1.8b}$$

where the surface energies $\gamma(\text{hkl})$ depend on the crystal orientation. This statement, known as the Wulff theorem, was first enunciated in 1901 (Herring 1951, 1953). If γ is isotropic, the form is a sphere in the absence of gravity, as wonderful pictures of water droplets from space missions have shown us. The sphere is simply the unique geometrical form which minimizes the surface area for a given volume. With gravity, for larger and more massive drops, the shape is no longer spherical, and the 'sessile drop' method is one way of measuring the surface tension of a liquid (Adamson 1990, section 2.9, Hudson 1992, chapter 3); before we all respected the dangers of mercury poisoning, this was an instructive high school experiment. For a solid, there are also several methods of measuring surface tension, most obviously using the zero creep method, in which a ball of material, weight mg, is held up by a fine wire, radius r, in equilibrium via the surface tension force $2\pi r \gamma$ (Martin & Doherty, 1976, chapter 4). But in fact, it isn't easy to measure surface tension or surface energy accurately: we need to be aware of the likelihood of impurity segregation to the surface (think soap or oxidation again), and as we shall see in section 1.3, not all surfaces are in true equilibrium.

The net result is that one needs to know $\gamma(\text{hkl})$ to deduce the equilibrium shape of a small crystal; conversely, if you know the shape, you *might* be able to say something about $\gamma(\text{hkl})$. We explore this in the next section within a simple model.

1.2.2 The terrace–ledge–kink model

Consider a simple cubic structure, lattice parameter a, with nearest neighbor (nn) bonds, where the surface is inclined at angle θ to a low index (001) plane; a two-dimensional (2D) cut of this model is shown in figure 1.3, but you should imagine that the 3D crystal also contains bonds which come out of, and go into, the page.

In this model, bulk atoms have six bonds of strength ϕ. The sublimation energy L, per unit volume, of the crystal is the $(6\phi/2)(1/a^3)$, where division by 2 is to avoid double

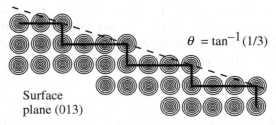

$$\theta = \tan^{-1}(1/3)$$

Surface
plane (013)

Figure 1.3. 2D cut of a simple cubic crystal, showing terrace and ledge atoms in profile. The tangent of the angle θ which the (013) surface plane makes with (001) is 1/3. The steps, or ledges, continue into and out of the paper on the same lattice.

counting: 1 bond involves 2 atoms. Units are (say) eV/nm^3, or many (chemical) equivalents, such as kcal/mole. Useful conversion factors are 1 $eV \equiv 11\,604$ K $\equiv 23.06$ kcal/mole; these and other factors are listed in Appendix C.

Terrace atoms have an extra energy e_t per unit area with respect to the bulk atoms, which is due to having five bonds instead of six, so there is one bond missing every a^2. This means

$$e_t = (6-5)\ \phi/2a^2 = \phi/2a^2 = La/6 \text{ per unit area.} \tag{1.9a}$$

Ledge atoms have an extra energy e_l per unit length *over terrace atoms*: we have four bonds instead of five bonds, distributed every a. So

$$e_l = (5-4)\ \phi/2a = La^2/6 \text{ per unit length.} \tag{1.9b}$$

Finally kink atoms have energy e_k *relative to the ledge atoms*, and the same argument gives

$$e_k = (4-3)\ \phi/2 = La^3/6 \text{ per atom.} \tag{1.9c}$$

More interestingly a kink atom has 3ϕ *relative to bulk atoms*. This is the same as $L/atom$, so adding (or subtracting) an atom from a kink site is equivalent to condensing (or subliming) an atom from the bulk.

This last result may seem surprising, but it arises because moving a kink around on the surface leaves the number of T, L and K atoms, and the energy of the surface, unchanged. The kink site is thus a 'repeatable step' in the formation of the crystal. You can impress your friends by using the original German expression 'wiederhohlbarer Schritt'. This schematic simple cubic crystal is referred to as a Kossel crystal, and the model as the TLK model, shown in perspective in figure 1.4. The original papers are by W. Kossel in 1927 and I.N. Stranski in 1928. Although these papers seem that they are from the distant past, my own memory of meeting Professor Stranski in the early 1970s, shortly after starting in this field, is alive and well. The scientific 'school' which he founded in Sofia, Bulgaria, also continues through social and political upheavals. This tradition is described in some detail by Markov (1995).

Within the TLK model, we can work out the surface energy as a function of (2D or 3D) orientation. For the 2D case shown in figure 1.3, we can show that

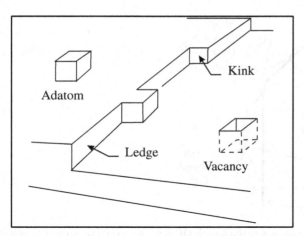

Figure 1.4. Perspective drawing of a Kossel crystal showing terraces, ledges (steps), kinks, adatoms and vacancies.

$$e_s = (e_t + e_1/na) \cos\theta. \tag{1.10a}$$

But $1/n = \tan|\theta|$. Therefore, $e_s = e_t \cos|\theta| + e_1/a \sin|\theta|$, or, within the model

$$e_s = (La/6)(\cos|\theta| + \sin|\theta|). \tag{1.10b}$$

We can draw this function as a polar diagram, noting that it is symmetric about $\theta = 45°$, and repeats when θ changes by $\pm 90°$. This is sufficient to show that there are cusps in all the six $\langle 100 \rangle$ directions, i.e. along the six $\{100\}$ plane normals, four of them in, and two out, of the plane of the drawing. The $|\theta|$ form arises from the fact that θ changes sign as we go through the $\{100\}$ plane orientations, but $\tan|\theta|$ does not. In this model is does not matter whether the step train of figure 1.3 slopes to the right or to the left; if the surface had lower symmetry than the bulk, as we discuss in section 1.4, then the surface energy might depend on such details.

1.2.3 Wulff construction and the forms of small crystals

The Wulff construction is shown in figure 1.5. This is a polar diagram of $\gamma(\theta)$, the γ-plot, which is sometimes called the σ-plot. The Wulff theorem says that the minimum of $\int \gamma dA$ results when one draws the perpendicular through $\gamma(\theta)$ and takes the inner envelope: this is the equilibrium form. The simplest example is for the Kossel crystal of figure 1.3, for which the equilibrium form is a cube; a more realistic case is shown in figure 1.5.

The construction is easy to see qualitatively, but not so easy to prove mathematically. The deepest cusps (C in figure 1.5) in the γ-plot are always present in the equilibrium form: these are *singular* faces. Other higher energy faces, such as the cusps H in the figure, may or may not be present, depending in detail on $\gamma(\theta)$. Between the singular faces, there may be rounded regions R, where the faces are *rough*.

The mathematics of the Wulff construction is an example of the calculus of variations; the history, including the point that the original Wulff derivation was flawed, is

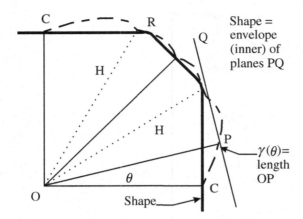

Figure 1.5. A 2D cut of a γ-plot, where the length OP is proportional to $\gamma(\theta)$, showing the cusps C and H, and the construction of the planes PQ perpendicular to OP through the points P. This particular plot leads to the existence of facets and rounded (rough) regions at R. See text for discussion

described by Herring (1953). There are various cases which can be worked out precisely, but somewhat laboriously, in order to decide by calculation whether a particular orientation is mechanically stable. Specific expressions exist for the case where γ is a function of one angular variable θ, or of the lattice parameter, a. In the former case, a face is mechanically stable or unstable depending on whether the surface stiffness

$$\gamma(\theta) + d^2\gamma(\theta)/d\theta^2 \text{ is } > \text{ or } < 0. \tag{1.11}$$

The case of negative stiffness is an unstable condition which leads to *faceting* (Nozières 1992, Desjonquères & Spanjaard 1996). This can occur at 2D internal interfaces as well as at the surface, or it can occur in 1D along steps on the surface, or along dislocations in elastically anisotropic media, both of which can have unstable directions. In other words, these phenomena occur widely in materials science, and have been extensively documented, for example by Martin & Doherty (1976) and more recently by Sutton & Balluffi (1995). These references could be consulted for more detailed insights, but are not necessary for the following arguments.

A full set of 3D bond-counting calculations has been given in two papers by MacKenzie *et al.* (1962); these papers include general rules for nearest neighbor and next nearest neighbor interactions in face-centered (f.c.c.) and body-centered (b.c.c.) cubic crystals, based on the number of broken bond vectors ⟨uvw⟩ which intersect the surface planes {hkl}. There is also an atlas of 'ball and stick' models by Nicholas (1965); an excellent introduction to crystallographic notation is given by Kelly & Groves (1970). More recently, models of the crystal faces can be visualized using CD-ROM or on the web, so there is little excuse for having to duplicate such pictures from scratch. A list of these resources, current as this book goes to press, is given in Appendix D.

The experimental study of small crystals (on substrates) is a specialist topic, aspects of which are described later in chapters 5, 7 and 8. For now, we note that close-packed faces tend to be present in the equilibrium form. For f.c.c. (metal) crystals, these are $\{111\}$, $\{100\}$, $\{110\}$ ... and for b.c.c. $\{110\}$, $\{100\}$ ...; this is shown in γ-plots and equilibrium forms, calculated for specific first and second nearest neighbor interactions in figure 1.6, where the relative surface energies are plotted on a stereogram (Sundquist 1964, Martin & Doherty 1976). For really small particles the discussion needs to take the discrete size of the faces into account. This extends up to particles containing $\sim 10^6$ atoms, and favors $\{111\}$ faces in f.c.c. crystals still further (Marks 1985, 1994). The properties of stereograms are given in a student project which can be found via Appendix D.

The effect of temperature is interesting. *Singular* faces have low energy and low entropy; *vicinal* (stepped) faces have higher energy and entropy. Thus for increasing temperature, we have lower free energy for non-singular faces, and the equilibrium form is more rounded. Realistic finite temperature calculations are relatively recent (Rottman & Wortis 1984), and there is still quite a lot of uncertainty in this field, because the results depend sensitively on models of interatomic forces and lattice vibrations. Some of these issues are discussed in later chapters.

Several studies have been done on the anisotropy of surface energy, and on its variation with temperature. These experiments require low vapor pressure materials, and have used Pb, Sn and In, which melt at a relatively low temperature, by observing the profile of a small crystal, typically 3–5 μm diameter, in a specific orientation using scanning electron microscopy (SEM). An example is shown for Pb in figures 1.7 and 1.8, taken from the work of Heyraud and Métois; further examples, and a discussion of the role of roughening and melting transitions, are given by Pavlovska *et al.* (1989).

We notice that the anisotropy is quite small (much smaller than in the Kossel crystal calculation), and that it decreases, but not necessarily monotonically, as one approaches the melting point. This is due to three effects: (1) a nearest neighbor bond calculation with the realistic f.c.c. structure gives a smaller anisotropy than the Kossel crystal (see problem 1.1); (2) realistic interatomic forces may give still smaller effects; in particular, interatomic forces in many metals are less directional than implied by such bond-like models, as discussed in chapter 6; and (3) atomistic and layering effects at the monolayer level can affect the results in ways which are not intuitively obvious, such as the missing orientations close to (111) in the Pb crystals at 320 °C, seen in figure 1.7(b). The main qualitative points about figure 1.8, however, are that the maximum surface energy is in an orientation close to $\{210\}$, as in the f.c.c. bond calculations of figure 1.6(b), and that entropy effects reduce the anisotropy as the melting point is approached. These data are still a challenge for models of metals, as discussed in chapter 6.

1.3 Thermodynamics versus kinetics

Equilibrium phenomena are described by thermodynamics, and on a microscopic scale by statistical mechanics. However, much of materials science is concerned with kinetics, where the rate of change of metastable structures (or their inability to change) is

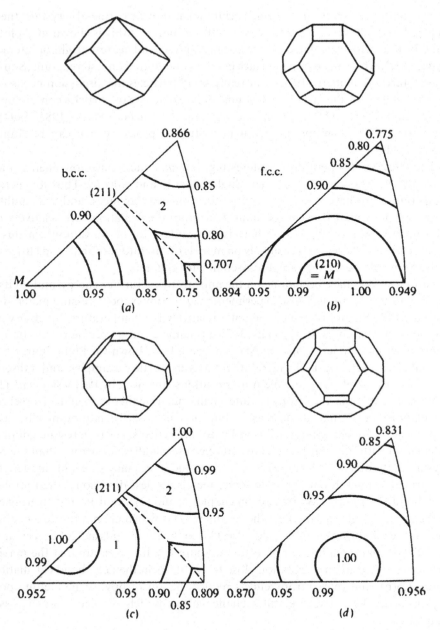

Figure 1.6. γ-Plots in a stereographic triangle (100, 110 and 111) and the corresponding equilibrium shapes for (a) b.c.c., (b) f.c.c., both with $\rho=0$; (c) b.c.c. with $\rho=0.5$, and (d) f.c.c. with $\rho=0.1$; ρ is the relative energy of the second nearest bond to that of the nearest neighbor bond (from Sundquist 1964, via Martin & Doherty 1976, reproduced with permission).

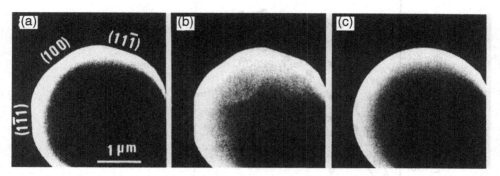

Figure 1.7. SEM photographs of the equilibrium shape of Pb crystals in the [011] azimuth, taken *in situ*: (a) at 300 °C, (b) at 320 °C, showing large rounded regions at 300 °C, and missing orientations at 320 °C; (c) at 327 °C where Pb is liquid and the drop is spherical (from Métois & Heyraud 1989, reproduced with permission).

dominant. Here this distinction is drawn sharply. An equilibrium effect is the vapor pressure of a crystal of a pure element; a typical kinetic effect is crystal growth from the vapor. These are compared and contrasted in this section.

1.3.1 Thermodynamics of the vapor pressure

The sublimation of a pure solid at equilibrium is given by the condition $\mu_v = \mu_s$. It is a standard result, from the theory of perfect gases, that the chemical potential of the vapor at low pressure p is

$$\mu_v = -kT \ln (kT/p\lambda^3), \qquad (1.12)$$

where $\lambda = h/(2\pi mkT)^{1/2}$ is the thermal de Broglie wavelength. This can be rearranged to give the equilibrium vapor pressure p_e, in terms of the chemical potential of the solid, as[1]

$$p_e = (2\pi m/h^2)^{3/2} (kT)^{5/2} \exp (\mu_s/kT). \qquad (1.13)$$

Thus, to calculate the vapor pressure, we need a model of the chemical potential of the solid. A typical μ_s at low pressure is the 'quasi-harmonic' model, which assumes harmonic vibrations of the solid, at its (given) lattice parameter (Klein & Venables 1976). This free energy per particle

$$F/N = \mu_s = U_0 + \langle 3h\nu/2 \rangle + 3kT\langle \ln(1 - \exp(-h\nu/kT)) \rangle, \qquad (1.14)$$

where the $\langle \rangle$ mean average values. The (positive) sublimation energy at zero temperature T, $L_0 = -(U_0 + \langle 3h\nu/2 \rangle)$, where the first term is the (negative) energy per particle in the solid relative to vapor, and the second is the (positive) energy due to zero-point vibrations.

---

[1] This result is derived in most thermodynamics textbooks but not all. See e.g. Hill (1960) pp. 79–80, Mandl (1988) pp. 182–183, or Baierlein (1999) pp. 276–278.

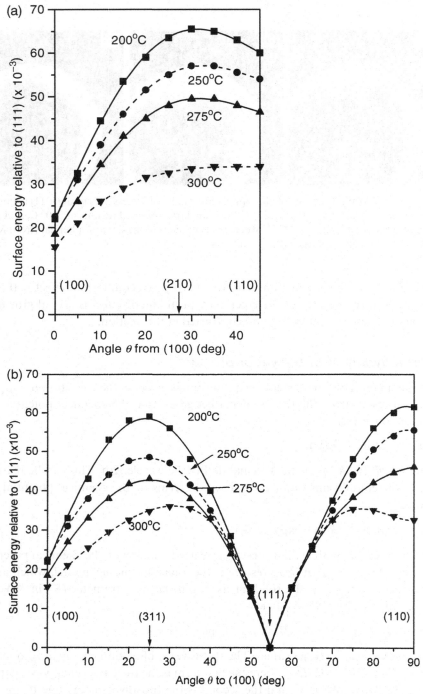

Figure 1.8. Anisotropy of $\gamma(\theta)$ for Pb as a function of temperature, where the points are the original data, with errors $\sim \pm 2$ on this scale, and the curves are fourth-order polynomial fits to these data: (a) in the $\langle 100 \rangle$ zone; (b) in the $\langle 110 \rangle$ zone. The relative surface energy scale is $(\gamma(\theta)/\gamma(111) - 1) \times 10^{-3}$, so 70 corresponds to $\gamma(\theta) = 1.070 \times \gamma(111)$ (after Heyraud & Métois 1983, replotted with permission).

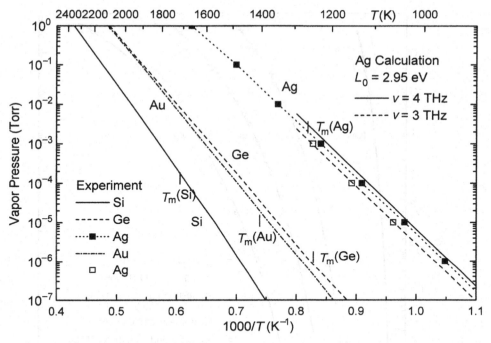

Figure 1.9. Arrhenius plot of the vapor pressure of Ge, Si, Ag and Au, using data from Honig & Kramer (1969). In the case of Ag, earlier handbook data for the solid are also given (open squares); the Einstein model with $L_0 = 2.95$ eV and $\nu = 3$ and 4 THz is shown for comparison with the Ag data.

The vapor pressure is significant typically at high temperatures, where the Einstein model of the solid is surprisingly realistic (provided thermal expansion is taken into account in U_0). Within this model (all $3N$ νs are the same), in the high T limit, we have $\langle \ln(1 - \exp(-h\nu/kT)) \rangle = \langle \ln(h\nu/kT) \rangle$, so that $\exp(\mu_s/kT) = (h\nu/kT)^3 \exp(-L_0/kT)$. This gives

$$p_e = (2\pi m \nu^2)^{3/2} (kT)^{-1/2} \exp(-L_0/kT), \tag{1.15}$$

so that $p_e T^{1/2}$ follows an Arrhenius law, and the pre-exponential depends on the lattice vibration frequency as ν^3. The absence of Planck's constant h in the answer shows that this is a classical effect, where equipartition of energy applies.

The $T^{1/2}$ term is slowly varying, and many tabulations of vapor pressure simply express $\log_{10}(p_e) = A - B/T$, and give the constants A and B. This equation is closely followed in practice over many decades of pressure; some examples are given in figures 1.9 and 1.10. Calculations along the above lines yield values for L_0 and ν, as indicated for Ag on figure 1.9. Values abstracted using the Einstein model equations in their general form are given in table 1.1. For the rare gas solids, vapor pressures have been measured over 13 decades, as shown in figure 1.10; yet this can still often be well fitted by the two-parameter formula (Crawford 1977). This large data span means that the sublimation energies are accurately known: the frequencies given here are good to

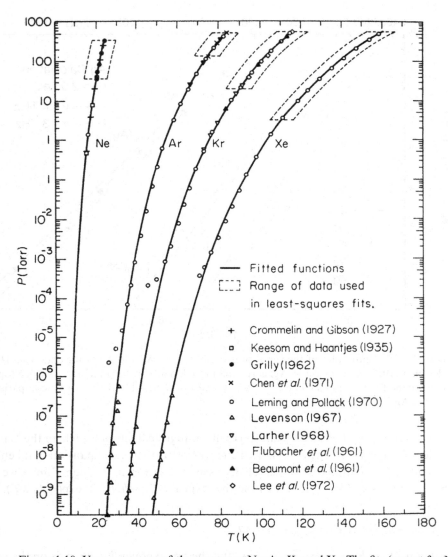

Figure 1.10. Vapor pressure of the rare gases Ne, Ar, Kr and Xe. The fits (except for Ne) are to the simplest two- parameter formula $\log_{10}(p_e) = A - B/T$ (from Crawford 1977, and references therein; reproduced with permission).

maybe ±20%, and depend on the use of the (approximate) Einstein model. These points can be explored further via problem 1.3.

The point to understand about the above calculation is that the vapor pressure does not depend on the structure of the surface, which acts simply as an intermediary: i.e., the surface is 'doing its own thing' in equilibrium with both the crystal and the vapor. What the surface of a Kossel crystal looks like can be visualized by Monte Carlo (MC) or other simulations, as indicated in figure 1.11. At low temperature, the terraces are

Table 1.1. *Lattice constants, sublimation energies and Einstein frequencies of some elements*

Element	Lattice constant (a_0) nm	Sublimation energy (L_0) eV or K		Einstein frequency ν (THz)
Metals				
Ag	0.4086 (f.c.c.) at RT	2.95 ± 0.01 eV		4
Au	0.4078	3.82 ± 0.04		3
Fe	0.2866 (b.c.c.)	4.28 ± 0.02		11
W	0.3165	8.81 ± 0.07		7
Semiconductors				
Si	0.5430 (diamond)	4.63 ± 0.04		15
Ge	0.5658	3.83 ± 0.02		6
Van der Waals				
Ar	0.5368 (f.c.c.) at 50 K	84.5 meV or 981 K		1.02
Kr	0.5692	120	1394	0.84
Xe	0.6166	167	1937	0.73

almost smooth, with few adatoms or vacancies (see figure 1.4 for these terms). As the temperature is raised, the surface becomes rougher, and eventually has a finite interface width. There are distinct roughening and melting transitions at surfaces, each of them specific to each {hkl} crystal face. The simplest MC calculations in the so-called SOS (solid on solid) model show the first but not the second transition. Calculations on the roughening transition were developed in review articles by Leamy *et al.* (1975) and Weeks & Gilmer (1979); we do not consider this phenomenon further here, but the topic is set out pedagogically by several authors, including Nozières (1992) and Desjonquères & Spanjaard (1996, section 2.4).

1.3.2 The kinetics of crystal growth

This picture of a fluctuating surface which doesn't influence the vapor pressure applies to the equilibrium case, but what happens if we are not at equilibrium? The classic paper is by Burton, Cabrera & Frank (1951), known as BCF, and much quoted in the crystal growth literature. We have to consider the presence of kinks and ledges, and also (extrinsic) defects, in particular screw dislocations. More recently, other defects have been found to terminate ledges, even of sub-atomic height, and these are also important in crystal growth. The BCF paper, and the developments from it, are quite mathematical, so we will only consider a few simple cases here, in order to introduce terms and establish some ways of looking at surface processes.

First, we need the ideas of supersaturation $S = (p/p_e)$, and thermodynamic driving force, $\Delta\mu = kT\ln S$. $\Delta\mu$ is clearly zero in equilibrium, is positive during condensation, and negative during sublimation or evaporation. The variable which enters into exponents is therefore $\Delta\mu/kT$; this is often written $\beta\Delta\mu$, with $\beta \equiv 1/kT$ standard notation in

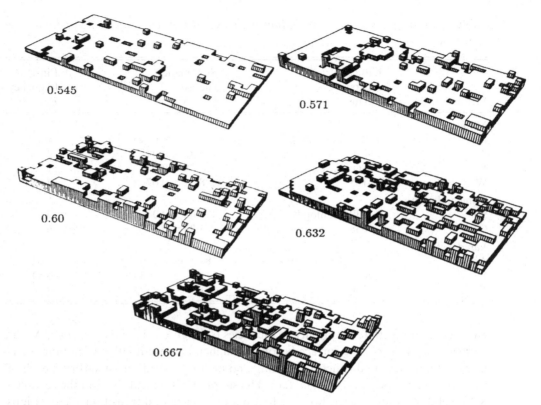

0.545

0.571

0.60

0.632

0.667

Figure 1.11. Monte Carlo simulations of the Kossel crystal developed within the solid on solid model for five reduced temperature values (kT/ϕ). The roughening transition occurs when this value is ~ 0.62 (Weeks & Gilmer 1979, reproduced with permission).

statistical mechanics. The deposition rate or flux (R or F are used in the literature) is related, using kinetic theory, to p as $R = p/(2\pi mkT)^{1/2}$.

Second, an atom can adsorb on the surface, becoming an adatom, with a (positive) adsorption energy E_a, relative to zero in the vapor. (Sometimes this is called a desorption energy, and the symbols for all these terms vary wildly.) The rate at which the adatom desorbs is given, approximately, by $\nu\exp(-E_a/kT)$, where we might want to specify the pre-exponential frequency as ν_a to distinguish it from other frequencies; it may vary relatively slowly (not exponentially) with T.

Third, the adatom can diffuse over the surface, with energy E_d and corresponding pre-exponential ν_d. We expect $E_d < E_a$, maybe much less. Adatom diffusion is derived from considering a random walk in two dimensions, and the 2D diffusion coefficient is then given by

$$D = (\nu_d a^2/4) \exp(-E_d/kT),\qquad (1.16)$$

and the adatom lifetime before desorption,

$$\tau_a = \nu_a^{-1} \exp(E_a/kT).\qquad (1.17)$$

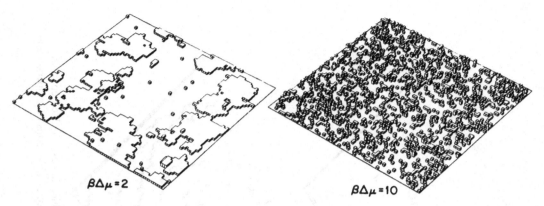

Figure 1.12. MC interface configurations after 0.25 monolayer deposition at the same temperature on terraces, under two different supersaturations $\beta\Delta\mu = 2$ and 10; the bond strength is expressed as $\phi = 4kT$ (Weeks & Gilmer 1979, reproduced with permission).

BCF then showed that $x_s = (D\tau_a)^{1/2}$ is a characteristic length, which governs the fate of the adatom, and defines the role of ledges (steps) in evaporation or condensation. It is a useful exercise to familiarize oneself with the ideas of local equilibrium, and diffusion in one dimension. Local equilibrium can be described either in terms of differential equations or of chemical potentials as set out in problems 1.2 and 1.4; diffusion needs a differential equation formulation and/or a MC simulation.

The main points that result from the above considerations are as follows.

(1) Crystal growth (or sublimation) is difficult on a perfect terrace, and substantial supersaturation (undersaturation) is required. When growth does occur, it proceeds through nucleation and growth stages, with monolayer thick islands (pits) having to be nucleated before growth (sublimation) can proceed; this is illustrated by early MC calculations in figure 1.12.
(2) A ledge, or step on the surface captures arriving atoms within a zone of width x_s either side of the step, statistically speaking. If there are only individual steps running across the terrace, then these will eventually grow out, and the resulting terrace will grow much more slowly (as in point 1). In general, rough surfaces grow faster than smooth surfaces, so that the final 'growth form' consists entirely of slow growing faces;
(3) The presence of a screw dislocation in the crystal provides a step (or multiple step), which spirals under the flux of adatoms. This provides a mechanism for continuing growth at modest supersaturation, as illustrated by MC calculations in figure 1.13 (Weeks & Gilmer 1979).

Detailed study shows that the growth velocity depends quadratically on the supersaturation for mechanism 3, and exponentially for mechanism 1, so that dislocations are dominant at low supersaturation, as shown in figure 1.14. Growth from the liquid and from solution has been similarly treated, emphasizing the internal energy change on melting L_m, and a single parameter α proportional to L_m/kT, where $\alpha < 2$ typical for melt growth of elemental solids corresponds to rough liquid–solid interfaces (Jackson

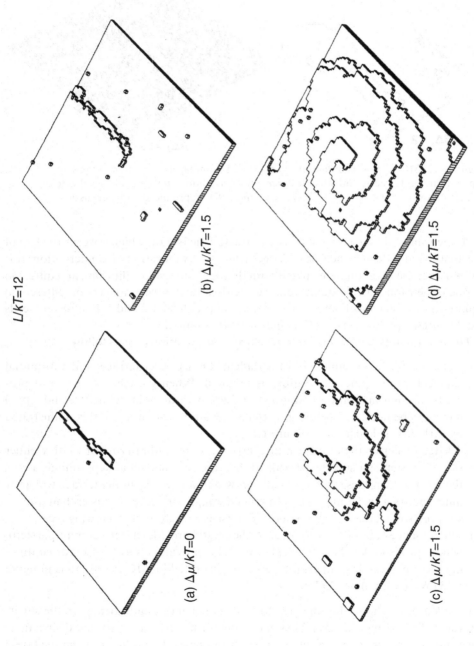

$L/kT=12$

(a) $\Delta\mu/kT=0$

(b) $\Delta\mu/kT=1.5$

(c) $\Delta\mu/kT=1.5$

(d) $\Delta\mu/kT=1.5$

Figure 1.13. MC interface configurations during deposition in the presence of a screw dislocation which causes a double step (a) in equilibrium, and (b)–(d) as a function of time under supersaturation $\beta\Delta\mu = 1.5$, for bond strength expressed in terms of temperature as $L/kT = 12$, equivalent to $\phi = 4kT$ (Weeks & Gilmer 1979, reproduced with permission).

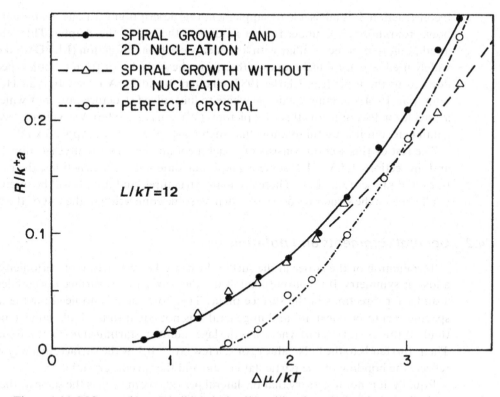

Figure 1.14. MC growth rates (R/k^+a) during deposition for spiral growth (in the presence of a screw dislocation) compared with nucleation on a perfect terrace as a function of supersaturation $\beta\Delta\mu$, for bond strength expressed in terms of temperature as $L/kT = 12$, equivalent to $\phi = 4kT$ (Weeks & Gilmer 1979, reproduced with permission).

1958, Jackson et al. 1967, Woodruff 1973). Growth from the vapor via smooth interfaces are characterized by larger α values, either because the sublimation energy $L_0 \gg L_m$, and/or the growth temperature is much lower than the melting temperature. Such an outline description is clearly only an introduction to a complex topic, and further information can be obtained from the books quoted, from several review articles (e.g. Leamy et al. 1975, Weeks & Gilmer 1979), or from more recent handbook articles (Hurle 1993, 1994). But the reader should be warned in advance that this is not a simple exercise; there are considerable notational difficulties, and the literature is widely dispersed. We return to some of these topics in chapters 5, 7 and 8.

1.4 Introduction to surface and adsorbate reconstructions

1.4.1 Overview

In this section, the ideas about surface structure which we will need for later chapters are introduced briefly. However, if you have never come across the idea of surface reconstruction, it is advisable to supplement this description with one in another text-book from those given under further reading at the end of the chapter. This is also a good point to become familiar with low energy electron diffraction (LEED) and other widely used structural techniques, either from these books, or from a book especially devoted to the topic (e.g. Clarke 1985, chapters 1 and 2). A review by Van Hove & Somorjai (1994) contains details on where to find solved structures, most of which are available on disc, or in an atlas with pictures (Watson *et al.* 1996). We will not need this detail here, but it is useful to know that such material exists (see Appendix D).

The rest of this section consists of general comments on structures (section 1.4.2), and, in sections 1.4.3–1.4.8, some examples of different reconstructions, their vibra-tions and phase transitions. There are many structures, and not all will be interesting to all readers: the structures described all have some connection to the rest of the book.

1.4.2 *General comments and notation*

Termination of the lattice at the surface leads to the destruction of periodicity, and a loss of symmetry. It is conventional to use the z-axis for the surface normal, leaving x and y for directions in the surface plane. Therefore there is no need for the lattice spacing $c(z)$ to be constant, and in general it is not equal to the bulk value. One can think of this as $c(z)$ or $c(m)$ where m is the layer number, starting at $m = 1$ at the surface. Then $c(m)$ tends to the bulk value c_0 or c, a few layers below the surface, in a way which reflects the bonding of the particular crystal and the specific crystal face.

Equally, it is not necessary that the lateral periodicity in (x, y) is the same as the bulk periodicity (a, b). On the other hand, because the surface layers are in close contact with the bulk, there is a strong tendency for the periodicity to be, if not the same, a simple multiple, sub-multiple or rational fraction of a and b, a commensurate structure. This leads to Wood's (1964) notation for surface and adsorbate layers. An example related to chemisorbed oxygen on Cu(001) is shown here in figure 1.15 (Watson *et al.* 1996). Note that we are using (001) here rather than the often used (100) notation to empha-size that the x and y directions are directions *in* the surface; however, these planes are equivalent in cubic crystals and can be written in general as {100}; similarly, specific directions are written [100] and general directions ⟨100⟩ in accord with standard crys-tallographic practice (see e.g. Kelly & Groves 1970).

But first let us get the basic notation straight, as this can be somewhat confusing. For example, here we have used (a, b, c) for the lattice constants; but these are not necessar-ily the normal lattice constants of the crystal, since they were defined with respect to a particular (hkl) surface. Also, several books use $a_{1,2,3}$ for the real lattice and $b_{1,2,3}$ for the reciprocal lattice, which is undoubtedly more compact. Wood's notation originates in a (2×2) matrix M relating the surface parameters (a, b) or $\mathbf{a}_s$ to the bulk (a_0, b_0) or $\mathbf{a}_b$. But the full notation, e.g. Ni(110)c(2×2)O, complete with the matrix M, is rather for-bidding (Prutton 1994). If you were working on oxygen adsorption on nickel you would simply refer to this as a c(2×2), or 'centered 2 by 2' structure; that of adsorbed O on Cu(001)-($2\sqrt{2} \times \sqrt{2}$)R45°-2O shown in figure 1.15 would, assuming the context were not confusing, be termed informally a $2\sqrt{2}$ structure.

(a)

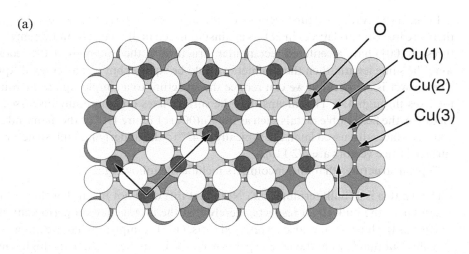

Cu(100)-(2√2x√2)R45°-2O (top view)   BALSAC plot

(b)

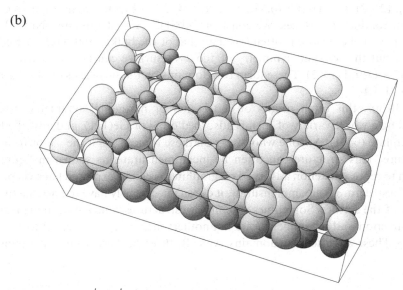

Cu(100)-(2√2x√2)R45°-2O (perspective)   BALSAC plot

Figure 1.15. Wood's notation, as illustrated for the chemisorbed structure Cu(001)-
$(2\sqrt{2} \times \sqrt{2})$R45°-2O in (a) top and (b) perspective view. The $2\sqrt{2}$ and the $\sqrt{2}$ represent the ratios
of the lengths of the absorbate unit cell to the substrate Cu(001) surface unit cell. The R45°
represents the angle through which the adsorbate cell is rotated to this substrate surface cell,
and the 2O indicates there are two oxygen atoms per unit cell. The different shading levels
indicate Cu atoms in layers beneath the surface (after Watson *et al.* 1996, reproduced with
permission).

From the surface structure sections of the textbooks referred to, we can learn that there are five Bravais lattices in 2D, as against fourteen in 3D. For example, many structures on (001) have a centered rectangular structure. If the two sides of the rectangle were the same length, then the symmetry would be square; but is it a centered square? The answer is *no*, because we can reduce the structure to a simple square by rotating the axes through 45°. This means that the surface axes on commonly discussed surfaces, e.g. the f.c.c. noble metals such as the Cu(001) of figure 1.15 or the diamond cubic {001} surfaces discussed later, are typically at 45° to the underlying bulk structure; the surface lattice vectors are $a/2\langle 110\rangle$.

Typical structures that one encounters include the following.

* (1×1): this is a 'bulk termination'. Note that this does not mean that the surface is similar to the bulk in all respects, merely that the average lateral periodicity is the same as the bulk. It may also be referred to as '(1×1)', implying that 'we know it isn't really' but that is what the LEED pattern shows. Examples include the high temperature Si and Ge(111) structures, which are thought to contain mobile adatoms that do not show up in the LEED pattern because they are not ordered.
* (2×1), (2×2), (4×4), (6×6), c(2×2), c(2×4), c(2×8), etc: these occur frequently on semiconductor surfaces. We consider Si(001)2×1 in detail later. Note that the symmetry of the surface is often less than that of the bulk. Si(001) is four-fold symmetric, but the two-fold symmetry of the 2×1 surface can be constructed in two ways (2×1) and (1×2). These form two domains on the surface as discussed later in section 1.4.4.
* $\sqrt{3}\times\sqrt{3}$R30°: this often occurs on a trigonal or hexagonal symmetry substrate, including a whole variety of metals adsorbed on Si or Ge(111), and adsorbed gases on graphite (0001). Anyone who works on these topics calls it the $\sqrt{3}$, or root-three, structure. This structure can often be incommensurate, as shown in figure 1.16, drawn to represent xenon adsorbed on graphite, as can be explored later via problem 4.1. If a structure is incommensurate, it doesn't necessarily have to have the full symmetry of the surface. Sometimes we can have structures which are commensurate in one direction and incommensurate in another: these may be referred to as striped phases. These will also form domains, typically three, because of the underlying symmetry.

1.4.3   Examples of (1x1) structures

These 'bulk termination' structures include some f.c.c. metals, such as Ni, Ag, Pt(111), Cu and Ni(001), and Fe, Mo and W(110) amongst b.c.c. metals. One may expect this list to get shorter with time, rather than longer, as more sensitive tests may detect departures from (1×1). For example, W and Mo(001) are 1×1 at high temperature, but have phase transitions to (2×1) and related incommensurate structures at low temperature (Debe & King 1977, Felter *et al.* 1977, Estrup 1994). Lower symmetries are more common at low temperature than at high temperature in general. This is a feature that surfaces have in common with bulk solids such as ferroelectrics. The interaction between the atoms is strongly anharmonic, leading perhaps to double-well interaction

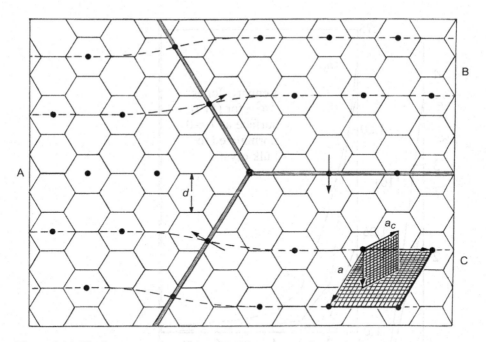

Figure 1.16. The incommensurate $\sqrt{3} \times \sqrt{3}R30°$ structure of adsorbed xenon (lattice parameter *a*) on graphite with a lattice parameter a_c. Note that the Xe adatoms approximately sit in every third graphite hexagon, close to either A, B or C sites; they would do so exactly in the commensurate phase. The arrows indicate the displacement, or Burgers, vectors associated with the domain walls, sometimes called misfit dislocations. On a larger scale these domain walls form a hexagonal network, spacing *d*, as in problem 4.1 (after Venables & Schabes-Retchkiman, 1978, reproduced with permission).

potentials. At high temperature, the vibrations of the atoms span both the wells, but at low temperature the atoms choose one *or* the other. There is an excellent executive toy which achieves the same effect with a pendulum and magnets . . . check it out!

The *c*-spacing of metal (1×1) surface layers have been extensively studied using LEED, and are found mostly to relax inwards by several percent. This is a general feature of metallic binding, where what counts primarily is the electron density around the atom, rather than the directionality of 'bonds'. The atoms like to surround themselves with a particular electron density: because some of this density is removed in forming the surface, the surface atoms snuggle up closer to compensate. We return to this point, which is embodied in embedded atom, effective medium and related theories of metals in chapter 6.

Rare gas solids (Ar, Kr, Xe, etc.) relax in the opposite sense. These solids can be modeled fairly well by simple pair potentials, such as the Lennard-Jones 6–12 (LJ) potential; they are accurately modeled with refined potentials plus small many-body corrections (Klein & Venables 1976). Such LJ potential calculations have been used to explore the spacings and lattice vibrations at these (1×1) surfaces (Allen & deWette 1969, Lagally 1975). The surface expands outwards by a few percent in the first two–three layers, more for the open surface (110) than the close packed (111), as shown

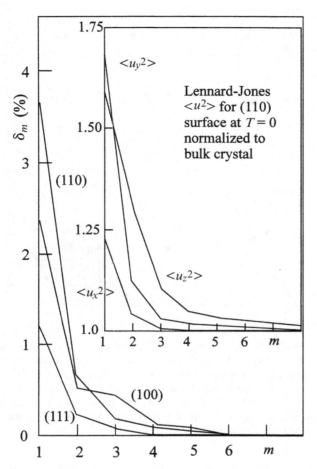

Figure 1.17. Static displacements δ_m, expressed as a percentage of the lattice spacing m layers from the surface, for (001), (110) and (111) surfaces of an f.c.c. Lennard-Jones crystal; right-hand inset). Ratios of mean square displacement amplitudes $\langle u^2 \rangle$, expressed as a ratio of the bulk value, for the (110) surface of an f.c.c. Lennard-Jones crystal approximating solid argon (after Allen & deWette 1969, Lagally 1975, replotted with permission).

in figure 1.17. The inset in figure 1.17 and table 1.2 explore the vibrations calculated for the LJ potential, and remind us that the lower symmetry at the surface means that the mean square displacements are not the same parallel and perpendicular to the surface; on (110) all three modes are different. Different lattice dynamical models have given rather different answers. This is because the vibrations are sufficiently large for anharmonicity to assume greater importance at the surface.

1.4.4   Si(001) (2 ×1) and related semiconductor structures

Let us start by drawing Si(001) 2×1 and 1×2. First, draw the diamond cubic structure in plan view on (001), labeling the atom heights as 0, 1/4, 1/2 or 3/4 (or equivalently

Table 1.2. *The ratios of the mean square displacements of surface atoms to those in the bulk for a Lennard-Jones crystal*

Surface	Component	SFC [1]	QH ($T_m/2$) [2]	MD ($T_m/2$) [3]
(001)	$\langle u_x^2 \rangle = \langle u_y^2 \rangle$	1.46	2.03	2.23 ± 0.17
	$\langle u_z^2 \rangle$	1.87	2.77	3.07 ± 0.15
(111)	$\langle u_x^2 \rangle = \langle u_y^2 \rangle$	1.30	1.45	1.27 ± 0.07
	$\langle u_z^2 \rangle$	1.86	2.85	3.48 ± 0.18
(110)	$\langle u_x^2 \rangle$	1.50	1.78	1.89 ± 0.20
	$\langle u_y^2 \rangle$	2.14	3.21	4.99 ± 0.65
	$\langle u_z^2 \rangle$	1.83	3.03	3.57 ± 0.25

Notes:
[1] Simple force-constant model, with force constants at the surface equal to those in the bulk.
[2] Quasi-harmonic approximation, changes in the surface force constants determined at $T_m/2$ (where T_m is melting temperature).
[3] Molecular Dynamics (MD) computer experiment at $T_m/2$, which includes anharmonicity.
Sources: After Allen & deWette 1969, Lagally 1975.

$-1/4$), three to four unit cells being sufficient, after the manner of figure 1.18. The surface can occur between any of these two adjacent heights. There are two domains at right angles, aligned along different $\langle 110 \rangle$ directions. The reconstruction arises because the surface atoms dimerize along these two [110] and [1$\bar{1}$0] directions, to reduce the density of dangling bonds, producing a unit cell which is twice as long as it is broad; hence the 2×1 notation. Once you have got the geometry sorted out, you can see that the two different domains are associated with different heights in the cell, so that one terrace will have one domain orientation, then there will be a step of height 1/4 lattice constant, and the next terrace has the other domain orientation. This is already quite complicated!

Listening to specialists in this area can tax your geometric imagination, because the dimers form into rows, which are perpendicular to the dimers themselves – dimer and dimer row directions are both along $\langle 110 \rangle$ directions, but are *not* along the same direction, they are at right angles to each other. Moreover, there are two types of 'single height steps', referred to as S_A and S_B, which have different energies, and alternate domains as described above. There are also 'double height steps' D_A and D_B, which go with one particular domain type. Then you can worry about whether the step direction will run parallel, perpendicular or at an arbitrary angle to the dimers (or dimer rows, if you want to get confused, or vice versa). The dimers can also be symmetric (in height) or unsymmetric, and these unsymmetric dimers can be arranged in ordered arrays, 2×2, c(2×4), c(2×8), etc.

With all the intrinsic and unavoidable complexity, it is sensible to ask yourself whether you really need to know all this stuff. Semiconductor surface structures, and the growth of semiconductor devices, are specialist topics, which we will return to later

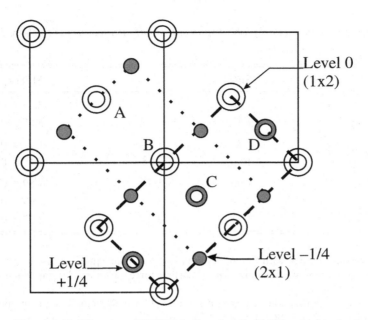

Figure 1.18. Diagram of Si(001) bulk unit cells (full lines), showing how the 2×1 and 1×2 domains arise. If the surface atoms are at level 0, atoms A and B move together, i.e. they dimerize, leading to the 2×1 cell given by the dotted line; but if the surface atoms are at level $+1/4$, atoms C and D dimerize, leading to the 1×2 cell shown by the dashed line.

in chapters 7 and 8. However, I am assuming that several of you really do need to know about these structures. To my way of thinking, they are remarkably interesting and important! Why? Because semiconductor technology has arrived at the point of growing devices with nanometer dimensions on clean surfaces, using MBE, CVD, ALE or whatever new technique is invented next year; the surface processes which take place at the monolayer level actually influence performance and reliability. This is an amazing fact of late twentieth century life, one which is set to be dominant for electrical and chemical engineering in the twenty-first century. As we will explore in chapters 2 and 3, we now have experimental techniques for producing, analyzing and visualizing these nanometer scale structures, often down to atomic detail. Thus, it is worth sticking with the topic for awhile.

Meanwhile, on the subject of surface reconstructions, we abstract three salient points.

(a) The existence of a particular type of structure, e.g. 2×1, does *not* determine the actual atomic arrangement. This typically has been determined by a detailed analysis of LEED Intensity–Voltage (I–V) curves, and an experiment–theory comparison in the form of a reliability or R-factor (Clarke 1985, chapter 7). For example, as shown in figure 1.19, three different models of Si(001) 2×1 were proposed before the dimer model (figure 1.19(a)) became widely accepted.

(b) The number of possible domains depends on the symmetry. For Si(111) with the

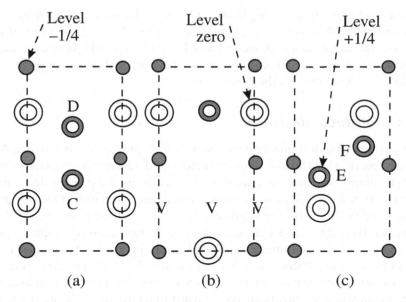

Figure 1.19. Models of the Si(001) 2×1 structure: (a) dimer model after Chadi (1979), with the dimerized atoms C and D as in the (1×2) cell in figure 1.18, but with a shifted origin. Higher order, e.g. c(4×2) structures are formed if C and D are different heights, and are ordered alternately; (b) vacancy model after Poppendieck et al. (1978), with surface vacancies in the two highest levels at V. Their c(4×2) structure also contained distortions in the third layer; (c) conjugated chain model after Jona et al. (1977), with atoms E and F forming the chain in a horizontal ⟨110⟩ direction, and considerable implied distortions between the two highest levels.

metastable 2×1 structure which is produced by cleaving, there are three domains. This 2×1 structure, and its electronic structure, are described by Lüth (1993/5), with the π-bonded chain model finding favor. There is also the possibility of anti-phase boundaries, between domains which are in the same orientation, but which are not registered identically to the underlying bulk structure.

(c) Major contributions to calculations and explanations of the surface, step and boundary structures have been made, as set out briefly by Chadi (1989, 1994). Several writers have discussed the physical reasons behind such reconstructions. We pursue this topic in chapter 7.

1.4.5 The famous 7×7 structure of Si(111)

This structure is described in many places and you cannot leave a course on surfaces, or put down a book, without having realized what this amazing structure is. The question of why nature chooses such a complex arrangement is absolutely fascinating, and we will look at how people have thought about this in chapter 7. It was determined by a combination of LEED, scanning tunneling microscopy (STM) and transmission high energy electron diffraction (THEED) (Takayanagi et al. 1985). It has three structural units, dimers, adatoms and stacking faults, and is hence known as a DAS model. The 7 $\times$ 7 is just one possible structure of this type, all of which have odd numbers of multiples between the surface and bulk meshes. The LEED or THEED patterns of the 7×7 structure contains 49 superstructure spots (or beams) of different intensity, which needed to be analyzed to solve the structure in detail.

1.4.6   Various 'root-three' structures

These structures arise in connection with metals and semi-metals (B, Cu, Ag, Au, In, Sb, Pb, etc.) on the (111) face of semiconductors, and adsorption of gases on hexagonal layer compounds such as graphite. Here again we have three domains, but they are positional, as well as sometimes orientational, in nature. One can put the atoms in three positions on the substrate, but if you put them on one lattice (A), the other two (B and C) are excluded, in the case of rare gases on graphite because of the large size of the adatoms, as indicated earlier in figure 1.16. Studies of such structures have a long history in statistical mechanics, as in the 'three-state Potts model', where the three equivalent positions leads to a degenerate ground state, and interesting higher temperature properties. Adsorption is discussed here in more detail in chapter 4.

Figure 1.20 shows the reported structure of Ag adsorbed on Si or Ge(111), which has been determined by surface X-ray diffraction (Howes *et al.* 1993), with the surface and bulk lattices indicated. The interesting point in the present context about this Ag-induced structure is to realize how much has to happen at the surface, to produce these structures. Deposition of metal atoms alone is not nearly enough to produce it starting from Si(111)7$\times$7 or Ge(111)2$\times$8. Substantial diffusion of both metal and semiconductor is required. The same consideration applies to producing Si(111) surfaces by cleavage, which results in the 2$\times$1 structure. This π-bonded structure, which does not require any long range atomic motion is, however, metastable. Heating to around 250 °C causes it to transform irreversibly into the 7$\times$7, which is the equilibrium structure below the reversible 7$\times$7 to '1$\times$1' transformation at 830 °C; these transformations involve major movement of atoms at the surface.

1.4.7   Polar semiconductors, such as GaAs(111)

When lower symmetry structures are combined with the lower symmetry of the surface, various curious and interesting phenomena can occur. For example, GaAs and related III–V semiconductors are cubic, but low symmetry ($\bar{4}$3m point group). Looked at along the [111] direction, the atomic sequence is asymmetric, as in (Ga, As, space) versus (As, Ga, space). This results in 'polar faces', with (111) being different from ($\bar{1}\bar{1}\bar{1}$). These are the A and B faces, and can have different compositions and charges on them. Atomic composition and surface reconstruction interact to cancel out long range electric fields. For 'non-polar' faces, e.g. GaAs (110), this composition/charge imbalance does not occur, and these tend to have (1$\times$1) surfaces. This

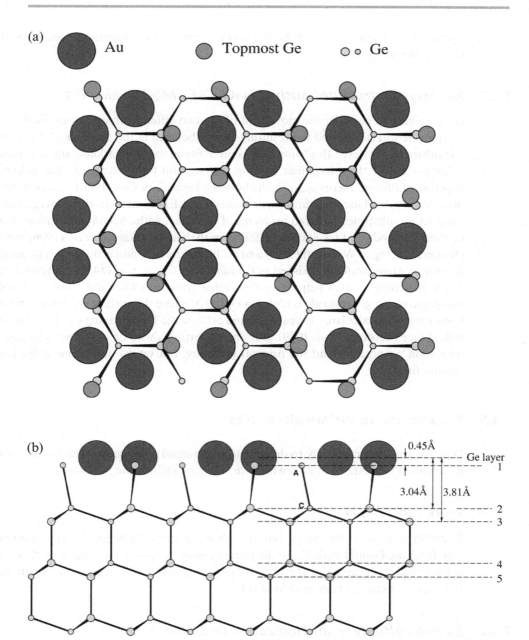

Figure 1.20. (a) Top and (b) side views of the Ag/Ge(111) root-three structure, as determined by surface X-ray diffraction, showing the spacings normal to the surface which have been determined (after Howes *et al.* 1993, reproduced with permission).

again is a specialist topic, combining surface structure with surface electronics, that we consider in chapter 7.

1.4.8 Ionic crystal structures, such as NaCl, CaF$_2$, MgO or alumina

Here we have to consider the movement of the two different charged ions, likely to be in opposite directions, and the resulting charge balance in the presence of the dielectric substrate. However, this 'rumpling' is often found to be remarkably small, typically a few percent of the interplanar spacing; a first point for a search of what is known experimentally and theoretically is the book by Henrich & Cox (1996). A recent development is to combine structural experiments (e.g. LEED) on ultra-thin films grown on conducting substrates, to avoid problems of charging, with *ab initio* calculation. Some of these methods and results can be found in the atlas of Watson *et al.* (1996), review chapters in King & Woodruff (1997) and a 1999 conference proceedings on *The Surface Science of Metal Oxides* published as *Faraday Disc. Chem. Soc.* **114**. Grazing incidence X-ray scattering is also helping to determine structures (Renaud 1998). A notable exception to the general rule is alumina (A1$_2$O$_3$), where the surface oxygen ion relaxations have been calculated to reach around 50% of the layer spacing on the hexagonal (0001) face (Verdozzi *et al.* 1999). But are we getting ahead of ourselves: you can see how soon we need to read the original literature, but we do need some more background first!

1.5 Introduction to surface electronics

Here we are concerned only to define and understand a few terms which will be used in a general context. The terms which we will need include the following.

1.5.1 Work function, ϕ

The work function is the energy, typically a few electronvolts, required to move an electron from the Fermi Level, E_F, to the vacuum level, E_0, as shown in figure 1.21(a). The work function depends on the crystal face {hkl} and rough surfaces typically have lower ϕ, as discussed later in section 6.1.

1.5.2 Electron affinity, χ, and ionization potential Φ

Both of these would be the same for a metal, and equal to ϕ. But for a semiconductor or insulator, they are different. The electron affinity χ is the difference between the vacuum level E_0, and the bottom of the conduction band E_C, as shown in figure 1.21(b). The ionization potential Φ is $E_0 - E_V$, where E_V is the top of the valence band. These terms are not specific to surfaces: they are also used for atoms and molecules generally, as the energy level which (a) the next electron goes into, and (b) the last electron comes from.

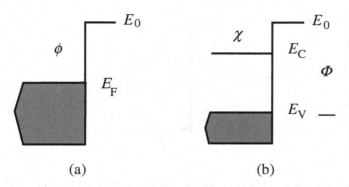

Figure 1.21. Schematic diagrams of (a) the work function, ϕ; (b) the electron affinity, χ and ionization potential Φ, both in relation to the vacuum level E_0, the Fermi energy E_F, and conduction and valence band edges E_C and E_V.

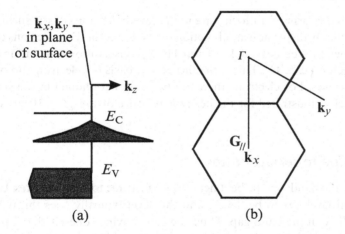

Figure 1.22. Schematic diagrams of: (a) a surface state defined by wave vector $\mathbf{k}_{//} = \mathbf{k}_x + \mathbf{k}_y$, and $\mathbf{k}_\perp = \mathbf{k}_z$; (b) the surface Brillouin zone and 2D reciprocal lattice vector $\mathbf{G}_{//}$ for the $\sqrt{3} \times \sqrt{3} R 30°$ structure, plotted in the same orientation as the real (xenon) lattice of figure 1.16.

1.5.3 Surface states and related ideas

A surface state is a state localized at the surface, which decays exponentially into the bulk, but which may travel along the surface. The wave function is typically of the form

$$\psi \approx u(r)\exp(-i k_\perp |z|) \exp(i k_{//} r), \tag{1.18}$$

where, for a state in the band gap, $k_\perp$ is complex, decaying away from the surface on both sides, as shown in figure 1.22(a). Such a state is called a resonance if it overlaps with a bulk band, as then it may have an increased amplitude at the surface, but evolves continuously into a bulk state. A surface plasmon is a collective excitation located at the surface, with frequency typically $\omega_p/\sqrt{2}$, where ω_p is the frequency of a bulk plasmon.

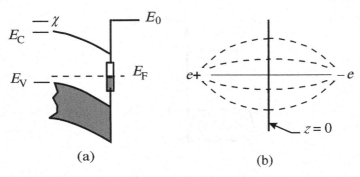

Figure 1.23. Schematic diagrams of: (a) band bending due to a surface state on a p-type semiconductor; (b) the E-field between an electron at position z and a metal surface is the same as that produced by a positive image charge at $-z$.

1.5.4 Surface Brillouin zone

A surface state takes the form of a Bloch wave in the two dimensions of the surface, in which there can be energy dispersion as a function of the $\mathbf{k}_{//}$ vector. For electrons crossing the surface barrier, $\mathbf{k}_{//}$ is conserved, $\mathbf{k}_{\perp}$ is not. The $\mathbf{k}_{//}$ conservation is to within a 2D reciprocal lattice vector, i.e. $\pm \mathbf{G}_{//}$. This is the theoretical basis of (electron and other) diffraction from surfaces. For electrons there are two states contained in the surface Brillouin zone, which is illustrated for the hexagonal lattice of the $\sqrt{3} \times \sqrt{3}R30°$ structure in figure 1.22(b).

1.5.5 Band bending, due to surface states

In a semiconductor, the bands can be bent near the surface due to surface states. Under zero bias, the Fermi level has to be 'level', and this level typically goes through the surface states which lie in the band gap. Thus one can convince oneself that a p-type semiconductor has bands that are bent downwards as the surface is approached from inside the material, as shown in figure 1.23(a). This leads to a reduction in the electron affinity. Some materials (e.g. Cs/p-type GaAs) can even be activated to negative electron affinity, and such NEA materials form a potent source of electrons, which can also be spin-polarized as a result of the band structure.

1.5.6 The image force

You will recall from elementary electrostatics that a charge outside a conducting plane has a field on it equivalent to that produced by a fictitious image charge, as sketched in figure 1.23(b). The corresponding potential felt by the electron, $V(z) = -e/4z$. For a dielectric, with permitivity ε, there is also a (reduced) potential $V(z) = -(e/4z)(\varepsilon-1)/(\varepsilon+1)$. It is often useful to think of metals as the limit $\varepsilon \to \infty$, and vacuum as $\varepsilon \to 1$. Typical semiconductors have $\varepsilon \sim 10$, with $\varepsilon = 11.7$ for Si and 16 for Ge; so semiconductors and metals are fairly similar as far as dielectric response is concerned, even though they are not at all similar in respect of electrical conductivity.

1.5.7 Screening

The above description emphasizes the importance of screening, in general, and also in connection with surfaces. We can also notice the very different length scales involved in screening, from atomic dimensions in metals, $(2k_F)^{-1}$, increasing through narrow and wide band gap semiconductors to insulators, and vacuum; there is no screening (at our type of energies!), unless many ions and electrons are present (i.e. in a plasma). In general, nature tries very hard to remove long range (electric and magnetic) fields, which contribute unwanted macroscopic energies. We will come back to this point, which runs throughout the physics of defects; in this sense, the surface is simply another defect with a planar geometry.

Further reading for chapter 1

Adamson, A.W. (1990) *Physical Chemistry of Surfaces* (Wiley, 5th Edn) chapters 2 and 3.

Blakely, J.W. (1973) *Introduction to the Properties of Crystal Surfaces* (Pergamon) chapters 1 and 3.

Clarke, L.J. (1985) *Surface Crystallography: an Introduction to Low Energy Electron Diffraction* (John Wiley) chapters 1, 2 and 7.

Desjonquères, M.C. & D. Spanjaard (1996) *Concepts in Surface Physics* (Springer) chapter 1.

Gibbs, J.W. (1928, 1948, 1957) *Collected Works, vol. 1* (Yale Univerity Press, New Haven); reproduced as (1961) *The Scientific Papers, vol. 1* (Dover Reprint Series, New York).

Henrich, V.E. & P.A. Cox (1996) *The Surface Science of Metal Oxides* (Cambridge University Press) chapters 1, 2.

Hudson, J.B. (1992) *Surface Science: an Introduction* (Butterworth-Heinemann) chapters 1, 3–5, 17.

Kelly, A. & G.W. Groves (1970) *Crystallography and Crystal Defects* (Longman) chapters 1–3.

Lüth, H. (1993/5) *Surfaces and Interfaces of Solid Surfaces* (3rd Edn, Springer) chapters 3, 6.1, 6.2.

Prutton, M. (1994) *Introduction to Surface Physics* (Oxford University Press) chapters 3 and 4.

Sutton, A.P. & R.W. Balluffi (1995) *Interfaces in Crystalline Materials* (Oxford University Press) chapter 5.

Problems for chapter 1

These problems test ideas of bond counting, elementary statistical mechanics, diffusion and surface structure. When set in conjunction with a course, they have typically not been done 'cold', but have been used to open a discussion on topics which are best attempted through problem solving rather than by lecturing. Note that there are further problems of a similar type in Desjonquères & Spanjaard (1996, chapters 2 and 3).

Problem 1.1. Bond counting and surface (internal) energies of a static lattice

Consider the (012) face on a Kossel (simple cubic) crystal with six nearest neighbor bonds.

(a) Use the analysis of section 1.2 to consider the surface energy of this crystal in terms of the sublimation energy L and the lattice parameter a. Find the ratio of the surface energy of the (012) and the (001) face.
(b) Repeat this exercise for the (012) face of a f.c.c. crystal with 12 nearest neighbor bonds. Compare your result with figures 1.6 and 1.8, and comment on the relative values.

Note: this problem can be done most readily by drawing the structure and counting bonds. There is a more general vector-based approach by MacKenzie et al. (1962), but this is not simple for a first try, or for complex structures. If the stereograms (figure 1.6) are not familiar, see Kelly & Groves (1970) or another crystallography book, or obtain web-based information via Appendix D.

Problem 1.2. Local equilibrium at the surface of a crystal at temperature T

Consider the (001) face of an f.c.c. crystal with 12 nearest neighbor bonds, and (small concentrations of) adatoms and vacancies at this surface. The sublimation energy is 3eV and the frequency factor is 10 THz. Use the appropriate formulations of section 1.3 to do the following.

(a) Construct a differential equation to describe the processes of arrival of atoms from, and re-evaporation into, the vapor, to find the equilibrium concentration of adatoms in monolayer (ML) units. Find the adatom concentration at $T = 1000$ K if the arrival rate $R = 1$ ML/s.
(b) Use the chemical potential formulation to express the local equilibrium between the bulk crystal and the surface adatoms, to obtain their equilibrium concentrations at the same temperature, ignoring arrival from, or sublimation to, the vapor. Hence decide whether the case (a) corresponds to under- or over-saturation, and calculate the thermodynamic driving force in units of kT.

Problem 1.3. Effects of vacancies and/or lattice vibrations on the sublimation pressure

Consider the model of the vapor pressure of a solid described in section 1.3, table 1.1 and figure 1.9. This model neglects the effects of vacancies, and the model of lattice vibrations is only a first approximation.

(a) How might you consider the effects of vacancies, which are expected to have an energy of 1 eV in Ag, and reduce the frequency of atomic vibration in the vicinity of the neighbors of the vacancy to 80% of the value in the bulk?

(b) How might you consider the effect of other lattice dynamical models, for example the cell model, discussed in more detail in chapter 4.2?

Note: this problem is useful for a discussion of points of principle and practicality, and could be expanded via detailed computation for a course project.

Problem 1.4. Crystal growth at steps and the condensation coefficient

Consider a surface consisting of terraces of width d, separated by monatomic height steps.

(a) Set up a one-dimensional rate-diffusion equation describing the diffusion of adatoms to the steps in the presence of both adatom arrival and desorption. Explain what boundary conditions you use at the steps.

(b) Show that the steady state profile of adatoms between the steps depends on the ratio $\cosh(x/x_s)/\cosh(d/2x_s)$, where x_s is the BCF length (section 1.3). Show that the fraction of atoms which get incorporated into the steps, the condensation coefficient, is given by $(2x_s/d)\tanh(d/2x_s)$. Evaluate the limits $(2x_s/d) \gg 1$ and $\ll 1$, and give reasons why these limits are sensible.

Problem 1.5. Surface reconstructions of particular crystals

Consider a surface structure in which you are interested. In metals this could be W and Mo(100) which have transitions below room temperature to 2×1, and 2×1-like structures, or in semiconductors the difference between 2×1, $c4 \times 2$ and $p\,2 \times 2$ superstructures on Si or GaAs(100). Use your chosen system to explore the relation between the structure, the symmetry and size of the surface unit cell, and the diffraction pattern, most obviously the LEED patterns in the literature.

2 Surfaces in vacuum: ultra-high vacuum techniques and processes

This chapter presents a practically oriented introduction to modern vacuum techniques, in the context of studying surface and thin film processes. The following sections review the science behind the technologies, and give a few worked examples, which we refer to later. Section 2.1 reviews the kinetic theory concepts on which vacuum systems are based; section 2.2 outlines the basic ideas involved in ultra-high vacuum (UHV) system design. The next section 2.3 deals with vacuum system hardware, in order to make sense of the large range of chambers, flanges, pumps and gauges which make up a complete system. It is assumed that the reader is already familiar with a basic vacuum system and its components, so that only a few figures of apparatus are needed. Section 2.4 describes the procedures used in performing experiments under UHV conditions, and discusses some of the challenges involved in scaling these procedures up to manufacturing processes. Finally, section 2.5 briefly lists some of the more commonly used thin film deposition techniques, and describes where more information on such processes can be found.

2.1 Kinetic theory concepts

2.1.1 Arrival rate of atoms at a surface

The arrival rate R of atoms at a surface in a vacuum chamber is related to the molecular density n, the mean speed of the molecules $\bar{v}$ and the pressure p, via the standard kinetic theory formulae (Dushman & Lafferty 1992, Hudson 1992)

$$R = n\bar{v}/4 \text{ per unit area} = p/(2\pi mkT)^{1/2} ; \tag{2.1}$$

you may also need $n = p/kT$ and $\bar{v} = (8kT/\pi m)^{1/2}$, which are required to connect the two versions of the above formula. The notation can be confusing: R is sometimes called the deposition flux, F, and $\bar{v}$ is sometimes written v, c or $\bar{c}$.

Now let us work through an example to find the molecular density n, the mean free path λ, and the monolayer arrival time, τ. We take, as a typical example, the residual gas in a vacuum system, which is often a mixture of CO, H_2 and H_2O; the following calculations are for carbon monoxide, CO, which has molecular weight 28. Then the molecular mass, $m = 28 \times 1.6605 \times 10^{-27}$ kg; Boltzmann's constant $k = 1.3807 \times 10^{-23}$ J/K; and $T = 293$ K (UK, if you're lucky), or 300 K (Arizona, ditto); we shall also need the molecular diameter of CO, $\sigma = 0.316$ nm.

The question of units, especially of pressure, is important. The SI unit is the pascal (Nm^{-2}). One bar $= 10^5$ Pa, and modern vacuum gauges are calibrated in millibar: 1 mbar $= 100$ Pa. The older unit torr (mm Hg) is named after the inventer of the mercury barometer, Torricelli, who worked with Galileo in the seventeenth century: 1 Torr $=$ 1.333 mbar (760 Torr $= 1013$ mbar $= 1$ atmosphere). These units and conversion factors are collected in Appendix C.

2.1.2 The molecular density, n

At low pressures $n = Ap$. With n per cm^3 and p in mbar, we have the constant $A = (100)/(kT \times 10^6)$. This gives

$$n = p/kT = 7.2464 \times 10^{18} p/T. \tag{2.2}$$

Roth (1990, chapter 1) for example, has suitable diagrams and tables which spell this relationship out for air at 25 °C. Don't forget that in all these equations, T is the absolute *Kelvin* temperature (K): $T(\text{K}) = T(°\text{C}) + 273.15$. For our example of CO, a typical number to get hold of is that at 10^{-6} mbar there are 2.42×10^{10} molecules/ cm^3 in Arizona and 2.47×10^{10} in the UK. Just checking: it's the temperature! There are still lots of molecules around, even in the best vacuum.

2.1.3 The mean free path, λ

The mean free path between molecular collisions in the gas phase is inversely related to the density n and the molecular cross section proportional to σ^2. The proportionality constant f in the equation $\lambda = f/n\sigma^2$ was solved by Maxwell in 1860 (for a historical account see Garber *et al.* 1986): $f = 1/\pi\sqrt{2} = 0.225$. Thus for CO, with σ^2 almost exactly equal to 0.1 nm^2, we have

$$\lambda = 2.25 \times 10^{14}/n \text{ (cm)}, \tag{2.3}$$

where n is expressed as in (2.2); the mean free path at 10^{-6} mbar is of order 100 m at room temperature. Thus λ is much greater than the typical dimensions of a UHV chamber, operating, say, below 10^{-9} mbar; the gas molecules will travel from wall to wall, or from wall to sample, without intermediate collisions.

Higher pressure gas reactors, operating at 10^{-3} mbar and above, start to run into gas collision and diffusion effects, but the UHV community largely ignore this, except for particle accelerators where particles circulate at close to the velocity of light for many hours. At a large installation, such as CERN or FermiLab, the accelerated particles are constrained to miss the walls, but of course they hit the residual gas molecules. There are other effects, such as high power (up to several kW/m of path) synchrotron radiation produced when the beam travels in a circle, which desorbs molecules from the walls;

on sufficiently long timescales, this initially bad effect can be turned to advantage, in the form of beam cleaning. One of the challenging aspects of the (late) superconducting supercollider was how to design a toroidal pipe some 80 km long and say 0.15 m in diameter with a vacuum everywhere better than 10^{-12} mbar. The large electron–positron (LEP) storage ring at CERN (*only* 26.7 km in circumference) has an only marginally less severe specification, and beam conditioning effects are an important aspect of the operation (Reinhard 1983, Dylla 1996). Vacuum design and procedures have to be taken rather seriously!

2.1.4   *The monolayer arrival time, τ*

If N_0 is the number of atoms in a monolayer (the ML unit) then $R\tau = N_0$. We have already had that $R = Cp$ with $C = (2\pi mkT)^{-1/2}$. Now we need the conversion from millibars to pascal, T and the other constants. For CO, R in atoms·m^{-2}·s^{-1}, and p in mbar, C is then 2.876×10^{24} for $T = 300$ K, or $R = 2.876 \times 10^{18}$ atoms·m^{-2}·s^{-1} at 10^{-6} mbar, i.e. of order 3×10^{14} at 10^{-10} mbar, a typically (good) UHV pressure. Watch out for whether cm^{-2} are used in place of m^{-2} as (area)$^{-1}$ units; factors of 10^4 are significant!

The definition of N_0 requires above all consistency. It can be defined in terms of the substrate, the deposit or the gas molecules, but it must be done consistently, and the ML unit needs definition, essentially in each paper or description: there is no accepted standard. For example, consider condensation on Ag(111), with a (1×1) structure. It is perfectly reasonable to define N_0 as the number of Ag atoms per unit area. With the bulk lattice parameter $a_0 = 0.4086$ nm, the surface mesh area is $(\sqrt{3}/2)a^2$, where the surface lattice constant $a = a_0/\sqrt{2}$. Thus $N_0 = 1.383 \times 10^{19}$ atoms·m^{-2}. With this definition, $\tau = 4.81$ s at 10^{-6} mbar (CO) and 13.4 hours at 10^{-10} mbar. This is, of course, the reason for doing experiments in UHV conditions; only at low pressures can one maintain a clean surface for long enough to do the experiment.

However, the above definition of the monolayer arrival time only makes sense if we have a well-defined substrate. If the substrate N_0 is ill-defined or irrelevant (e.g. the inside of a stainless steel vacuum chamber, or for an incommensurate deposit), then a definition in terms of the deposit makes more sense. In our case we might use a close-packed monolayer of condensed CO; with $a = 0.316$ nm, the corresponding values of $\tau = 4.02$ s at 10^{-6} mbar (CO) and 11.2 hours at 10^{-10} mbar. Although these are of the same order, they are not the same. Thus for quantitative work, it is important *either* to define the ML unit explicitly, *or* to work with a value of R expressed in atoms·m^{-2}·s^{-1}, rather than in ML/s. Note also that had the deposit been something other than CO, and we wanted to track the result in terms of pressure, then we have to use the correct m and T in the constant C.

To summarize: (1) the density n is still high even in UHV; (2) the mean free path $\lambda \gg$ apparatus dimensions; (3) the monolayer arrival time τ is greater than 1 h only for $p < 10^{-9}$ mbar; and for good measure (4) the monolayer (ML) unit, if used, needs to be defined consistently. The various quantities calculated in this section are displayed in figure 2.1.

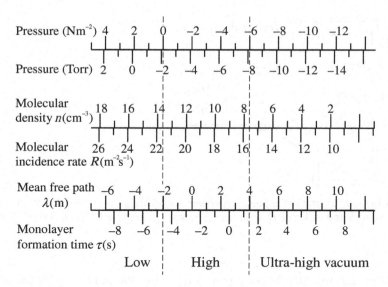

Figure 2.1. Plot of n (cm^{-3}), R (atoms·m^{-2}·s^{-1}), λ (m) and τ (s) for CO at temperature $T = 300$ K, as a function of pressure p, on a logarithmic scale in units of 1 mbar = 100 Pascal or Nm^{-2}, and the older unit 1 Torr = 1.333 mbar. The division into low, high and ultra-high vacuum regimes are approximate terms based on usage.

2.2 Vacuum concepts

2.2.1 System volumes, leak rates and pumping speeds

The system to be pumped has a system volume, V, measured in liters, at pressure p (mbar or torr), as indicated schematically in figure 2.2. It is pumped by a pump, with a pumping speed, S (liter/s). The pump-down equation for a constant volume system, with a leak rate Q into the system, is then:

$$pS = -V dp/dt + Q. \tag{2.4}$$

The leak rate is composed of two elements: $Q = Q_1 + Q_o$, where Q_1 is the *true* leak rate (i.e. due to a hole in the wall) and Q_o is a *virtual* leak rate. A virtual leak is one which originates inside the system volume; it can be caused by degassing from the walls, or from trapped volumes, which are to be avoided strongly.

The solution of the pump-down equation, assuming everything except the pressure is constant, separates into:

(a) a short time limit: $p = p_0 \exp(-t/\tau)$, with $\tau = V/S$, (2.5a)

where the leak rate is negligible. This stage will be essentially complete in 10τ. Typical values 10×50 liter/ 50 liters/s = 10 s. It isn't quite this short in practice, but it is short;

(b) a long time limit: p_u, the ultimate pressure = Q/S. (2.5b)

When the true leak rate is negligible, $Q \rightarrow Q_o$, which depends on the surface area A, material and the treatment of the surface. For example, if the system volume

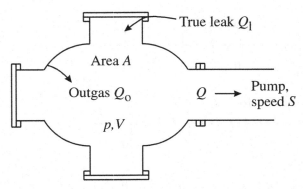

Figure 2.2. Schematic diagram of a pumping system, comprising the volume V, internal area A, pumping speed S and leak rate Q, comprising outgas Q_o and true leaks Q_1. See text for discussion.

$V = 50$ liter (e.g. $50 \times 20 \times 50$ cm^3), then A is ~ 1 m^2. We can take $Q_o = qA$, so with a typical (good) value for q around 10^{-8} mbar·liter·m^{-2}·s^{-1}, we deduce from (2.5b) that $p_u = 2 \times 10^{-10}$ mbar.

This value is a typical pressure to aim for after bakeout. The bakeout is required to desorb gases, particularly H_2O, from the walls. Water is particularly troublesome because it is always present and desorbs so slowly; it is essentially impossible (with standard stainless steel/glass systems) to achieve pressures below 10^{-8} mbar in a sensible time without baking. The role of water vapor in vacuum systems has been reviewed by Berman (1996). Practical bakeout procedures are indicated by Lüth (1993/5 Panel I) and Yates (1997) with comments here in section 2.3.

It is worth remembering, throughout this chapter, that both design and preparation procedures are lengthy; they can be disastrous, and very expensive, if they are not thought through or go wrong. Thus we should give even the simple models described here due respect! It is also important not to use these simple calculations blindly, and to check with experts who have a feel for the points which are difficult to quantify. An example is the following.

In applying the pump-down equation (2.4), there is some possibility of confusion, as it can be used too uncritically, and used to deduce answers which run counter to practical experience. To deduce, via (2.5a) above, that you can get down to 10^{-6} mbar, say, in a minute or so, is not correct. It is, however, correct to deduce that in that time the term $-V \mathrm{d}p/\mathrm{d}t$ becomes smaller in magnitude than $+Q$; but Q itself varies (decreases) with time, as the walls outgas, and S is also, in general, a function of pressure. This means that for almost all UHV situations we are interested in the long time limit (2.5b), but with variable Q, depending on the bakeout and other treatments of the vacuum system, and with variable S, depending on the type of pump and the pressure. Manufacturers' catalogues typically give a plot of how S varies with p. As the pump approaches its ultimate pressure limit, the speed S drops off to an ineffective value.

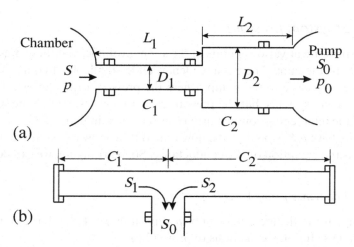

Figure 2.3. The effect on pumping speeds S of pipe conductances C_i: (a) in series, and (b) in parallel with a pump of speed S_0.

2.2.2 The idea of conductance

The pumping speed of the pump is reduced by the high impedance, or low conductance, of the pipework between the pump and the vacuum chamber, as shown in figure 2.3. The conductance of a pipe is defined as the flow rate through it, divided by the pressure difference between the two ends. But, of course, large pipes increase both the system volume and the internal surface area. So, one needs to take care in the design of the system, to avoid obvious pitfalls. Individual pipes have a conductance C_i, and several of these, with different lengths and diameters, may be in series with the pump. Then with the pump speed as S_0, we have the effective pumping speed S at the chamber given by

$$S^{-1} = \Sigma_i C_i^{-1} + S_0^{-1}, \tag{2.6}$$

where the C_i are measured in liters/s. Thus we need to choose the C_i large enough so that S is not $\ll S_0$; or equivalently, if S is sufficient, we can economize on the size (S_0) of the pump. As with all design problems, we need to have enough in hand so that our solution works routinely and is reliable. On the other hand, over-provision is (very) expensive. We consider actual values of C in section 2.3 and Appendix F.

Sometimes, if high pumping speed is essential, or if the geometrical aspect ratio is unfavorable (as in the accelerator examples cited in section 2.2.1), we would use multiple pumps distributed along the length of the apparatus. In this case the conductances are distributed in parallel, and

$$S = \Sigma_i S_i, \ C = \Sigma_i C_i. \tag{2.7}$$

Whether this is a good solution should be clear from geometry. Obviously, a solution involving one UHV pump is simpler, if possible. Sometimes we use more than one pump because different pumps have different characteristics, as described in section 2.3.

2.2.3   Measurement of system pressure

One can measure pressure at various points in the system volume, and these pressures will not necessarily be the same. There is a tendency, understandable of course, to put the pressure gauge very close to the pump, where the pressure (p_0) is lowest. This is useful for tests on small systems, but the pressure p in a large system volume will be worse because of the intervening conductance C, as indicated in figure 2.3.

By continuity, we have $p_0 S_0 = pS = Q$, the flow rate. But the flow rate is also equal to $C(p - p_0)$, as this is the definition of the conductance. So, rearranging, we can deduce that

$$p - p_0 = p_0 S_0 / C \text{ and hence } p = p_0 (1 + S_0 / C). \tag{2.8}$$

Thus there is a big error in the measurement of p at position p_0, if S_0 is large and/or C small. One can also use the above relations to prove (2.6).

Note that in general both S and C can be functions of pressure. In the molecular flow regime, at low p where the gas molecules only collide with the walls, and where we are not near the ultimate pressure of the pump, then they are, in fact, both constant. The parameter most commonly used to distinguish different flow regimes is the Knudsen number K_n, which is simply the ratio of the the mean free path λ to the the pipe diameter D. Molecular flow, considered here, is valid for $K_n > 1$. The viscous flow regime, valid at high pressures, arises when $K_n < 0.01$. There is an intermediate regime, named after Knudsen, when $0.01 < K_n < 1$, in which the flow is turbulent. Backing pumps may operate in the turbulent regime during the initial pump down phase from atmospheric pressure, with $Q > 200D$ (D in cm), but otherwise this regime is unimportant for UHV systems (Roth 1990, Delchar 1993).

2.3   UHV hardware: pumps, tubes, materials and pressure measurement

2.3.1   Introduction: sources of information

It is not really possible to do justice to the subjects of hardware and experimental design practices in a book; it takes too much space, so let me first comment on additional sources of information. Of the general surface science textbooks, Lüth (1993/5) has used his Panel I to convey the feel of UHV equipment. Detailed books on vacuum technology exist, including O'Hanlon (1989), Roth (1990) and Dushman & Lafferty (1992), and there are several concise summaries including Delchar (1993) and Chambers *et al.* (1998). A highly regarded general text on experimental design is that by Moore *et al.* (1989), which has a chapter on vacuum and a short section on UHV design. A useful compendium of designs and know-how for experimental surface science has been compiled by Yates (1997).

Manufacturers' catalogues are useful, assuming that you know that they are attempting to get you to buy something (in the long run). Although all such catalogues provide detailed information about the products, the Leybold-Heraeus catalogue has traditionally included a tutorial section which helps one understand what the products are doing, and what choices the purchaser needs to make. Relatively small performance improvements in vacuum components can cause quite a commercial stir. So one always needs to consider what the latest model is really doing. The physical principles on which these devices are based are emphasized here, in the hope that these do not change too fast.

2.3.2 Types of pump

There are many types of pump, but the ones used to create UHV conditions are typically one or more of the following: turbomolecular, diffusion, ion or sputter ion, sublimation or getter, or cryo-pumps. In choosing a pump for a system, you need to know, first of all, its general characteristics.

Turbopumps are extremely useful general purpose pumps, with high throughput, and produce a pressure ratio between their input and output ends. They are poor for low mass molecules, especially hydrogen, because they work by giving an additional velocity, in the required direction, to the molecules, and thus are less effective when the molecular speed is high. The ultimate pressure depends on the backing pressure, and so p_u can be improved using two pumps in series. There are newer versions with magnetic levitation bearings which make the pumps contamination free and much quieter than earlier versions. The rotor of a small pump typically turns at over 100 000 revs/min, with tip speeds in excess of 250 m/s; these high speeds means that the lightness and tensile strength, as well as the geometric form of the rotor blades are important materials parameters (Becker & Bernhardt 1983, Bernhardt 1983). Turbopumps are used extensively in semiconductor manufacturing facilities, the 'Fabs' of the silicon age. UHV pumps constitute a major cost of these facilities. There is an active current effort (Helmer & Levi 1995, Schneider *et al.* 1998) in modeling the performance of such pumps, with the goal of making less expensive (rather than simply more powerful) turbopumps for future facilities.

Diffusion pumps are the workhorses of standard high vacuum systems. For UHV use, they are always fitted with a liquid nitrogen cooled trap, in order to stop oil entering the vacuum chamber. This trap is situated behind a valve that can be sealed off should the trap need to be warmed up, or if any disaster occurs. One of the claims in favor of diffusion pumps is that the cost for a given pumping speed is lower than for other types of pump; they also pump hydrogen and helium well.

Ion, sputter-ion, sublimation, getter and cryo-pumps are characterized as capture pumps, since they trap the gas inside the system (Welch 1994). Thus they are not good if there is a heavy gas load, but can be very good for a static vacuum under clean conditions. Chemical pumps comprise those capture pumps which work primarily via chemical reactions at the internal surfaces; these pumps are poor for rare gases. Getters are chemical pumps which have been traditionally been used in static vacua such as lamp bulbs, cathode ray and TV tubes, and they are also used in accelerators such as LEP (Reinhard 1983, Ferrario 1996).

Cryopumps have very high speed, but produce vibration from the closed cycle displacer motor used for refrigeration, and are quite expensive. Specific characteristics of all these pumps can perhaps best be assessed by visiting a laboratory or facility, or by visiting a trade show at a conference. The important points to understand in advance are the principles of their operation.

2.3.3 Chambers, tube and flange sizes

The second type of decision concerns pumping speed and flange sizes. These design requirements affect the size, weight and cost, and via these factors, the viability of the apparatus. Tubes and flange sizes are standard, as can be seen from the manufacturers' catalogues. The standard sizes are in Appendix F, and their conductance per meter length, and with a 10 cm end into the chamber, is given. The conductance calculations are then sufficient to make estimates which will enable you to sketch a reasonable design. Then, one typically needs to discuss it with someone who has done a design previously; it may be the most important factor in your experiment, and should not be done blind.

Useful formulae for conductances in the molecular flow regime are the following. For an aperture, diameter D (cm),

$$C = 2.86 \, (T/M)^{1/2} \, D^2 \text{ liter/s,} \tag{2.9}$$

with M the molecular weight, and T (K). For a long or short pipe, with length L in cm also,

$$C = 3.81 \, (T/M)^{1/2} \, D^3/(L + 1.33D) \text{ liter/s.} \tag{2.10}$$

The flanges are typically made of stainless steel, and are sealed with copper gaskets. They are loosely referred to as conflat flanges, though Conflat® is a trade mark of Varian, Inc.; they are available from many suppliers. These tubes/flanges are referred to as ports on the central chamber. Even if one has access to a good machine shop, it is not particularly cost-effective to try to make one's own vacuum components: there are several specialist firms who make tubes, flanges and chambers on a routine basis in both North America and Europe, at least. What we then have to do is to 'pick and mix' accessories for our needs, typically around a special chamber which has been designed for the job in hand, and made by one of these firms. Many of the accessories relate to particular measurement or sample handling techniques, which are the subject of section 2.4.

We need to match pump speeds to pipe dimensions and conductances, as set out in Appendix F, table F2. A rule of thumb is that you need 1 liter/s of pump speed for every 100 cm^2 of wall area. This can be seen by taking a value for q (see section 2.2.1) = 2×10^{-12} mbar·liter·s^{-1}·cm^{-2} or 2×10^{-8} mbar·liter·s^{-1}·m^{-2}, which is a reasonably conservative design figure.

Both sublimation and cryopump designs can trap a large fraction of the gas which enters the throat of the pump; in practice certainly greater than a quarter. This means that $S > C/4$, where C is the aperture conductance, which can be high, e.g. for 4–8 inch diameter pipes in the range 400–3000 liter/s, depending on the precise flanging arrangements. A titanium sublimation pump (TSP) chamber can be designed relatively easily

Table 2.1. *Manufacturer's quoted information for TSP pumping speeds (liter·s⁻¹·cm⁻²)*

Wall T	H_2	N_2	O_2	CO	CO_2	H_2O	CH_4/Inert
20 °C	3	4	2	9	8	3	0
77 K	10	10	6	11	9	14	0

for these needs, using data from table 2.1. For example, a tube 20 cm long and 20 cm (8 inch) diameter has an internal area of about 1200 cm². If we take a pessimistic view that, with the wall at 20 °C, the average for all relevant gases is 2 liter·s⁻¹·cm⁻², this still gives us a pumping speed $S = 2 \times 1200 = 2400$ liter/s, which is quite large enough to be greater than, or around, $C/4$ for a reasonable pump aperture.

But we should note some other points too. A TSP outgasses when it is being 'fired', and the pressure therefore goes up before coming down; if the walls are too close to the hot filament, this problem is worse. Second, these pumps do not pump unreactive gases at all well. Cooling the walls with liquid nitrogen helps; even water cooling is quite effective in improving the performance, but it still doesn't pump unreactive gases. The result of such concerns means that you should not economize on the wall area, and you should use a somewhat larger diameter tube than you might calculate on the simplest basis.

2.3.4 Choice of materials

UHV experiments require the use of materials with low vapor pressures, and it is helpful to have your own notes and diagrams which give you easy access to such information (see Appendix G for some pointers). Since the outgas leak rate $Q_o = qA$, we should use low q materials, and minimize the area A of the design. As materials and accessories have improved there is a tendency to put more and more equipment into the vacuum system. This may make life more difficult in the long run: to try to do everything often means you may achieve nothing.

There are lists of q (sometimes called q_d to represent desorption) values for different materials and treatments in vacuum books and review articles, but they need to be treated as general guides only. If you need to make measurements of q for particular materials, this is not without pitfalls (Redhead 1996), and is rarely done for small scale applications. The main materials, stainless steel, copper, aluminum, ceramics, all produce values below 2×10^{-12} mbar·liter·s⁻¹·cm⁻² after a modest bakeout at around 200 °C for 12–24 h. These values are satisfactory for most purposes, and the trend is to avoid more stringent bakeouts at higher temperatures or over longer times.

It is imperative that you know what is in your system before you bakeout, or this important stage in your experiment may cause irreversible damage, and repairs may be very expensive. In particular, *do not* bakeout your system to temperatures which seem routine from the research literature (Redhead *et al.* 1968, Hobson 1983, 1984)! Some equipment contains materials, particularly high temperature plastics, e.g. for insulating electrical

Table 2.2. *Classification of vacuum gauges*

Physical property involved	Kind of gauge	Kind of pressure recorded
(1) Pressure exerted by the gas	Bourdon, capacitance	Total pressure, all gases
	McCleod (gas compression)	Pressure, non-condensable gas
(2) Viscosity of the gas	Spinning rotor	Total, depends on gas
(3) Momentum transfer	Radiometer, Knudsen gauge	Total, ~independent of gas
(4) Thermal conductivity	Pirani, thermocouple gauge	Total, depends on gas
(5) Ionization	Bayard–Alpert gauge	Total, depends on gas
	Partial pressure analyzers	Partial pressure

wires, which are very sensitive to the exact bakeout temperature, say between 150 and 220°C. Despite this caution, the availability of such plastics, coated wires, and even electric motors which work under UHV, has made surface science techniques much more widespread and routine.

2.3.5 Pressure measurement and gas composition

As with pumps, the practitioner needs to know what the different types of gauge can do, and what principles they are based on. There are three general purpose gauges for accurate pressure measurement: the ion gauge, the Pirani gauge and the capacitance gauge. The ion gauge works by ionization of the gas molecules, and the fine wire collector reduces the low pressure limit due to X-ray emission of electrons, which mimics an ion current. It should only be used below 10^{-1} mbar, works well below 10^{-3} mbar, and has a lower limit typically below 10^{-11} mbar, depending on the design. The cold cathode (Penning) gauge also works by ionizing the gas molecules, and works over the range $5\cdot10^{-8}$ to 10^{-2} mbar; but it also functions as a sputter ion pump to some extent, and so the pressure tends to be underestimated.

The Pirani gauge utilizes the thermal conductivity of the gas molecules, and works over a range from about 10^{-3} to 10^2 mbar; it typically is used for semi-quantitative monitoring of the fore-vacuum. A capacitance gauge is extremely precise above 10^{-4} mbar, but requires different heads for different pressure ranges. This is sometimes referred to as a baratron, but (spelt with a capital B) this is the trade name of a company making such equipment; these gauges are used very widely in all aspects of pressure measurement, process and flow control, for example in chemical vapor deposition (CVD) reactors. An outline description of such process equipment is given by Lüth (1993/5, section 2.5); more details are given in various sections of Glocker & Shah (1995).

A list of such gauges is given in table 2.2. There are some relatively new ones, including the spinning rotor gauge, based on gas viscosity, which has been developed and marketed over the last ten years. To find out more about such a development requires a two-pronged approach. One needs manufacturer's catalogues to find out what is actually available commercially. The second line of enquiry is to search the vacuum

Table 2.3. *Typical ion-gauge sensitivities relative to nitrogen*

H_2	D_2	He	H_2O	CH_4	Ne	CO	N_2	C_2H_6	O_2	Ar	CO_2	Kr	Xe
0.6	0.4	0.25	0.86	1.4	0.29	1.1	1.0	2.8	0.8	1.4	1.45	1.86	2.7

Note: True pressure = indicated pressure divided by sensitivity quoted.

journals: further development of the spinning rotor guage is described by Bentz *et al.* (1997) and Isogai (1997). The basic high and ultra-high vacuum gauge is still the ionization gauge, developed originally by Bayard & Alpert (1950) as described, for example, by Redhead *et al.* (1968). Commercial gauges are typically calibrated for N_2. Other gases have different sensitivities, as set out in table 2.3.

The determination of gas composition is also very important, and is typically done with a compact mass spectrometer known as a residual gas analyzer, or RGA. This produces a characteristic mass spectrum, as in the example shown in figure 2.4(a), taken from an American Vacuum Society educational monograph (Drinkwine & Lichtman 1979). A more recent example at better pressure after bakeout is shown in figure 2.4(b). It is helpful to record such spectra, and to store examples of when your system is working well, as the spectrum when you have a real leak is typically quite different from if you have performed an inadequate bakeout, or have let unwanted or corrosive gases into your system. As we have implied in section 2.1, the vacuum composition for a well outgassed system is typically dominated by H_2, CO and H_2O, very different from the atmosphere (see Table 1.3 in Roth 1990). With a real leak, the O_2 peak at mass 32 is much higher than in these examples, where it is very, or unmeasurably, small. The second spectrum also shows that some peaks around mass 62 are due to reactions with the hot filament in the ion source of the mass spectrometer, in this case Re^{3+} ions.

Most of the less expensive RGAs are based on a quadrupole mass spectrometer, or QMS, whose principle is explained by Lüth (1993/5, Panel 4) and by Moore *et al.* (1989, section 5.5). Higher mass resolution is obtained in more specialized magnetic sector or time of flight instruments (Duckworth *et al.* 1986), which are typically attached to specialist facilities for cluster research, secondary ion mass spectrometry, atom probe microanalyis, or isotope dating (e.g. in archaeology). In these latter cases, the mass spectrometer represents a major fraction of the overall cost of the equipment.

2.4 Surface preparation and cleaning procedures : *in situ* experiments

2.4.1 *Cleaning and sample preparation*

There are two aspects of cleaning: (a) cleaning of sample chambers, pieces of equipment; and (b) sample cleaning. The first is a rather obvious combination of dirt removal, degreasing, ultrasonic rinsing, use of solvents, etc. This requires care, and is

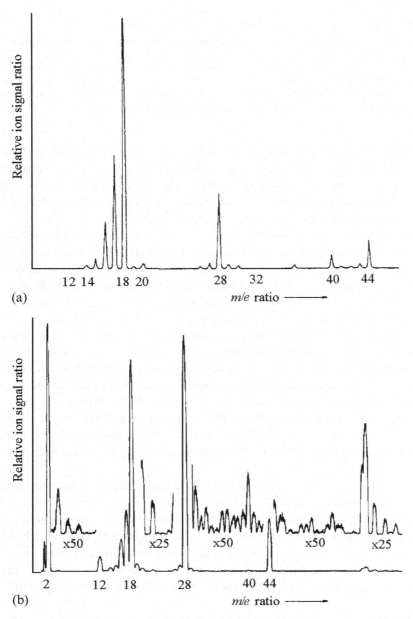

Figure 2.4. QMS spectra of (a) a 20 liter laboratory UHV system, pumped with an ion pump. The pressure $p \sim 3 \times 10^{-7}$ torr before bakeout, with large water derived peaks (16–18), plus $CO + N_2$ (28), CH_4 (16), CO_2 (44), Ar (40) and Ne (20) the next most prevalent gas phase species (after Drinkwine & Lichtman 1979, replotted with permission); (b) a larger multichamber system shown in figure 2.5, pumped with ion and Ti-sublimation pumps at $p = 5 \times 10^{-11}$ torr. This spectrum, after several days' bakeout at up to 180 °C, has peaks at 2(H_2), 16–18 and 28. The high-end peaks with mass numbers in the 60s are from reactions in the ion source of the mass spectrometer, in this case Re^{3+} ions (from Zeysing & Johnson 1999, reproduced with permission).

time-consuming; it is a clear candidate for 'more haste less speed', since it is essential to be systematic; thinking that this is a 'low-level' activity which you should be able to race through does not help. Cultivate high level thought in parallel, but *concentrate on the details*. A discussion of possible sets of prescriptions is given in Appendix H.

The second type of cleaning is very specific to the material concerned, and to the experiment to be performed. Indeed it may be most helpful to think of it as the first stage of the experiment itself, rather than as a separate cleaning operation. For example, in semiconductor processing under UHV conditions, where there are many such cleaning and preparation stages, 'clean' means 'good enough so that the next stage is not messed up'. Thus, acting quickly, transferring under inert gas, or any trick that will work (i.e. increase throughput/reliability), all count under this heading; there is no absolute standard.

For research purposes the criteria are remarkably similar. Thus a cleaning process which is good enough for one experiment or technique, may not be sufficient for a more refined technique. An example is that the surface has to be reasonably clean at the sub-ML level to give a sharp LEED pattern; however it does not have to be particularly flat. Once people began to examine surfaces by a UHV microscopy technique, it became clear that many of the cleaning treatments employed (e.g. high temperature oxidation followed by a 'flash' anneal) did not produce flat surfaces at all, so it was necessary to reconsider options carefully. Some systems are 'known to be difficult'. This means that a large part of one's (e.g. thesis) time can be taken up with such work, and that the results may well depend on satisfactory resolution of such problems.

The various possibilities for sample cleaning include the following: heating, either resistive, using electron bombardment or laser annealing; ion bombardment; cleaving; oxidation; *in situ* deposition and growth. These may be applied singly, or more often in combination or in various cycles. Typically, the first time a sample is cleaned, the procedure is more lengthy, or more cycles are required. Thereafter, relatively simple procedures are needed to restore a once-cleaned surface, provided it has been kept under vacuum.

Two examples will be sufficient to give the flavor of such UHV preparation treatments, which typically follow specific external treatments including cutting, X-ray orientation, diamond, alumina and/or chemical polishing and degreasing.

(i) W and Fe(110)

The b.c.c., close-packed, W(110) substrate has been used many times because it was possible to clean it reproducibly. Fe(110), which is arguably more interesting, is more difficult because of its reactivity and internal impurities. Both substrates can be cleaned on a holder equipped for electron bombardment of the rear side of the sample. Tungsten is typically cleaned by heating in oxygen at around 10^{-6} mbar at 1400–1500 °C for around an hour (to convert C and impurities into oxides), alternated with flash heating to 2000 °C to desorb and/or decompose the oxides. Only electron bombardment heating can readily deliver sufficient power density to reach such temperatures.

However, Fe cannot be heated to anywhere near such temperatures, since there is a crystal phase transition (b.c.c. to f.c.c.) at $T = 911$ °C, and one might also be nervous about going above the (ferro- to para-) magnetic phase transition at 770 °C. The solution is typically to use ion bombardment at room temperature, followed by annealing at moderate temperature $T \sim 5$–600 °C. This removes C and O, but promotes surface segregation of sulfur, which is a major impurity in Fe; so a lengthy iterative process is required to reduce S to an acceptable level. This cleaning process is typically monitored by Auger Electron Spectroscopy (AES), as discussed in chapter 3.

(ii) Si and Ge(111)

These semiconductor substrates can be prepared in various ways, and it is known that the equilibrium reconstruction of Si(111) at moderate temperatures is the 7×7 structure (see section 1.4). But temperatures above 900 °C are needed to clean the surface by (resistive or focused high power lamp) heating, and this is above the 7×7 to '1×1' transition at 837 °C. Thus the procedure is typically to heat to say 1000 °C at $< 10^{-9}$ mbar until clean, then cool slowly through the phase transition to allow large domains of 7×7 to grow, followed by a more rapid cool to room temperature. By contrast, the Ge(111) surface, which has the $c2 \times 8$ to '1×1' transition at 300 °C, and has a much more 'mobile' surface, is quite a lot easier to clean. It is less reactive to oxygen, and can be cleaned by heating at 500–600 °C after an initial light ion bombardment, or by cycles of ion bombardment and annealing at around 400 °C.

The above Fe(110) example is described at greater length by Noro (1994) and Noro *et al.* (1995), and there are many other examples locked up in doctoral theses around the world, and in recipes (patented or not), fiercely guarded by firms whose livelihood depends on similar tricks. Discussion often does not appear in article or book form; for this reason, conference proceedings on the topic can offer useful insights (e.g. Nemanich *et al.* 1992, Higashi *et al.* 1993, 1997).

2.4.2 Procedures for in situ *experiments*

Most surface experiments are performed *in situ*, i.e. without breaking the vacuum. The progress of such experiments and manufacturing processes proceeds along the following lines.

(a) Degassing components during and after bakeout. This may apply to masks for deposition, evaporation sources, gauge and TSP pump filaments. The main point is that such equipment will degas during use, worsening the pressure, often directly in the neighborhood of the sample; prior degassing will lessen, but rarely eliminate, these effects. A typical procedure is to leave evaporation sources (say) powered up during the later stages of bakeout, but at a low enough level so as not to cause significant evaporation.

(b) Cleaning the sample and characterizing it for surface cleanliness, typically with AES, for surface crystallography, e.g. by LEED or Reflection High Energy Electron Diffraction (RHEED), and maybe on a microscopic scale using, say Scanning Electron (SEM) or Scanning Tunneling (STM) Microscopy.

(c) Performing the treatment or experiment: deposit/anneal, react with gases, bend the sample, implant a million computer chips, whatever is your field of interest, or current task. *And finally:*

(d) examine the sample with the techniques at your disposal!

One can see why it is useful to think of the cleaning the sample (b above) as the first stage of the experimental *process*, because what you can characterize is determined by what you have bolted onto the system. Even if you have the particular equipment, you might decide not to use it because it takes too long, or doesn't answer the question you are currently asking. And, as implied at the beginning of section 2.3.4, it is helpful not to have too many accessories bolted on to the system at the same time. Not only will the pressure be worse than it might be; *none of the accessories will actually be working when you need them!*

Of course, as in all design problems, the real situation is a balance between competing tendencies; if you change the vacuum and accessory configuration too frequently, you pay a large price in inconvenience, down time and loss of output, measured in either scientific results or material products. But if you don't build in the possibility for configurational changes at the design stage, you risk wasting a very large investment.

2.4.3 Sample transfer devices

Increasingly, UHV-based *in situ* techniques are being applied in engineering and manufacturing situations. Given the availability of quite complex sample transfer devices, whole sequences of surface engineering processes can be performed on samples, as for example in molecular beam epitaxy (MBE) and other (commercial) equipment. Transfer of samples between equipment with a UHV device was first demonstrated by Hobson & Kornelsen (1979), including showing that it was possible to transport the equipment by air across the Atlantic at pressures below 10^{-9} mbar. However, this is still not a routine, nor a necessarily desirable, procedure.

Sample transfer is done on a regular basis when samples are to be examined at major facilities, such as a synchrotron radiation laboratory. It can be much more efficient to prepare the sample in a dedicated surface science or MBE chamber, and then transport the sample, typically pumped using a moderately sized ion pump, to the measuring station. One such design is indicated in figure 2.5, which is specialized for X-ray diffraction measurements at the HASYLab synchrotron in Hamburg (Johnson 1991).

This design consists of a small chamber, built inside the ion pump housing itself, with a flange on the end capped with a thin hemispherical beryllium window, through which the X-rays can pass in and out. The sample sits at the center of a two-axis diffractometer, and can be heated or cooled during the experiment; the whole transfer chamber is mounted on a rigidly engineered rotatable goniometer stage. Transfer to and from the preparation chamber is effected by closing the gate valve, shown in figure 2.5(a), unbolting the assembly from the goniometer, and proceding *cautiously*. When bolted to the other chamber, the sample can then be withdrawn into the sample preparation position using a transfer shaft fixed to that chamber.

There are many designs for such transfer shafts which maintain full UHV conditions internally. A common design is to use a shaft with a magnetic slug, coupled to a strong permanent magnet external to the vacuum chamber. A schematic drawing of a set of chambers connected with several such magnetic transfer devices is shown in figure 2.5(b). Indicated also are the entry locks, and the place where the X-ray transfer chamber is attached. The fact that it looks like a space station is not entirely coincidental – it *is* a space station with deep space on the *inside*.

2.4.4   *From laboratory experiments to production processes*

I have noted here that UHV-based experimental research and manufacturing technologies are quite capital-intensive, and have argued the case that adequate thought must be put into both the design and operation stages. Once one considers scale-up from the laboratory to production, these points have even greater weight. The dollar figures involved in the semiconductor industry are quite astounding, and control of contamination during the surface processing involved is a major concern and expense.

For example, Ouellette (1997) reports that 'an estimated 80% of equipment failures in silicon wafer process lines arise from contamination related defects. Since most wafer fabrication lines average an 80% yield, as much as 16% of the total loss of yield may be due to contamination'. She goes on to estimate that a single Fab-line can lose $15M/month from contamination-related defects. The definition of defects is suitably wide: anything from peeling paint, worn bearings, bits of PTFE seals, particles, right down to the individual atomic defects incorporated into the materials themselves. As pointed out by O'Hanlon (1994), the major drive for UHV in the semiconductor industry comes from the need to control the purity of reacting gases at the parts per billion level. This is understandable, given the predominance of chemical vapor deposition systems using good high vacuum, rather than UHV technology. This is a problem that simply won't go away.

The other important industry is based even more directly on chemistry. Estimates for the catalytic industry (Bell 1992, Ribiero & Somorjai 1995) suggest that 17% of all manufactured goods go through at least one step involving catalytic processes. Rabo (1993) reported that the yearly catalyst market was projected to be $1.8 billion in 1993, with auto emissions catalysts the fastest growing component. With sensors also a growing market, and environmental concerns growing all the time, these industrial applications are becoming rapidly more important. Most of this activity involves heterogeneous catalysis, in which gases react over a surface.

There are three major types of catalyst which are the subject of intense study: these are (single crystal) metal and oxide catalysts, and supported metal catalysts, where small metal particles are suspended, typically on oxide surfaces. In all these cases, the properties of the catalyst may be dependent on point defects or steps on the surface, and may be very difficult to analyze. In the case of supported metal catalysts, the properties are very dependent on the dispersion of the metal, i.e. on the size and distribution of the small metal particles (SMPs). There is more surface area associated with a given volume of metal if the particle size is small, and additionally the reactivity of the

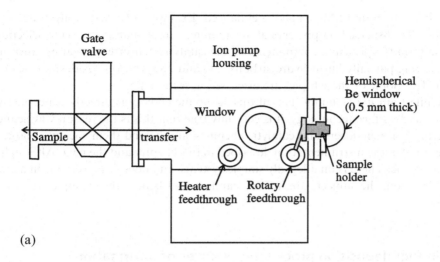

(a)

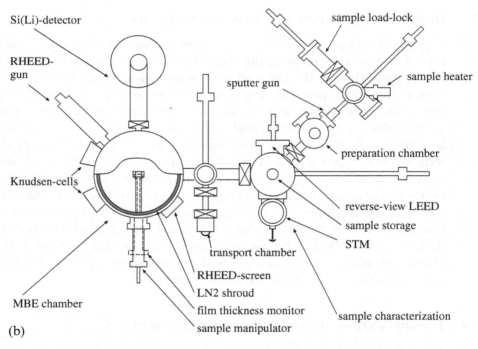

(b)

Figure 2.5. (a) Transfer chamber in use at HASYLab for studies of surface X-ray diffraction (after Johnson 1991); (b) sample preparation and analysis chambers connected via sample transfer shafts, showing sample load locks and transfer chamber docking (from Falkenberg & Johnson 1999, reproduced with permission).

less strongly bound SMP's may be enhanced. Examples of SMP catalysts are Pt, Pd and/or Rh dispersed on polycrystalline alumina, zirconia and/or ceria; a selection of these form the principal components of the catalytic converters in car exhaust pipes, converting partially burnt hydrocarbons, CO and NO_x (nitrous oxides) into CO_2, N_2 and H_2O. Some of these topics are discussed in section 4.5.

While I am *not* claiming that all this economic activity is directly concerned with UHV and surface technology, it is certainly true that this is the reason why semiconductor device engineers and catalytic chemists, and behind them society at large, are interested in the instrumentation described in this chapter and the next. Although our primary focus here is on allegedly simple systems, and doesn't go very far in a chemical direction, the subjects are in fact seamless, as I believe the examples chosen will show.

2.5   Thin film deposition procedures: sources of information

2.5.1   *Historical descriptions and recent compilations*

Thin films have been prepared ever since vacuum systems first became available, but deposition as a means of producing films for device purposes is a development of the past 40 years. Thin metallic film coatings on glass or plastic were among the first to be exploited for optical purposes, ranging from mirrors to sunglasses, and this still continues as a major, typically high vacuum, high throughput business. Most such films are examples of polycrystalline island growth; models of island growth are described here in some detail in chapter 5. An early survey of laboratory-based production methods of single crystal *epitaxial* metallic films on a range of single crystal substrates is given in the articles in *Epitaxial Growth, part A* (Matthews 1975). As thin film deposition processes have developed very rapidly over the past 25 years, particularly in the context of semiconductor devices, processes have become highly specialized, and have been described in textbook form (Smith 1995), and in updateable compilations such as the *Handbook of Thin Film Process Technology* (Glocker & Shah 1995), where the individual sections have themselves been edited and have multiple authors. The following sections describe some of these developments in outline.

2.5.2   *Thermal evaporation and the uniformity of deposits*

This technique is the simplest conceptually, corresponding to raising the temperature of the source material, either in an open boat, suspended on a wire, or by any other convenient means so that the material evaporates or sublimes onto the substrate. The boat/wire is typically chosen as a high temperature material such as W or Mo, and must not react adversely with the evaporant. Unless particular precautions are taken, the evaporant will be deposited all over the inside of the vacuum system, and will therefore be both inefficient in the use of the source material, very messy for the vacuum system, and will not produce a uniform deposit.

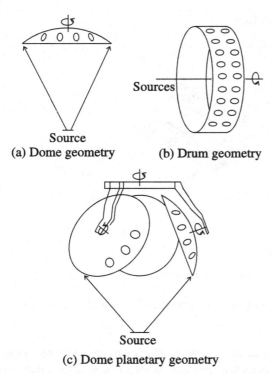

(a) Dome geometry   (b) Drum geometry

(c) Dome planetary geometry

Figure 2.6. The use of substrate rotation to produce more uniform films: (a) substrates mounted on a dome, which is rotated about the axis; (b) substrates on a drum with the sources placed along the center line, with the drum rotating about it; (c) use of planetary motion in a dome geometry for ultimate uniformity (after Graper 1995, redrawn with permission).

The production of uniform deposits with high throughput is a requirement for industrial processes, and this usually means that the substrate has to be rotated. This movement is required because the emission from the source is more or less peaked in a particular (forward) direction, so that the deposited films are thinner at the edges. Three examples of using substrate rotation to alleviate this effect are indicated in figure 2.6 (Graper 1995). Planar solutions are often used because of their simplicity, but the drum solution is preferred for highest throughput. Planetary solutions are required for best thickness uniformity, especially in the dome geometry as illustrated in figure 2.6(c). These are used for this reason even at the expense of throughput and reliability.

The vapor pressure of the source material is exponentially dependent on the temperature, as described in section 1.3.1, and shown for some elements in figures 1.9 and 1.10. The deposition rate is determined by the source area and temperature, and by the distance between the source and substrate. One should note that different materials have very different relations between the vapor pressure and the melting point, so that a satisfactory deposition rate may only be obtained if the material is liquid, which may drip off a wire or inclined boat.

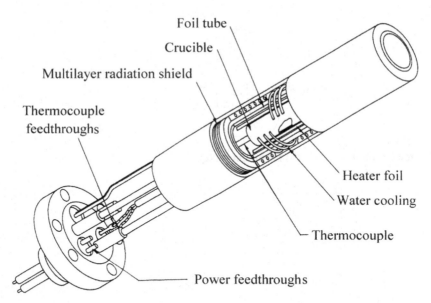

Figure 2.7. A small effusion source using a PBN oven (after Davies & Williams 1985, reproduced with permission).

To avoid such problems, one can use an oven, which can easily be mounted in an orientation such that the liquid does not spill out. With an oven or crucible, it is relatively simple to construct it so that the evaporant does not evaporate in all directions, but comes out in a more or less well directed beam, which can be further collimated so that the source material is directed preferentially onto the substrate. The sources can be characterized as effusion sources, with a relatively large area opening, or as Knudsen sources, where a small hole is used; in the latter case the model is that the material coming through the hole samples the vapor pressure of the source material inside, and that standard kinetic theory formulae are applicable. Small sources have been developed for use with graphite (Kubiyak et al. 1982) or pyrolytic boron nitride (PBN) ovens, as illustrated in figure 2.7 (Davies & Williams 1985). The design of such an evaporation source can be explored via problem 2.2. In practice considerable thought and effort is required to achieve a uniform temperature enclosure, via careful design of the crucible, of heater windings, radiation shields and water cooling, and by the use of anticipatory electronic control of heater currents based on thermocouple measurements. Such sources and controllers are now commercially available, and a pilot plant system (e.g. to deposit multilayers) may have many of them in action at any one time.

In order to deposit high temperature materials, or materials which interact with the crucible, electron beam evaporation is required. The design typically includes a heavy duty filament to emit many milliamps of current, and several kilovolts of high voltage in order to deliver the necessary power. The electron beam is directed onto the sample surface by a shaped magnetic field, typically using an inbuilt permanent magnet (Graper 1995). The heating so produced is very localized, and care is required to be sure that it is localized where it should be; this is also the case when using pulsed ultra-violet eximer lasers. Laser ablation, sometimes called pulsed laser deposition (PLD), has typically been used to deposit ceramic materials, including high temperature superconductors (Dijkamp *et al.* 1987, Venkatasan *et al.* 1988, Morimoto & Shimizu 1995); it produces very rapid deposition in which whole chunks of material can break off and be deposited during the immense peak powers which typically last for 10–20 ns. A particular advantage of rapid evaporation is the control of stoichiometry, since the different species do not have time to segregate to the surface during the evaporation phase.

2.5.3 Molecular beam epitaxy and related methods

The large scale use of such evaporation sources, especially for depositing semiconductor or metallic multilayers, has become known as molecular beam epitaxy (MBE). This acroymn has spawned several sub-cases, such as GSMBE (Gas Source MBE), and MOMBE (Metal–Organic MBE) which is sometimes called CBE (chemical beam epitaxy). Thus MBE really spans a range of techniques which range from being merely a fancy name for thermal evaporation, to much more chemically oriented flow process techniques such as chemical vapor deposition (CVD) and MOCVD which we describe briefly in section 2.5.5. MBE is covered in several books (e.g. Parker 1985, Tsao 1993, Glocker & Shah 1995) and in many review articles and ongoing conference series. Only a few general points can usefully be made here, but this preparation method lies behind much of the science discussed later in chapters 5–8.

The growth end of such a chamber used for GSMBE or MOMBE is shown schematically in figure 2.8 (Abernathy 1995). The halfspace in front of the substrate is a bank of effusion cells and various forms of gas injector cells, which transport reacting chemicals to the substrate. The growth of III–V compounds such as GaAs or AlGaAs has been pursued using such techniques, starting from metal–alkyl compounds such as triethylaluminum (TEA) and triethylgallium (TEG) and hydrides such as AsH_3. The group five hydrides are injected via cracker cells which convert them catalytically into As_2 and hydrogen, which then impinge on the substrate to react with the alkyl beam to produce the growing film (Panish & Sumski 1984, Abernathy 1995).

These cells are surrounded by liquid nitrogen cooled shrouds, which both condense unwanted evaporants and improve the vacuum in the sample region, which is monitored by the mass spectrometer. One of the advantages of MBE and related methods is the relative ease by which *in situ* diagnostic tools can be incorporated into the vacuum system. The most widely used are RHEED, shown in figure 2.8 and described in section 3.2.2, plus various optical techniques, some of which can also work in higher pressure environments. A compendium of these 'real-time diagnostics', particularly in the context of semiconductor growth, is given by several authors in Glocker & Shah (1995, part D).

2.5.4 Sputtering and ion beam assisted deposition

There are many uses for ions in connection with the production of thin films. Sputtering using relatively low energy (100 eV–2 keV) ions is often used for cleaning samples, while higher energy (5–200 keV) ions can be used for doping layers with

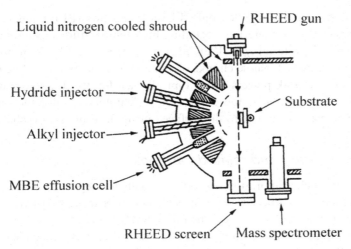

Figure 2.8. GSMBE or MOMBE growth chamber containing effusion and cracker cells, and RHEED diagnostics (after Abernathy 1995, reproduced with permission). See text for discussion.

electrically active impurities. Ions can also be used for deposition, where individual ions, or charged clusters can be deposited at a range of energies. Clearly, the fact that the ions are charged allows extra control, and there are various methods by which this can be done. Directed ion beams from an accelerator form the most obvious possibility, but plasma and magnetron sources are also widely used (Bunshah 1991, Vossen & Kern 1991, Shah 1995, Smith 1995).

One of the main claims for IBAD (Ion Beam Assisted Deposition) or IBSD (Ion Beam Sputter Deposition) is that better quality deposits can be obtained at lower substrate temperatures, thus avoiding large scale interdiffusion which results from high temperature processing. However, the deposited films adhere well to the substrate because of the localized limited mixing caused by the ion impacts (Itoh 1989, 1995, Marton 1994). This is one example of device engineers trying to reduce the 'thermal budget' and so produce sharper interfaces between dissimilar components. Another possibilty is to produce clusters in a supersonic expansion source, and to ionize them, controlling their flight during cluster depostion with applied voltages. This has been termed ICB (Ionized Cluster Beam) technology by the inventers (Takagi 1986, Takagi & Yamada 1989). Yet another possibility is to use ion beams to react with the substrate and grow compounds such as oxides or nitrides at and near surfaces (Herbots *et al.* 1994).

All of these procedures are inherently complex; for almost all purposes the question is whether they produce 'better' films for particular applications, i.e. whether they give a large enough improvement over existing methods to justify the considerable investment involved. Although our aim here is to give an outline description, we will not pursue models of how of these methods work in any detail; they are all rather specific to the combination of material and ion beam technique, and whether the processes are reactive, in the sense of involving (ion) chemistry, or physical, meaning that only collisions and clustering are involved. Nonetheless, ion beam assisted methods are very

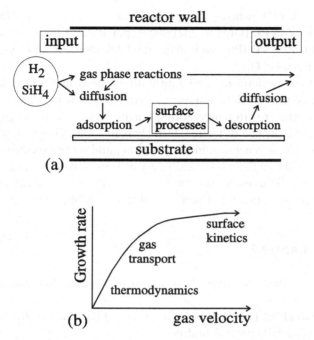

Figure 2.9. (a) Schematic diagram of reactions involved in the CVD of silicon from silane (SiH$_4$) and hydrogen (H$_2$); (b) typical variation of the growth rate on gas velocity with rate limiting steps indicated (after Vescan 1995, reproduced with permission).

widespread and are of great economic importance; sputter deposition in particular has high throughput for the production of (textured) polycrystalline thin films of a wide variety of materials (Bunshah 1991, Vossen & Kern 1991, Shah 1995). The language used to describe such ion based processes (see Greene 1991) necessarily starts from a similar vantage point to that used to describe thermal evaporation; the latter topic is treated here in detail in chapter 5.

2.5.5 Chemical vapor deposition techniques

The growth of semiconductor layers for device production is most frequently done by a variant of CVD (Chemical Vapor Deposition). This is usually implemented as a flow technique in which the reacting gases pass over a heated substrate, as indicated in figure 2.9(a) for growth of silicon from silane and hydrogen. CVD reactors are either of the hot wall or cold wall type, and are surrounded by an extended gas handling and pumping system with rigorous control of the (often highly toxic) gases (Carlsson 1991, Vescan 1995). Because the pressure may be up to an atmosphere in certain cases (APCVD: Atmospheric Pressure CVD) the growth rate can be high, and UHV technology is not an absolute necessity; but control of impurities is a dominant problem, as pointed out in section 2.4.4 (O'Hanlon 1994). Thus UHV-CVD is becoming more widespread, where the total pressure is <1 mbar. However, most commercial processing corresponds to

LPCVD (Low Pressure CVD) where the pressure is 0.1–1 mbar. MOCVD (Metal–Organic CVD) or OMVPE (Organo–Metallic Vapor Phase Epitaxy) is also a widely used technique, as is PECVD (Plasma Enhanced CVD); there are many variants on this theme (Vossen & Kern 1991).

The question of reaction mechanisms and rate limiting steps in CVD is highly complex. Under LPCVD conditions diffusion processes in the gas are typically not the dominant effect, so that at the growth temperatures, kinetic processes on the growing surface are rate limiting, as indicated in figure 2.9(b) (Vescan 1995). However, we can see from figure 2.9(a) that all the reactions, in the gas phase and on the surface of the growing film, are in series and that there is typically very little information on the intermediate states of the reaction. Thus understanding CVD in atomic and molecular terms is very much an ongoing research project, which we will return to later in section 7.3.

Further reading for chapter 2

Dushman, S. & J. Lafferty (1992) *Scientific Foundations of Vacuum Technique* (John Wiley).

Glocker, D.A. & S.I. Shah (Eds) (1995) *Handbook of Thin Film Process Technology* (Institute of Physics), especially parts A and B.

Hudson, J.B. (1992) *Surface Science: an Introduction* (Butterworth-Heinemann) chapters 8 and 9.

Lüth, H. (1993/5) *Surfaces and Interfaces of Solid Surfaces* (3rd Edn, Springer) chapters 1 and 2.

O'Hanlon, J.F. (1989) *A Users Guide to Vacuum Technology* (John Wiley).

Matthews, J.W. (Ed.) (1975) *Epitaxial Growth, part A* (Academic).

Moore, J.H., C.C. Davis & M.A. Coplan (1989) *Building Scientific Apparatus* (2nd Edn, Addison-Wesley) chapters 3 and 5.

Roth, A. (1990) *Vacuum Technology* (3rd Edn, North-Holland).

Smith, D.L. (1995) *Thin-Film Deposition: Principles and Practice* (McGraw-Hill).

Tsao, J.Y. (1993) *Materials Fundamentals of Molecular Beam Epitaxy* (Academic).

Problems for chapter 2

These problems are to practice and test ideas about vacuum systems, design problems and surface preparation techniques.

Problem 2.1. Design of vacuum systems for specific purposes

Use your knowledge of (and appendices on) conductances of standard size tubes, and the characteristics of vacuum pumps, to suggest (and justify semi-quantitatively) design choices in the following situations.

(a) Pumping an approximately spherical chamber of diameter 0.5 m. The chamber is to be let up to air infrequently, and we want to achieve as good a pressure as possible ($<10^{-10}$ mbar).

(b) Pumping a cylindrical chamber of length about 10 m and diameter 0.1 m. The chamber is to be periodically flooded with rare gases up to about 10^{-3} mbar pressure, and the important point is to be able to achieve pressures below 10^{-8} mbar quickly and economically.

(c) Pumping a state of the art particle accelerator from sections of pipe of length about 50 m and diameter 0.1 m, with a total length in excess of 50 km (kilometers) at a pressure of $<10^{-11}$ mbar.

Problem 2.2. Design of a Knudsen source for depositing elemental metal films

A Knudsen source is an evaporation furnace which relies on the establishment of the vapor pressure above a solid or liquid source material. A small hole in the furnace above the source material, plus collimating holes in front of the source, allow a beam of the source material to be directed at the sample. Use your knowledge of vapor pressures and kinetic theory to design a source which will deposit one monolayer per minute on a sample held 0.15 m away from the exit of the source, will be uniform on the sample within 1% for the central 0.01 m diameter, and will not deposit any material on the sample outside a radius of 0.02 m. Do this in stages, with discussion, as follows.

(a) Consider the formula $R = nv/4$ for the number of atoms hitting unit area per second of an enclosure, and how this formula applies to a Knudsen source. Derive the formula by considering the relevant integrations over angles and the Maxwell–Boltzmann velocity distribution.

(b) Consider the geometry of the design, and the constraints on the uniformity and area of the deposit. Show that this will limit the size of the hole in the furnace, and suggest a suitable size for holes in both the furnace and the collimator.

(c) Choose an elemental metal of interest to you, and find out the formula for the vapor pressure as a function of furnace temperature. Using the relationship between density n and pressure p for this material, coupled with your design from part (b) work out the temperature at which the source will have to operate to satisfy the deposition rate requirement, explaining your assumptions.

(d) If you actually want to design a real source for this material, consider carefully the materials of which the source can be made, whether you should be using a Knudsen or some other type of source, and how to power the furnace to achieve sufficient temperature uniformity, etc.

Note: short descriptions of most possible deposition techniques are given by Smith (1995) and by Glocker & Shah (1995); some specific designs are in Yates (1997).

Problem 2.3. Some questions on surface preparation and related techniques

Questions about surface preparation are always very specific to the materials concerned, but here are a few which may be relevant and which spring from the text of this section.

(a) Why should one either cool the sample slowly through a surface phase transition (e.g. as in the case of Si(111)), or not anneal the sample above a bulk phase transition (e.g. in preparing b.c.c. Fe surfaces)?

(b) What is the main reason why Si(111) produces a 2×1 reconstruction after cleavage, when the equilibrium surface structure is the 7×7?

(c) Device engineers always grow a 'buffer layer' on Si(100) before attempting to grow a device, e.g. by molecular beam epitaxy. Why is this precaution taken, and how does it improve the quality of the devices grown on such surfaces?

(d) Mass spectrometry shows a range of mass numbers (M/e ratios) for the contents of the vacuum system, but they don't seem to be simply related to the molecules, e.g. O_2, N_2, CO, H_2O, CO_2 which are present. What range of processes are responsible for this discrepancy?

(e) GaAs often evaporates to leave small liquid Ga droplets on the surface. Why does this happen, and how can it be prevented?

3 Electron-based techniques for examining surface and thin film processes

This book presumes that the reader is interested in experimental techniques for examining surface and thin film processes; however, there are many books devoted to surface physics and chemistry techniques, some of which are given as further reading at the end of the chapter. There are even several books which are just about *one* technique, such as Pendry (1974) or Clarke (1985), both on low energy electron diffraction (LEED) in relation to surface crystallography. By the mid-1980s it was already stretching the limits of the review article format to compare the capabilities of the available surface and thin film techniques (Werner & Garten 1984).

Since then, the various sub-fields have proliferated, so we cannot be comprehensive, or give all the latest references here. In section 3.1 we discuss ways of classifying the large number of techniques which exist, and thereafter the chapter is restricted to techniques based on the use of electron beams. Section 3.2 discusses the most widely used (elastic scattering) diffraction techniques used for studying surface structure. Section 3.3 discusses forms of electron spectroscopy based on inelastic scattering, which are used to obtain composition and chemical state information. Individuals can look in more detail into a particular technique. Students have been asked to present a talk to the class, followed by questions and discussion; some of the topics considered in this way are listed, along with selected problems, at the end of the chapter. As examples, some case studies are given in section 3.3 on Auger electron spectroscopy, in section 3.4 on quantification of AES, and in section 3.5 on the development of secondary and Auger electron microscopy. Stress is placed on the relationship between microscopy and analysis: *micro*structure and *micro*analysis. The frontier is at *nano*structures and *nanometer resolution* analysis.

3.1 Classification of surface and microscopy techniques

3.1.1 Surface techniques as scattering experiments

Most physics techniques can be classified as scattering experiments: a particle is incident on the sample, and another particle is detected after the interaction with the sample. Surface physics is no exception: we can think of an incident probe and a response as set out in table 3.1.

Table 3.1. *Particle scattering techniques*

Probe {	Electrons (E_0, $\mathbf{k}_0$, ... spin s) Radiation (ω_0, $\mathbf{k}_0$, ... polarization) Atoms (E_0, $\mathbf{k}_0$, atomic number Z) Ions (E_0, $\mathbf{k}_0$, Z, charge state $\pm n$)		*Response* {	Electrons (E, $\mathbf{k}$, ... spin s) Radiation (ω, $\mathbf{k}$, ... polarization) Atoms (E, $\mathbf{k}$, atomic number Z') Ions (E, $\mathbf{k}$, Z', charge state $\pm n'$)	

The probe will be formed from a particular type of particle, and typically will have a well defined energy E_0, and often a well defined wave vector $\mathbf{k}_0$, or equivalently momentum $\mathbf{p}_0 = \hbar\mathbf{k}_0$. The response can either be the same or a different particle, and, depending on the detection system, its energy E and/or its wave vector (momentum) $\mathbf{k}(\mathbf{p})$, and maybe other attributes, can be measured. If we understand the nature of the scattering process, then we can interpret the experiment and deduce the corresponding characteristics of the sample. It is easy to see from table 3.1 how the number of techniques, and the corresponding acronyms, can be very large, especially once one realizes that any probe particle can give rise to several responses, and that we may have different names for essentially the same technique used at different energy, and different wave-vector (momentum or angular) regimes.

3.1.2 Reasons for surface sensitivity

The next question is 'which techniques are actually useful for studying surfaces?'. There are two cases. In the first case, the emergent (i.e. response) particle or the probe particle has a short mean free path, λ. This leads to a useful 'single surface' technique.

Examples are Auger electron spectroscopy (AES), where the emerging electrons in the energy range 100–2000 eV have λ for *inelastic* scattering in solids typically in the range 0.5–2.5 nm. Using an energy analyzer to measure only those Auger electrons which have not lost energy, attenuates the signal from subsurface layers strongly. A cruder form of energy discrimination is used in observing LEED patterns, where both the incident and the emergent electrons have short mean free paths for energy loss processes. In SIMS (secondary ion mass spectrometry), the emergent ions have a very high probability of being neutralized if they do not originate very near the surface. ICISS (impact collision ion scattering spectroscopy) is surface sensitive because the incident ion will be neutralized, and thereby not detected, if the probe particle penetrates the solid. An introduction to ion based techniques can be found in Feldman & Mayer (1986), in Rivière (1990), and in several chapters contained in Walls (1990); many other books can be unearthed via the web.

In the second case, the sample has a large surface to volume ratio. This condition allows us to extract surface information from techniques which are not particularly surface sensitive. We can perform heat capacity or other thermodynamic measurements, or study structures and dynamics by X-ray or neutron scattering. Here we need to know the signal from the bulk, and maybe subtract it in a differential measurement. Much of physical chemistry work on surfaces has been done this way, on powdered, or exfoliated, samples.

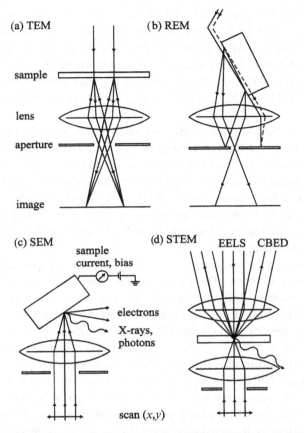

Figure 3.1. Schematic geometries for TEM, STEM, SEM and REM (after Venables *et al.* 1987, redrawn with permission).

3.1.3 *Microscopic examination of surfaces*

Microscopy can be categorized into fixed beam, scanned beam and scanned probe techniques. A typical fixed beam technique is transmission electron microscopy (TEM); the same instrument can often be used for reflection electron microscopy (REM). These techniques are illustrated schematically in figures 3.1(a) and (b). Examples will be given later which show that it is not essential to have these instruments operating at UHV in order to produce useful surface related information: UHV experiments followed by *ex situ* examination can be very informative, provided the final samples are not too reactive in air.

The central element of TEM or REM is the objective lens, a cylindrically symmetric strong magnetic field positioned just after the sample. As the equivalent of the first stage of an amplifier in electronics, this element is critical to the performance of the microscope, and the aberrations and phase transfer characteristics of this lens determine both the resolution and contrast that are seen in the image. A particularly useful feature is the use of the aperture situated in the back focal plane of the objective lens to select individual diffraction features and to relate these features to the image. In the case of REM, the image is strongly foreshortened, but this does not mean that the images are particularly difficult to interpret, as we experience the same sort of image foreshortening when we look ahead driving a car along the road. Both TEM and REM have been able to image surface steps, particularly on high atomic number relatively inert materials such as Au(111) and Pt(111), without requiring that the surfaces were truly clean.

There are many books on electron microscopy, and TEM in particular has the reputation for being difficult to understand, primarily due to the need for a dynamical theory of electron diffraction to interpret the images of crystalline samples. For an overview of the field, see Buseck *et al.* (1988), which includes a chapter on surfaces (Yagi 1988); recent surveys of high resolution (HR)-TEM, describing the approach to atomic resolution at surfaces and interfaces, are given by Smith (1997) and Spence (1999), both with extensive references. I have attempted a ten-minute sketch of the various techniques in Venables *et al.* (1987).

A few groups have converted their instruments to, or constructed instruments for, UHV operation, and *in situ* experiments. These instruments, which can also be used for transmission high energy electron diffraction (THEED) and reflection high energy electron diffraction (RHEED), have produced highly valuable information on surface studies, as reviewed, for example, by Yagi (1988, 1989, 1993). More recently low energy electron microscopy (LEEM) has been developed, which can be combined with LEED, and is making a major contribution (Bauer, 1994). This instrument can also be used for photoemission microscopy (PEEM), which has been developed in several different versions. A specialist form of microscopy with a venerable history is field ion microscopy (FIM), which is especially useful for studying individual atomic events such as diffusion and cluster formation, as discussed by Bassett (1983), Kellogg (1994), Ehrlich (1991, 1994, 1995, 1997) and Tsong & Chen (1997).

The great virtue of fixed beam techniques is that the information from each picture element (pixel) is recorded at the same time, in parallel. This leads to relatively rapid data acquisition, and the ability to study dynamic events, often in real time, e.g. via video recording. In contrast, data in a scanned beam technique, such as scanning electron microscopy (SEM) or scanning transmission electron microscopy (STEM), is collected serially, point by point, with the sample placed *after* the objective lens as illustrated in figures 3.1(c) and (d).

This configuration means that multiple signals (not just electrons at the probe energy as in TEM or REM) can be used, which makes the instruments very versatile. It also makes them ideally adapted for computer control and computer-based data collection, but can have a corresponding disadvantage: the need to concentrate a very high current density into a small spot means that not all forms of information can be obtained rapidly, that there will be substantial signal to noise ratio (SNR) problems, and that the beam can cause damage to sensitive specimens. Nonetheless SEM and STEM form the basis of a very useful class of techniques; UHV-SEM has been developed in several laboratories, including the University of Sussex, and UHV-STEM especially at Arizona State University. We examine particular developments in section 3.5.

The above techniques have been available for several decades, and have been substantially developed in an evolutionary sense, year by year. By contrast, the scanned probe techniques burst upon the scene in the early 1980s, in the revolutionary development of first scanning tunneling microscopy (STM) (Binnig *et al.* 1982), followed in quick succession by atomic force microscopy (AFM), scanning near-field optical microscopy (SNOM), and related spectroscopies. The first two techniques as illustrated in figure 3.2.

The central feature of STM operation is the tip, which has to be brought extremely close to the sample in a controlled manner to effect tunneling, as shown schematically in figure 3.2(a). In the simplest mode of operation, the z-motion is used to keep the tunneling current I_T constant. In addition, one has to be able to move the tip and sample relative to each other, both as a shift to find out where you are, and as a scan in x and y to produce the image. There are many different STM designs, but all successful designs have been based on piezo-electric elements and have paid due regard for design symmetry, which is necessary to minimize thermal drift.

A particularly appealing design is the 'beetle' STM developed by Besocke (1987) and Frohn *et al.* (1989) in which the sample rests on three piezo-tube 'legs' and probes the sample with the tip mounted on a piezo-tube 'feeler'. This design is shown in figure 3.2(b) in the version developed by Voigtländer & Zinner (1993) and Voigtländer (1999) for *in situ* deposition experiments. Coarse approach of the sample is effected by a special holder, in which a rotational motion is translated via a shallow ramp into z-motion; once the sample and tip can 'feel' each other, the feeler piezo takes over and STM proper can begin. Coarse movement in x and y uses the leg piezo drives in stick-slip motion; a fast jerk on the legs causes them to slip and the stage to move relative to the leg and tip, but a slow movement translates the stage and legs together. By repeated alternating stick and slip motions, stage translation can be made remarkably reproducible; the design will work either way up, though not on its side; it *uses* gravity. Either the sample holder or the tip holder assembly can be readily withdrawn for sample preparation.

The AFM also comes in many forms, and has the great advantage that it can be used on insulators as well as conductors. A key element here was the development of sensitive cantilever arms, whose deflection is typically monitored by a low powered He–Ne laser reflected onto a position sensitive diode array detector, as shown in figure 3.2(c) (Meyer & Amer 1988). These arms are usually made of lithographically etched silicon, with silicon nitride (Si_3N_4) as the tip material. Such an arm will have a characteristic resonant frequency, so that, in addition to steady (d.c.) measurements of tip displacement, many a.c. and phase sensitive measurement schemes are possible. Figure 3.2(d) shows a close up SEM view of such a Si_3N_4 tip (Albrecht *et al.* 1990).

Of the many recent books on scanned probe microscopy, arguably the best to start from are Chen (1993) and Wiesendanger (1994). There are several multi-author texts, including Stroscio & Kaiser (1993) and many review and specialist articles. Indeed, there are now a large number of techniques for studying surfaces on a microscopic scale: a description of these techniques and their applications would take a very long time. It is not possible to do justice to the full range of extraordinary possibilities offered by these techniques here, but several examples are given throughout the book which show how valuable they are in particular cases.

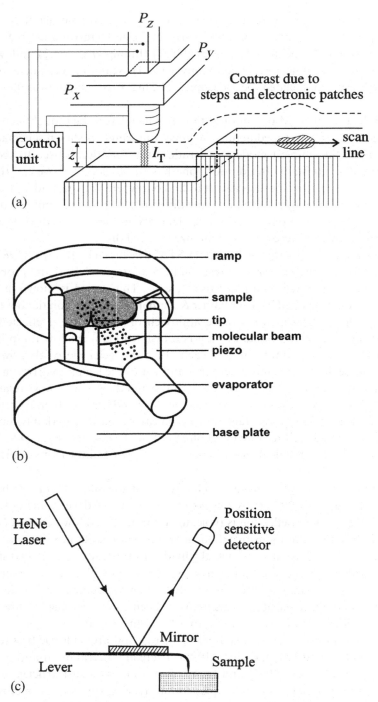

Figure 3.2. Scanned probe techniques: (a) principles of STM operation, indicating x, y and z piezo-elements P_x, P_y and P_z and contrast due to steps and electronic effects (after Binnig *et al.* 1982); (b) the 'beetle' STM design of Besocke (1987) and Frohn *et al.* (1989), as used for *in situ* deposition experiments by Voigtländer & Zinner (1993) and Voigtländer (1999);

(d)

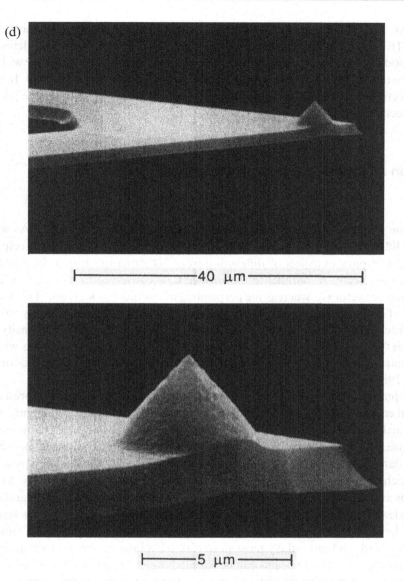

Figure 3.2. (*cont.*) (c) a widely used AFM design, with a He–Ne laser to detect the deflection of the cantilever on which the tip is mounted (after Meyer & Amer 1988); (d) close up SEM view of a Si_3N_4 AFM tip with a nominal tip radius <30 nm, which has achieved atomic resolution (after Albrecht *et al.* 1990; diagrams reproduced or redrawn with permission).

3.1.4 Acronyms

Acronyms are defined at various places throughout this book and are summarized in Appendix B. So far we have met and defined in the text: vacuum and electronics terms UHV, RGA, QMS, TSP, SNR; surface and crystal growth terms ML, TLK, BCF, CVD, MBE and others in section 2.5; diffraction techniques LEED, RHEED, THEED; chemical analysis and ion scattering techniques AES, SIMS, ICISS; microscopy types

TEM, REM, LEEM, PEEM, SEM, STEM, STM, AFM, SNOM. We also have two techniques (EELS, CBED) indicated on figure 3.1, which have not yet been defined. Now is a good time to check you know what these acronyms stand for, since we will be adding more to the list as we start to study individual techniques in more detail. In the following sections, we emphasize a common case, where electrons are both the probe and the detected (response) particle.

3.2   Diffraction and quasi-elastic scattering techniques

3.2.1   LEED

The common electron-based diffraction techniques are LEED and RHEED. As with all surface diffraction techniques, the analysis is based in terms of the surface reciprocal lattice. An important aspect of diffraction from 2D structures is that the component of the wave vector $\mathbf{k}$, parallel to the surface $\mathbf{k}_{//}$ is conserved to within a surface reciprocal lattice vector $\mathbf{G}_{//}$, whereas the perpendicular component $\mathbf{k}_{\perp}$ is not. This leads to the idea of reciprocal lattice *rods*; they express the fact that $\mathbf{k}_{\perp}$ can have any value, so that diffraction can take place at any angle of incidence. However, the intensity of diffraction is typically not constant at all values of $\mathbf{k}_{\perp}$, but is modulated in ways which reflect the partial 3D character of the diffraction (Lüth 1993/5 section 4.2, Woodruff & Delchar 1986 section 2.3).

The equipment for both diffraction techniques is simple, involving a fluorescent screen, with energy filtering in addition in the case of LEED, as indicated in figure 3.3, to remove inelastically scattered electrons. There are three types of LEED apparatus in regular use. The normal-view arrangement has the LEED gun and screen mounted on a UHV flange, typically 8 inches (200 mm) across, and the pattern is viewed past the sample, which therefore has to be reasonably small, or it will obscure the view. Most new systems are of the reverse-view type, where the gun has been miniaturized, and the pattern is viewed through a transmission screen and a viewport. This enables larger sample holders to be used, which helps for such operations as heating, cooling, straining, etc. The third, and potentially most powerful, technique is where, in addition to viewing the screen, the LEED beams can be scanned over a fine detector using electrostatic deflectors and focusing, in order to examine the spot profiles, which can be sensitive to surface steps and other forms of disorder at surfaces. This technique, which has been perfected by the Hannover group (Henzler 1977, 1997, Scheithauer *et al.* 1986, Wollschläger 1995), is now known as SPA-LEED, emphasizing the capability for spot profile analysis.

There are two aspects to electron diffraction techniques. The first, and simplest, is that the positions of the spots give the symmetry and size of the unit mesh, i.e. the surface unit cell. The common use of electron diffraction is primarily, often solely, in this sense. The second effect is that the positions of atoms in the mesh is not determined by this qualitative pattern (see discussion in section 1.4), but requires a quantitative analysis of LEED intensities. Application of dynamic theory has so far 'solved' several

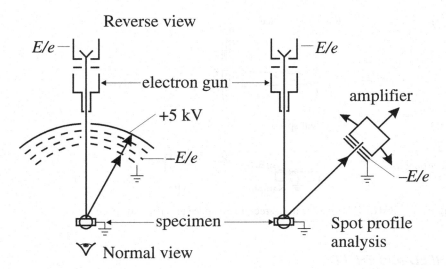

Figure 3.3. LEED apparatus types, illustrating schematically the configurations of normal and reverse view LEED, and spot profile analysis. The +5 kV is applied to a fluorescent screen, which for reverse view must be transparent (after Bauer 1975, redrawn with permission).

hundred surface structures (Watson *et al.* 1996). This is impressive, but it pales besides the number of bulk structures solved by X-rays, using (developments of) kinematic theory.

Experimentally, the intensities are typically collected in the form of so-called *I–V* or *I(E)* curves, where the size of the Ewald sphere is varied by varying the probe energy (say from 20–150 eV), and the intensity data are obtained by 'tracking' along an individual reciprocal lattice rod. Various computer controlled, and frame grabbing, schemes have been developed to do this. A description of standard experimental and theoretical methods is given by Clarke (1985); useful updates are reviews by Heinz (1994, 1995). Lüth (1993/5, chapter 4) or Prutton (1994, chapter 3) are other starting points for LEED, as they introduce the basic idea of multiple scattering in a relatively short space.

The difference between LEED and X-ray structure analysis is that a kinematic diffraction theory has limited usefulness, because the scattering is very strong, as explored in problem 3.2. Averaging different *I–V* curves at constant momentum transfer was once a promising method in the attempt to get around this problem, and some successes were recorded, particularly in obtaining the distances between lattice planes parallel to the surface, and surface vibration amplitudes (Webb & Lagally 1973, Lagally 1975). However, dynamical theory is constantly being developed, e.g. via adoption of the latest computational and approximation methods, which are closely related to band theory and have similar constraints (Pendry 1994, 1997); LEED is still the main method of surface structure analysis, now complemented by surface X-ray diffraction using synchrotron radiation (Feidenhans'l 1989, Johnson 1991, Robinson & Tweet 1992, Renaud 1998).

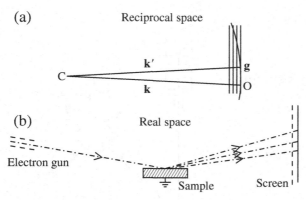

Figure 3.4. RHEED geometry in (a) reciprocal space and (b) real space.

3.2.2 RHEED and THEED

The basis of RHEED is very similar in LEED, but the language used is rather different, being similar to that used for THEED. The discussion can again be separated into geometry and intensities. The glancing angle geometry of RHEED means that the reciprocal lattice rods are closely parallel to the Ewald sphere near the origin, as shown in figure 3.4. This means that the low angle region often consists of streaks, rather than spots. This part of the pattern corresponds to the zero order Laue zone (ZOLZ) in THEED patterns. The higher angle parts of a RHEED pattern are then equivalent to higher order (HOLZ) rings in THEED patterns.

The apparatus for RHEED can consist of a simple 5–20 kV electrostatically focused gun, for instance to monitor the surface crystallography in an MBE experiment, where the glancing angle geometry has many practical advantages over LEED, especially in the ease of access around the sample. Or it can utilize an electron gun approaching electron microscope quality, operating at higher voltages, and produce finely focused diffraction spots over a wide angular range.

Several workers have perfected this technique especially in Japan (Ino 1977, 1988, Sakamoto 1988, Ichikawa & Doi 1988), and two examples of Si(111) 7×7 in different azimuths are seen in figure 3.5. Ino's group in particular have also developed a curved screen, centered on the sample, which could be viewed in two directions at right angles. In the normal view, from the right of figure 3.4, we see spots distributed on a series of arcs as shown for the Si(111)$\sqrt{3} \times \sqrt{3}$R30° Ag structure in figure 3.6(a); the same screen, viewed via a mirror in the perpendicular direction along the reciprocal lattice rods is seen to be an undistorted view of the reciprocal lattice (as in LEED) in figure 3.6(b). We can note that the (111)7×7 and (111)$\sqrt{3}$ patterns are strikingly different in a qualitative sense.

As in LEED, the question of intensities is much more detailed, involving multiple scattering and inelastic processes, and there are many discussions/assertions in the literature about whether streaks or spots constitute evidence for good (i.e. well prepared, flat) surfaces. Some general remarks are made in the next section.

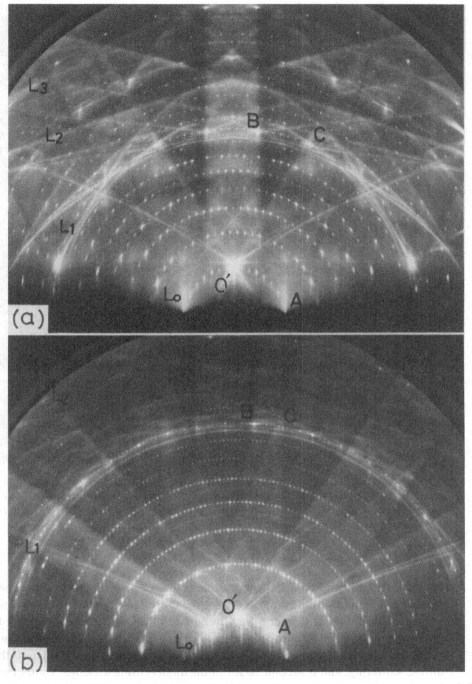

Figure 3.5. RHEED patterns (20 kV) of Si(111) with the 7×7 reconstruction along: (a) [$\bar{1}2\bar{1}$] and (b) [$01\bar{1}$] incidence. Note the reciprocal lattice unit cell O′ACB, and the six superlattice spots in each direction between these fundamental spots (from Ino 1977, reproduced with permission).

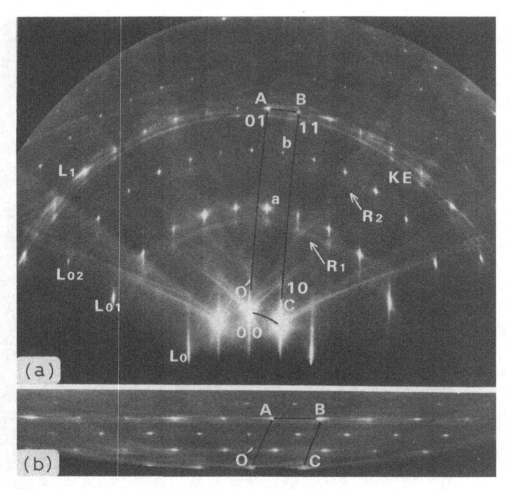

Figure 3.6. RHEED patterns (20 kV) of Si(111) in the '√3' structure associated with the ML phase of Ag deposited at 500 °C: (a) in the normal view, (b) in the perpendicular view, showing reciprocal lattice unit cell O'ACB, and the two spots between these spots (from Gotoh & Ino 1978, reproduced with permission).

3.2.3   Elastic, quasi-elastic and inelastic scattering

Models of LEED and RHEED concentrate on elastic scattering, where the energy of the outgoing electron is the same as that of the incoming electron. But experimentally we cannot discriminate in energy very well in a typical diffraction apparatus. LEED grids/screens are able to remove plasma loss electrons ($\sim$10–20 eV loss), but the intensities measured include phonon scattering ($\sim$25 meV losses and gains). This is thermal diffuse scattering, and is accounted for in the models using a Debye–Waller factor, as in standard X-ray treatments. At higher temperature, the intensities in the Bragg peaks fall off exponentially, as

$$I/I_0 = \exp - (K^2 \langle u^2 \rangle / 2), \text{ with } \langle u^2 \rangle = 3\hbar^2 T/(mk\theta_d^2), \tag{3.1}$$

θ_d being the Debye temperature, and **K** the scattering vector. This means that intensity measurements as a function of temperature measure $\langle u^2 \rangle$, and several such studies have been done with LEED. An interesting feature of such experiments is that the value of $\langle u^2 \rangle$ decreases towards the bulk value as the incident energy is increased, reflecting the increased sampling depth of the electrons (Lagally 1975, Woodruff & Delchar 1986,1994 chapter 2.7).

In a typical RHEED setup there is no energy filtering, other than that caused by the fact that higher energy electrons produce more light from phosphor screens. Yet the geometry is such that plasmons, especially surface plasmons, will be produced very efficiently. Because plasmon excitation produces only a small angular deflection, the diffraction pattern is not unduly degraded. A few groups have studied energy-filtered RHEED (Marten & Meyer-Ehmsen 1988, Ichimiya *et al.* 1997, Weierstall *et al.* 1999), but it is difficult to construct filters which work over a large angular range.

LEED (especially SPA-LEED) and RHEED, and the corresponding microscopies (LEEM and REM) have been shown to be very sensitive to the presence of surface steps and other types of defects, including domain structures. Some of these effects are due to the extra diffraction spots associated with particular domains; some are due to exploiting the difference between in-phase and out-of-phase scattering between terraces separated by steps; some again depend on the small static distortions and rotations produced by surface steps, and the increase in diffuse scattering (Yagi 1988, 1989, 1993, Henzler 1977, 1997, Bauer 1994, Wollschläger 1995). SPA-RHEED has also been demonstrated (Müller & Henzler 1995).

The basic reason for the surface sensitivity of LEED is the short inelastic mean free path for the excitation of plasmons (and other forms of electron–electron collision); this means that information from deeper in the crystal is effectively filtered out. One of the few calculations which is straightforward (see problem 3.2) is the pseudo-kinematic case, where one has single scattering and exponential attenuation. This calculation shows that the attenuation causes only a few layers at the surface to be sampled, which give rise to modulated reciprocal lattice rods, the width of the modulations being inversely proportional to the *imfp*.

In the full dynamical LEED calculations, the attenuation effect is included by an imaginary potential, V_{0i}. This is similar to the high energy case of RHEED and THEED, but the language is a little different. In TEM, imaginary potentials (V_{0i} and V_{gi}) are used to describe contrast in images caused by inelastic scattering; but these are dependent on the aperture size used, and are typically due to the scattering of phonons and defects. In contrast to plasmons, these scattering events cause a wide angular spread, and very little energy loss. Calculating RHEED intensities is a suitable combination of layer slicing, as in LEED, and high energy forward scattering as in THEED; reviews of these methods have been given by Peng *et al.* (1996) and Maksym (1977, 1999).

There are new electron diffraction techniques emerging, such as DLEED (Diffuse-LEED) (Heinz 1995) and electron holography (Saldin 1997), and continuous development of related theoretical methods (Pendry 1997). The above (outline) discussion has concentrated on the effect of inelastic processes on the interpretation of elastic

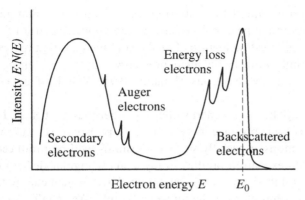

Figure 3.7. Electron energy spectrum, showing secondary, Auger, energy loss and backscattered electrons.

scattering processes. In the next two sections we are concerned with the understanding and use of the inelastic processes in their own right.

3.3   Inelastic scattering techniques: chemical and electronic state information

3.3.1   Electron spectroscopic techniques

If we bombard a sample with electrons or photons, electrons will be emitted which have an energy spectrum, shown schematically in figure 3.7 for the case of electron bombardment. The most well-known historical example is the photoelectric effect, and the modern version in UHV is called photoemission (Cardona & Ley 1978, Bonzel & Kleint 1995). Electron emission is commonly used; for example, secondary electrons are the signal normally used to form an image in the scanning electron microscope (SEM), and AES uses Auger electrons to determine surface chemical composition. Ion emission is also known, but is less widely used.

The problem of measuring the energy spectrum is non-trivial, and is discussed in many books (Bauer 1975, Ibach 1977, Walls 1990, Briggs & Seah 1990, Rivière 1990, Smith 1994); introductions are given by Prutton (1994, chapter 2) and Woodruff & Delchar (1986/1994, section 3.1). The field also supports specialist publications such as the *Journal of Electron Spectroscopy*, and *Surface and Interface Analysis*. There are various possible geometries for the analyzers and the measurements can be performed in an angle-integrated or angle-resolved (AR) mode. Thus we have a profusion of acronyms, e.g. UPS, ultra-violet photoelectron spectroscopy; ARUPS, the angular resolved version of the same technique, which is used to study band structure and surface states; XPS, X-ray photoelectron spectroscopy, also known as ESCA, electron spectroscopy for chemical analysis, so named by Siegbahn *et al.* (1967). The massive body of work by this Swedish group resulted in the Nobel Prize being awarded to Kai Siegbahn in

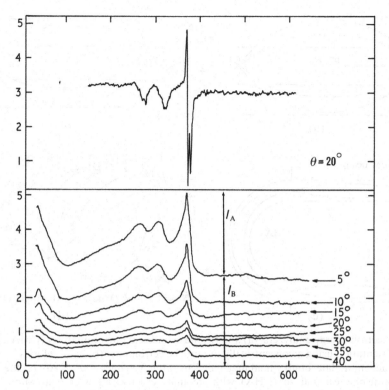

Figure 3.8. $N(E)$ (bottom) and $dN(E)/dE$ (top) spectra as a function of glancing incidence angle for a bulk sample of Cd (Janssen *et al.* 1977, reproduced with permission).

1981. Finally there is electron energy loss spectroscopy, which comes in two varieties (EELS and (high resolution) HREELS), the latter being used primarily for studies of surface and adsorbate vibrational structure (Ibach and Mills 1982, Ibach 1994, Lüth 1993/5).

The various analyzers also have acronyms. The magnetic sector spectrometer may be familiar from analysis using EELS in conjunction with TEM or STEM. It has very good energy resolving power, but collects electrons only over a small angular range; this is well suited to the strongly forward peaked scattering which occurs at TEM energies (>100 keV), as illustrated schematically here in figure 3.1(d). The retarding field analyzer (RFA) is the same arrangement as used for LEED, figure 3.3; the only difference is that one ramps, and may modulate the retarding voltage V on the grid (or the sample), collecting all the electrons with energy $E > eV$, i.e. the RFA is a high pass filter. The advantage of the RFA is simplicity and availability, plus the very large angular collection range; the disadvantage lies in the poor signal to noise ratio inherent in differentiating the collected signal (once or twice) to get the spectrum of interest.

One can appreciate these points by drawing a spectrum such as figure 3.7 and convincing yourself that the signal intensity collected, I, collected by an RFA corresponds to

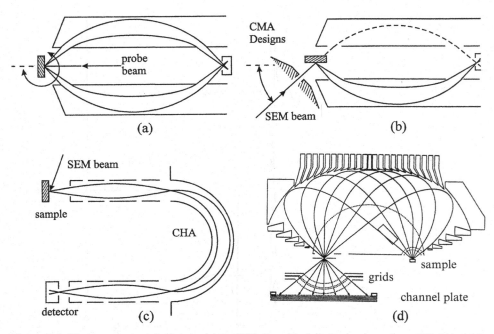

Figure 3.9. Electron energy analyzers: (a) and (b) the cylindrical mirror analyzer (CMA), (a) in the normal orientation using a concentrically mounted electron gun; (b) in an off-axis geometry to accommodate a bulky final magnetic lens (after Venables *et al.* 1980); (c) the concentric hemispherical analyzer (CHA), with pre-analyzer lenses allowing retardation and variable energy resolution, and/or multichannel detection; (d) a display analyzer used for synchrotron radiation (SR) research, with a multi-channel plate (MCP) parallel recording detector (after Daimon *et al.* 1995, reproduced with permission).

$$I = \int_E^{E_0} N(E)\,\mathrm{d}E, \qquad\qquad (3.2)$$

so that an $N(E)$ spectrum corresponds to differentiating the signal once, and a $\mathrm{d}N/\mathrm{d}E$ spectrum to differentiating twice. These two types of spectrum are shown in figure 3.8, ignoring the difference between $N(E)$ and $E{\cdot}N(E)$, which is discussed next.

The electron energy analyzers in common use are the cylindrical mirror analyzer (CMA) and concentric hemispherical analyzer (CHA), shown in figure 3.9. These are both band pass filters, passing a band of energy (ΔE) at a pass energy E, typically by adjusting a slit width (w) to change the energy resolution $\Delta E/E$. These analyzers can be operated with retardation, so that the pass energy is less than the energy of the electron being analyzed; this is easier for the CHA, with retarding lenses in front of the analyzer, and can lead to a high energy resolution in the resulting spectrum. If the analyzer is retarded to a constant pass energy, then the spectrum reflects $N(E)$, which is often peaked at low energies, since secondary electron emission is strong. If there is no retardation, or if the pass energy is a constant fraction of the analyzed energy, then the spectrum reflects $E{\cdot}N(E)$.

The goal of energy analysis is to combine high energy resolving power $\rho = (E/\Delta E)$,

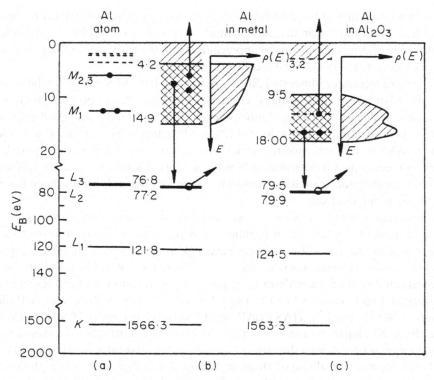

Figure 3.10. Energy levels and density of states of aluminum as (a) an atom, (b) a metal and (c) an oxide (after Bauer 1975, and Chattarji 1976; reproduced with permission).

with a high collection solid angle Ω. It is not very surprising that nature doesn't like you doing that: it suggests getting something for nothing. So the various analyzers have been optimized by all the tricks one can think of, such as second order focusing, where changing the angle of incidence to the optic axis α, produces aberrations of order α^2 or higher. The net result is that the energy resolution looks like

$$\Delta E/E = A(w/L) + B\alpha^n, \tag{3.3}$$

where L is a characteristic size of the analyzer, and A, B and $n \sim 2$ are constants for the equipment (Roy & Carette 1977, Moore *et al.* 1989). If one can detect neighboring energies in parallel, so much the better; this can be done relatively easily with the CHA, but is more difficult with the CMA.

3.3.2 Photoelectron spectroscopies: XPS and UPS

A comparison of the three main analytical techniques which use electron emission can be understood in relation to figure 3.10. UPS uses ultra-violet radiation as the probe, and collects electrons directly from the valence band, whereas XPS excites a core hole with X-rays. The core line is often split by spin-orbit interactions, whereas the valence line is wider because of band broadening, indicated by the $N(E)$ distributions in figures 3.10(b,c). An outline of photoemission models is given by Lüth (1993/5 chapter 6). The wide range of applications can be appreciated from the early case studies compiled by Ley & Cardona (1979) and the text by Hüfner (1996).

The third technique illustrated directly in figure 3.10 is AES, which can be excited by (X-ray) photons or, more usually, electrons. The basic Auger process involves three electrons, and leaves the atom doubly ionized. In general, XPS and AES are used for species identification, and core level shifts in XPS can also give chemical state identification. AES is routinely used to check surface cleanliness. UPS, especially ARUPS, is the main technique for determining band structure (of solids, not just the surface) and can also identify surface states. The surface sensitivity depends primarily on the energy of the outgoing electron.

Some details about the X-ray sources and monochromators used are given by Lüth (1993/5, panel 11) where the importance of synchrotron radiation sources to current research is emphasized. These sources have high intensity over a range of energies and very well defined direction, so that they are well suited to AR-studies; such studies form a substantial part of the wide-ranging programs at synchrotron facilities such as the Advanced Light Source (ALS) in the USA, the Daresbury Synchrotron Radiation Source (SRS) in the UK, HASYLAB or BESSY in Germany, the ESRF in France or the SPring 8 in Japan, to name only a few. Much useful information on these and other programs can be obtained directly via the internet, as indicated in Appendix D.

Until one has visited one of these installations, it is difficult to grasp the scale and complexity of the operation. Although the end product research overlaps strongly with that coming out of small scale laboratories, the tradition derives more from large budget particle physics, with the consequent need for substantial long range planning and technical backup. By the time one gets to the individual researcher/user, who is typically based at a university or industrial laboratory located some distance away, and who has a limited amount of 'beam time' allotted on a particular 'station', one is into social structures in addition to science. Safety training, where to (and how much) sleep, group organization and continuity are all extremely important factors influencing whether good work is produced. Stress is important as a spur to achievement in science, but sometimes it can get out of hand. As one who has never actually worked in such a facility, I can imagine that it takes a bit of getting used to, and strategies for effective working need to be thought about explicitly. Nonetheless, the upside is that all this wonderful equipment, and expert help, is *available* to help you produce the results you need!

There have been several attempts to develop display analyzers, where the angular information is displayed in parallel at a given energy, which is swept serially (or vice versa). All these analyzers are technically demanding attempts to utilize the beam time and low counting rate efficiently, and have typically been constructed for a synchrotron environment (Eastman *et al.* 1980, Leckey *et al.* 1990, Daimon 1988, Daimon *et al.* 1995). This last spectrometer is shown in figure 3.9(d), indicating that several finely fashioned grids are required to keep the fields in the different regions of the spectrometer isolated from each other. Designing a usable wide angle, gridless analyzer design remains quite challenging (e.g. Huang *et al.* 1993), but there are always some projects in train to try and improve analyzer performance, typically aiming to take advantage of parallel recording in either energy or angle.

There are many examples of ARUPS results in the literature; some from Au (111) and (112), Al (100) and (110) and other metals are described by Lüth; we can see that details of the band structure can be mapped out from such data, but it is quite difficult to separate surface and bulk states from a single set of spectra. The extraction of the surface state part of the spectrum for Si(100)2×1 is also described by Lüth, where it is seen that the data agree best with the dimer (pairing) model of the reconstruction. Thus detailed analysis of the electronic structure can in principle provide atomic structural information. These studies of electronic structure have occupied some of the best experimentalists and theorists over a substantial period (Siegbahn *et al.* 1967, Cardona and Ley 1979, Himpsel 1994, Bonzel & Kleint 1995, Hüfner 1996).

3.3.3 Auger electron spectroscopy: energies and atomic physics

In this and the following two sections, Auger electron spectroscopy (AES), its use as a quantitative analysis tool, and on a microscopic scale, is treated in more detail. The Auger process was discovered by Pierre Auger in 1925–26, when he observed tracks of constant length in a cloud chamber, and thereby inferred that the particles had constant energy; however, the technique was not used to study surfaces until the late 1950s and 1960s. Auger electron energies are closely related to the corresponding X-ray energy, and most usually are described in X-ray notation. For example, figure 3.11 shows the level scheme associated with the Si $KL_1L_{2,3}$ transition, one of a set of KLL transitions, the strongest of which is experimentally observed at ~ 1620 eV. The final state contains two holes in the L shell; the Auger process is a competing way to fill the initial core hole, and so is an *alternative* to Si K X-ray emission.

These Auger energies have historically been given by approximate formulae, e.g. by Chung & Jenkins (1970):

$$E(KL_1L_2) = E_K(Z) - 0.5\,\{E_{L1}(Z) + E_{L2}(Z)\} - 0.5\{E_{L1}(Z+\Delta) + E_{L2}(Z+\Delta)\}, \quad (3.4)$$

where the use of the average energy is due to being unable to distinguish which electron filled the core hole. The Δ has been used to indicate that the final emission is from an ion, not a neutral atom, which shifts the final energies downwards slightly. For practical surface analysis, one needs to know that these energies are known, and are typically displayed on a chart in every surface science laboratory. The energy measured does, however, depend on whether you are measuring in the $dN(E)/dE$ or N' mode, where the negative-going peak is typically quoted, or in the $N(E)$, or $E \cdot N(E)$ mode, where the positive-going peak is quoted (see figure 3.8). These can be separated by several electronvolts; the width of the Auger peak is typically 1–2 eV, due to a rather short lifetime before Auger emission. The Auger peak width can be further broadened by overlapping peaks, by wide valence bands, or by analyzers which are set to increase sensitivity at the expense of resolution. In addition, to quote the absolute energy relative to the vacuum level of the element, a negative correction to (3.4) equal to the work function of the analyzer (4.5 eV typically) is needed (McGilp & Weightman 1976);

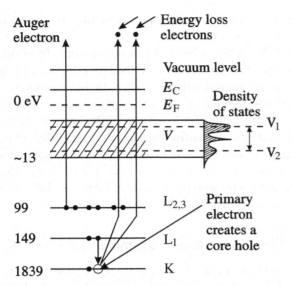

Figure 3.11. Si KLL Auger scheme, after Chang (1974). See text for discussion.

however, if one quotes energies relative to the Fermi level, this correction to (3.4) is not required (Seah & Gilmore 1996).

The theory of Auger emission is rooted in atomic physics, and is only modified slightly by solid state and surface effects. Thus the calculation needs to take into account whether the atom in question is L–S or j–j coupled, illustrated for the KLL series in figure 3.12. As atomic calculations have improved, formulae which explicitly acknowledge the energy relaxation due to the final state ion, and the spectroscopic configuration of the two holes have been developed (Shirley 1973, Weightman 1982); for example the main KLL transition in silicon has a $^1D_2$ final state. Si KLL is a primarily a case of L–S coupling, since the atomic number is small, whereas Ge KLL is in the intermediate coupling regime. A high energy resolution spectrum shows these effects, as for Ge LMM in figure 3.13 and for Si KLL and Ag MNN later in figure 3.25. What one can see in the spectrum depends on both energy resolution and signal to noise ratio, and if X-ray excitation is used, the secondary electron background can be much reduced over electron excitation. So the peak to background ratio, while very useful for analysis as we shall see later, does depend on the analyzer settings, the mode of excitation, and the excitation energy.

The other point which is clear from the atomic physics is that X-ray and Auger emission are alternatives: either/or, not both. For low energy transitions, Auger emission is strongly favored. The Auger efficiency $\gamma = 1 - \varpi$, where the X-ray fluorescence yield ϖ is shown in figure 3.14, taken from Burhop (1952). It is noticeable that this work was done a long time ago in the context of atomic, not surface physics. Figure 3.14 shows that the proportion of K-shell Auger emission is greater than 0.5 up to about $Z = 30$ (zinc). So, typically, one switches which transition is used as we move up the periodic table: KLL transitions for light elements, LMM after that, and then MNN.

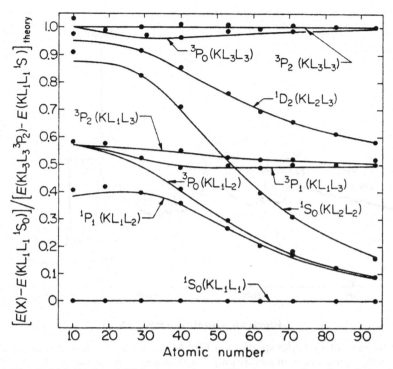

Figure 3.12. Reduced KLL Auger energy differences as a function of atomic number, corresponding to the transition form L–S coupling at low atomic number to j–j coupling at high atomic number. The nine components are labelled on the plot, with theoretical curves as full lines and points from experiment (after Shirley 1973, reproduced with permission).

The Auger effect is due to the Coulomb interaction between the core hole and the ejected electron which leaves the other hole behind. The transition rate for this process was solved very early in the history of quantum mechanics by Wentzel in 1927, who calculated transition matrix elements (Fermi's golden rule) between the initial (atomic) and the final (continuum) states. This is set out in detail for both relativisitic and non-relativistic cases by Chattarji (1976). One can use simple arguments which bring out what is basically going on as follows. The Auger transition rate,

$$b_n = (2\pi/\hbar)|\langle \chi_f \psi_f | e^2/|(\mathbf{r}_1 - \mathbf{r}_2)| | \chi_i \psi_i \rangle|^2, \tag{3.5}$$

where the wavefunctions ψ_f = emitted electron (continuum) and χ_f = electron in the lower band. On the other hand, the X-ray fluorescence yield is the one-electron dipole matrix element, namely

$$a_n = (2\pi/\hbar)k|\langle \chi_f | e\mathbf{r} | \chi_i \rangle|^2, \tag{3.6}$$

where the constant $k = 4/3 \, (\omega/c)^3$, with the radiation frequency $\omega/2\pi$ given by $\hbar\omega = E_K - E_L$. Now if we assume Bohr-like atoms $E_K \sim -Ze^2/r$ and $r \sim a_0/Z$, with the

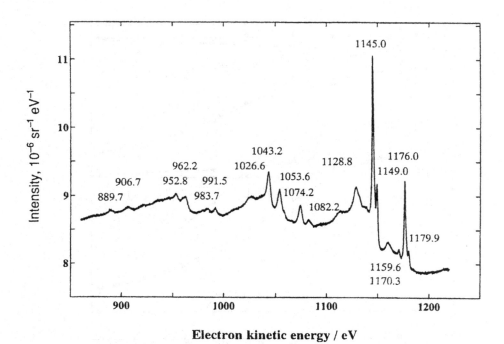

Figure 3.13. High resolution AES spectrum of Ge LMM for 5 keV incident energy (after Seah & Gilmore 1996, reproduced with permission). The strongest peaks, within the $L_2 M_{4,5} M_{4,5}$ series at 1145 and $L_3 M_{4,5} M_{4,5}$ series at 1180 eV have $^1G_4$ symmetry; the smaller peaks at higher E consist of three overlapping lines with ^3F symmetry (McGilp & Weightman 1976, 1978).

Bohr radius $a_0 = \hbar^2/(me^2)$, we can deduce that $\omega_K \sim Z^2$, and $\omega^3 r^2 \sim Z^4$. So the X-ray fluorescence yield

$$\varpi = \Sigma a_n / (\Sigma a_n + \Sigma b_n) = 1/(1 + a_K Z^{-4}), \tag{3.7}$$

where a_K is a constant for K-shell emission, as shown in figure 3.14. To recap: these models all depend on atomic and ionic wavefunctions for energies, transition rates, fluorescence yields, and are not specific to solids or surfaces. Note that several books on surface analysis pitch straight into the technical details without mentioning any of this at all; but atomic physics lies behind many of the finer effects.

3.3.4 AES, XPS and UPS in solids and at surfaces

The state of matter affects the lineshape, and causes energy shifts. If the transitions involve the valence band, then we refer to LVV etc, or more generally to core–valence–valence (CVV) transitions. For example, Si LVV has a transition at ~90 eV; these transitions are sensitive to chemical state, as shown in figure 3.15, and Al LVV has a different lineshape in metallic Al and in aluminum oxide (figure 3.10(c)). Many authors have studied core level shifts in UPS and XPS. An account of this history in the context of semiconductor surfaces is given by Himpsel (1994); the use of spectral shapes as 'fingerprints' of particular chemisorbed states is discussed by Menzel (1994).

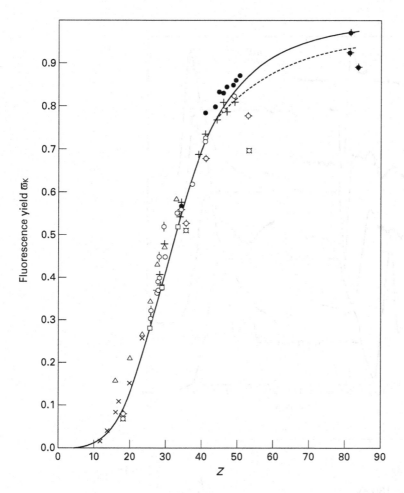

Figure 3.14. Fluorescence yield ϖ_K for K-shell X-rays compared with non-relativistic (full line) and relativistic (dashed line) models, compared with experimental data measured between 1926 and 1950 (after Burhop 1952, reproduced with permission).

The precise shape of an Auger or photoelectron line is an ongoing topic which is researched in a few specialist groups worldwide. In the simplest case, the UPS spectrum gives this density of states directly, and the lineshape of LVV Auger transitions reflects a self-convolution of the valence band density of states. Both of these shapes are different in a compound containing the element, from that in the element itself. Early Auger spectra for Si in different chemical environments are compared with optimized UPS spectra in individual oxidation states at the Si–silicon dioxide interface in figure 3.15 (Madden 1981, Himpsel *et al.* 1988, Himpsel 1994). The shift to lower energies in the compounds is apparent from these figures.

For both UPS and AES, many-body final state effects can cause significant changes in the lineshape, if the valence hole in the final state feels the effect of the core hole in the initial state. This depends on the energy cost of localizing two holes on the same

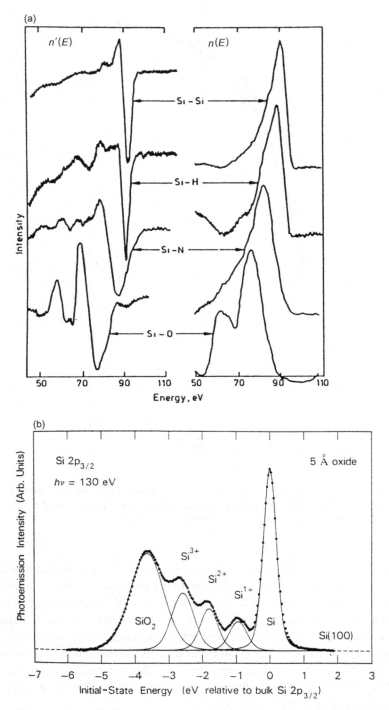

Figure 3.15. (a) Si Auger spectra in various environments (after Madden 1981); (b) Si plus thin oxide, as observed in synchrotron radiation UPS by Himpsel *et al.* (1988), showing a shift to lower energies with increasing oxidation state (both diagrams reproduced with permission).

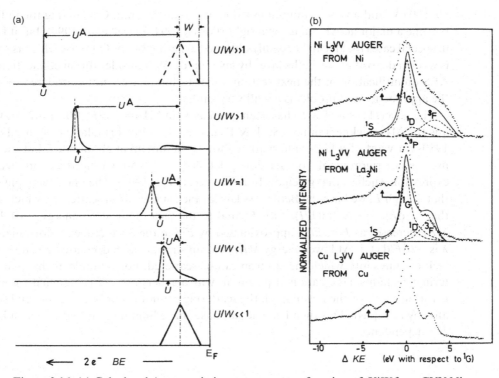

Figure 3.16. (a) Calculated Auger emission spectrum as a function of U/W for a CVV Ni spectra; (b) comparison of Ni L_3VV in Ni and in La$_3$Ni with Cu L_3VV Auger spectra, including calculated (full and dashed) lines based on appropriate values of U for the different atomic states (after Bennett *et al.* 1983, reproduced with permission).

atom (the correlation energy U, often associated with the 'Hubbard' model), versus the valence band width, W (Verdozzi & Cini 1995). If U/W is $\ll 1$, we see an unshifted self convolution (self fold) of the valence band, corresponding to a delocalized, band-like final state; in the other limit ($W/U \ll 1$) we see a shifted, but atomic-like, line. The Auger CVV transitions then show multiplet structure where the details depend on the variation of the matrix elements with the orbital character of the final state.

As a result of these considerations the lineshape can switch from one type to the other across a series of alloys. This effect happens particularly for nearly filled d-shells, as in Pd or Ni alloys and silicides. A schematic diagram of this effect plus an example of Ni LVV in Ni-based alloys (Fuggle *et al.* 1982, Bennett *et al.* 1983) is shown in figure 3.16, but there is ongoing discussion about how such models apply to unfilled bands, where many different possibilities for screening exist. This type of work is reviewed by Weightman (1982, 1995) and Hüfner (1996), and is becoming more important as the finer aspects of electron spectroscopy, including spin-polarized electron spectroscopy of magnetic materials as discussed in chapter 6, can be interpreted in terms of the electronic structure.

All these electron spectroscopies derive their surface sensitivity from the low inelastic mean free path (imfp) of electrons, which has a minimum in the neighborhood of

50–100 eV, and a typical minimum value of around 0.5 nm. Calculating the imfp has become a major activity in its own right (Powell 1994, Powell *et al.* 1994), but it is only in special cases that it can be readily obtained experimentally. One case which is soluble is when a deposit grows in the layer by layer mode. We consider this in connection with AES quantification in the next section; other techniques are not considered here, but the approaches and problems are all very similar.

The general form of $\lambda(E)$ has been given as λ (in ML) $= 538/E^2 + 0.41$ $(aE)^{1/2}$, where a is the ML thickness in nm (Seah & Dench 1979). This formula, widely used in the 1980s for want of a better alternative, shows a minimum at about 50eV, but the form used is not rigorous; if you are doing detailed work you will need a more accurate expression, such as that given by Tanuma *et al.* (1991, 1993). The underlying physics is that at high E, we have the Bethe loss law for inelastic electron scattering, which shows that $(- \mathrm{d}E/\mathrm{d}x) \sim E^{-1}\ln(E/I)$, with I equal to the mean ionization energy ~ 10–15 eV. $\lambda(E)$ increases as $E/\ln(E/I)$, approximated by $E^{1/2}$ in the Seah & Dench formula, since λ is $\sim (\mathrm{d}E/\mathrm{d}x)^{-1}$. At lower energy $\lambda(E)$ goes up strongly as E decreases, because of the lack of states into which the electron can be scattered. For example, if the main scattering mechanism is creation of plasmons with an energy of ~ 15 eV, then if the energy is within 15 eV of the Fermi level, the scattering cannot occur because the final state is already occupied. This then becomes a phase space limitation at low E, which has an E^{-2} dependence.

3.4   Quantification of Auger spectra

3.4.1   *General equation describing quantification*

The general equation governing the Auger electron current, I_A caused by a probe current I_p can be written down straightforwardly, but is not immediately transparent, and really needs to be explained using a schematic drawing, such as figure 3.17. The incoming electron causes an electron cascade below the surface, whose spatial extent is typically much greater than the imfp. For example, the spatial extent is about 0.5 μm at an incident energy $E_0 = 20$ keV, but also depends on the material and the angle of incidence, θ_0. As a result Auger electrons can be produced by the incoming primary electron beam, and also by the backscattered electrons as they emerge from the sample; the Auger signal intensity thus contains the backscattering factor, R, which is a function of the sample material, E_0 and θ_0.

The ratio I_A/I_p can be expressed as a product of terms describing the production and detection of the Auger electrons, as first developed by Bishop & Rivière (1969). The Auger yield Y is the number of Auger electrons emitted into the total solid angle ($\Omega = 4\pi$ sterad). It is therefore not dependent on the details of the analyzer. The detection efficiency D of the analyzer can be written as $(T \cdot \varepsilon)$, where T is a function $f(\Omega_a/4\pi)$, Ω_a being the solid angle collected by the analyzer, and ε is $f(\Delta E/E)$, most simply $\varepsilon = (\Delta E/E)$. Thus

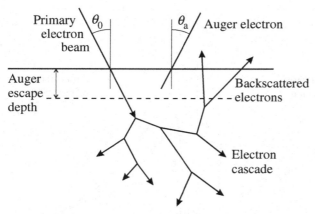

Figure 3.17. Schematic diagram of electron scattering in a solid, indicating the incident and detected angles, θ_0 and θ_a, plus the role of backscattered electrons in determining the Auger signal strength. The escape depth is qualitatively the thickness of the region from which most of the detected Auger electrons originate, of the same order as the imfp discussed in the text.

$$I_A/I_p = Y \cdot D = [\sigma \gamma R] \cdot \sec\theta_0 \cdot N_e(T \cdot \varepsilon). \tag{3.8}$$

Here we have Y expressed as the cross section for the initial ionization event (σ), the Auger efficiency (γ), discussed in outline in section 3.3, and the factor R. The $\sec\theta_0$ term describes the extra ionization path length caused by having the primary beam at an angle to the sample normal. Finally N_e is the effective number of atoms/unit area contributing to the (particular) Auger process.

What we actually want to know is: given a measured signal I_A, how many A-atoms are there on the surface? Typically there is not a unique answer to such a simple question, because the signal depends not only on the number of atoms but also on their distribution in depth. There are two cases which can be solved uniquely, which are instructive in showing how such analyses work. The first is when all the atoms are in the surface layer: then $N_e = N_1$, and if one knew all the other terms in the equation, we could determine N_1.

The second case is when the atoms are uniformly distributed in depth: in this case we can show that $N_e = \lambda N_m$, where N_m is the bulk (3D) density of A-atoms. The proof of the second case is as follows. We work out

$$(N_e \cdot T) = (1/4\pi)\int\int\int N(z) \exp(-z/\lambda\cos\theta_a) \cdot \sin\theta_a d\theta_a d\phi_a dz, \tag{3.9}$$

where the integral is over the two analyzer angles (θ_a, ϕ_a) and depth z. The path length in the sample in the direction of the analyzer is $z/\cos\theta_a$, so we are assuming exponential attenuation without change of direction; this is reasonable for inelastic scattering of the outgoing electrons. Because $N(z) = N_m$ is constant, we can do the integration over z first. This gives

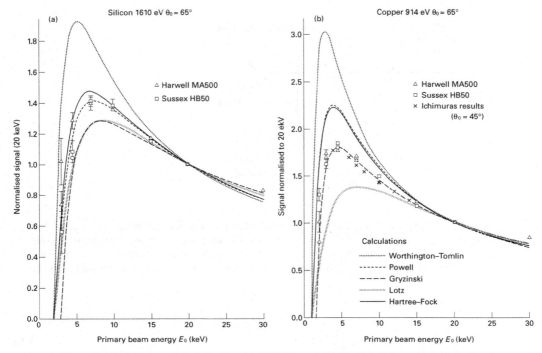

Figure 3.18. Auger peak heights for (a) Si KLL and (b) Cu L_3MM as a function of primary energy E_0, normalized to 20 keV, measured on Auger microprobe instruments at Harwell and Sussex, compared to the product (σR) calculated with the cross sections indicated (after Batchelor *et al.* 1989, reproduced with permission).

$$(N_e \cdot T) = \lambda N_m \cdot (1/4\pi) \iint \cos\theta_a \cdot \sin\theta_a \, d\theta_a \, d\phi_a = \lambda N_m \cdot f(\theta_a, \phi_a). \tag{3.10}$$

The angular double integral is just the cosine electron emission distribution, integrated over the solid angle of the analyzer; so $f = T$, $N_e = \lambda N_m$, as required.

A detailed experimental and computational study of the Auger signals from bulk elements has been performed, amongst others, by Batchelor *et al.* (1988, 1989). These studies show that the dependence of the Auger peak height on primary beam energy E_0 explores the variation of (σR) as shown in figure 3.18 for Si KLL and Cu L_3MM; the variation with θ_0 is determined by $(R\sec\theta_0 \cdot T)$. The dependence on atomic number Z is complicated, since all the above material variables and the Auger energy are properties of the individual element in question.

To make these comparisons we took a particular ionization cross section, and calculated the Auger backscattering factor $R = 1 + r$ with this cross section, where r is defined as

$$r = (1/\sigma(E_0) \sec\theta_0) \iint \sigma(E)(d^2\eta/dEd\theta)\sin\theta \, d\theta \, dE, \tag{3.11}$$

where the normal electron backscattering factor η is given by

$$\eta = \iint (d^2\eta/dEd\theta)\sin\theta \, d\theta \, dE. \tag{3.12}$$

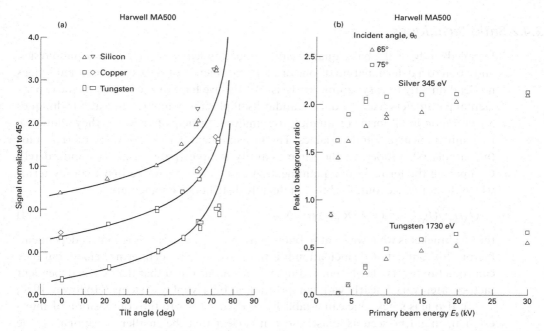

Figure 3.19. Variation of Auger peak to background ratio for (a) Si, Cu and W with incidence angle θ_0, compared with model calculation based on (3.13); (b) Si and Cu with primary beam energy E_0 at two incidence angles $\theta_0 = 65$ and $75°$ (after Batchelor *et al.* 1989, reproduced with permission).

This means that the Auger backscattering factor can be written as $R = 1 + \beta\eta$, where the factor β is typically greater than 1, and depends on the energy and angular distribution of backscattered electrons, and the Auger energy in relation to the beam energy. Comparison of the energy dependences of the cross sections and other checks on absolute values has advanced knowledge of which cross sections are reliable (Batchelor *et al.* 1988, Seah & Gilmore 1996).

The peak to background ratio (P/B or PBR) can be measured accurately and is very useful as an approximate compensation for topographic effects. The angular dependence is rather flat as shown in figure 3.19(a) for Cu, Si and W Auger electrons measured in two different instruments. The lines are a simple model calculation based on

$$P/B \propto \sec(\theta_0)R(\theta_0)/\eta(\theta_0), \tag{3.13}$$

and it is seen that this model, normalized at normal incidence, gives a good fit out to at least $\theta_0 = 70°$. At high incidence angle, there is much less variation in the back scattering factor as a function of atomic number, and the background becomes dominated by secondary electrons, which have similar excitation mechanisms to Auger electrons. This is mirrored in the PBR as a function of E_0, shown in figure 3.19(b). Above 15–20 keV both Cu L_3MM and Si KLL show no dependence on E_0; this effect sets in at lower E_0 for lower energy transitions, where the background comprises secondary rather than backscattered electrons (Batchelor *et al.* 1989).

3.4.2 Ratio techniques

Detailed study of the basic quantification equation leads to some interesting physics, and often to a determination of one or more parameters of the experiment, but it does not lead directly to easy sample analysis. For this we have to keep many of the experimental parameters fixed, and use standard samples for comparison. Ratio techniques are common in all forms of quantitative analysis, principally because they allow one to eliminate instrumental variables. The measurement is then the value of (I_A/I_p) for the sample (s), ratioed to the same quantity for the (pure element) standard (el). Comparing the terms in the quantification equation, we can see that the only terms which do not cancel out, for bulk, uniformly distributed samples, are

$$(I_A/I_p)_s/(I_A/I_p)_{el} = (R_s\lambda_s/R_{el}\lambda_{el})\cdot(N_s/N_{el}); \qquad (3.14)$$

the last bracket is what we want to know, and the previous term is a 'matrix dependent' factor. Without detailed calculation it is not obvious how such terms behave, but they can sometimes vary slowly. For example, it has been shown that the matrix dependent factors often vary linearly with composition in binary alloy systems (Holloway 1977, 1980). If one is stuck, one can establish standards closer to the composition of interest. This might be seen as a last resort: it is clear that the smaller extrapolation one makes, the more accurate the result is likely to be.

Many authors have developed programs for studying such effects, and national standards organizations (National Institute for Standards and Technology (NIST), National Physical Laboratory (NPL), etc.) are involved on an ongoing basis. The author most concerned in the USA is C.J. Powell of NIST, who has written extensively on this topic; his UK counterpart is M.P. Seah, who works for NPL; he has published an interesting commentary on the needs of, and competing pressures in this field (Seah 1996), and they have written several updates on the whole field (e.g. Powell & Seah 1990). It is no accident that they have concerned themselves with these aspects of quantitative analysis, and indeed run conferences with the title *Quantitative Surface Analysis* (QSA) at which such matters are discussed in great detail (Powell 1994); QSA-10 was held in 1998. Round-robins, which use standard samples to be tested in the different laboratories, are one of the methods used to find out where there is common ground and where there are difficulties. By characterizing the analyzers carefully, the spectra can be calibrated on an absolute scale, such as the example given earlier for Ge in figure 3.13 (Seah & Gilmore 1996).

There are also major consulting businesses based on such analyses, since the services and expertise are often very expensive to maintain in-house; the best known may be C.A. Evans and Associates. These topics are not discussed further here, but the flavor of this work can be obtained from the specialist books, especially Smith (1994); he too worked for NPL, and gives a list of significant papers by the above authors and others. The brochures and web-sites of such organizations are an increasing resource, which can be accessed via Appendix D.

A typical 'surface science' application of quantitative AES is to distinguish layer by layer from other types of growth. Layer growth is a case which one can work out easily if we disregard (the minor) changes in the backscattering factor; the experimental ratio is (I_A/I_p) for a multilayer divided by (I_A/I_p) for a ML. Here we assume that there are N_1 atoms in the first layer, N_2 in the second and so on, and their spacing is d. Then we can work out the signals at a coverage θ between n and $n+1$ ML from both the layers and the substrate, as sums which take into account the attenuation. For example, if $n=1$, and we neglect attenuation within the first layer

$$N_e = N_1[(2 - \theta) + (\theta - 1)\exp(-d/\lambda\cos\theta_e)] + N_2, \tag{3.15}$$

where the first (second) term in square brackets correspond to the proportion of the first layer which is uncovered (covered) by the second layer, and so on. This relation leads to a series of straight lines, often plotted as a function of deposition time, from which the λ can be deduced in favorable cases.

The simplest case, where there are the same numbers of atoms in each completed ML, can be worked out explicitly. In that case, the slope of the second ML line ratioed to that of the first ML gives $\exp(-d/\lambda\cos\theta_e)$, from which d/λ can be extracted if the effective analyzer angle θ_e is known. This effective angle is given by

$$\cos\theta_e = (1/\Omega_a)\int\int\cos\theta_a\cdot\sin\theta_a\,d\theta_a\,d\phi_a, \tag{3.16}$$

with the integrals taken over the analyzer acceptance. For detailed studies, it is advisable to construct computer programs which take the analyzer geometry into account, and then perform the angular integration numerically. There has been much discussion over what λ really is in such experimental comparisons; it is now accepted to be the attenuation length (AL), which is shorter than the imfp due to elastic (wide angle) scattering (Dwyer & Matthew 1983, 1984, Jablonski 1990, Matthew *et al.* 1997, Cumpson & Seah 1997). However, as pointed out by the last authors, unless the integration of (3.16) is performed separately (as implied here), the AL also depends on the type of analyzer used and the angular range accepted; thus in some papers the AL is not a material constant, and one should beware of using published values uncritically.

There are many layer growth analyses in the literature, in some cases with a large number of data points showing relatively sharp break points at well-defined coverage, such as for Ag/W(110) (Bauer *et al.* 1977). Experiments with fewer data points can still lead to firm conclusions, using the comparison with a layer growth curve of the type reported here. Such an analysis is shown in figure 3.20, which shows both the Auger spectra for a series of Ag deposits on Si(001) and the corresponding Auger curves as a function of coverage at both room and elevated temperature (Hanbücken *et al.* 1984, Harland & Venables 1985). This is a case where growth more or less follows the 'layer plus island', or Stranski–Krastanov (SK) mode, discussed in more detail in chapter 5. From the Auger curves we see that at room temperature, layer growth is followed for approximately 2 ML, but then experiment diverges from the model, indicating roughening or islanding. The high temperature behavior is much more extreme, as the first layer is ≤ 0.5 ML thick, and islands grow on top of, and in competition with, this dilute layer (Luo *et al.* 1991, Hembree & Venables 1992, Glueckstein *et al.* 1996).

The Ag/Si(111) and Ag/Ge(111) systems have also been studied, with the $\sqrt{3}$ reconstructed layer at high temperatures having a thickness of around 1 ML. The exact

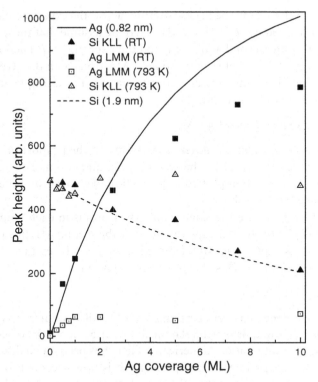

Figure 3.20. Auger $E \cdot N(E)$ peak intensities versus coverage for Ag/Si(001) at deposition temperatures $T = 293$ K (closed squares for Ag and triangles for Si) and at $T = 773$ K (open symbols). The lines are numerical layer growth calculations using the actual analyzer geometry and for an inelastic mean free path $\lambda = 0.82$ nm for Ag (355eV) in Ag (solid line) and $\lambda = 1.90$ nm for Si (1620 eV) in Ag (dashed line) (after Hanbücken *et al.* 1984, replotted with permission).

coverage of such layers has been quite controversial (Raynerd *et al.* 1991, Metcalfe & Venables 1996, Spence & Tear 1998), because the layers are only slightly more stable than the islands, and consequently kinetic effects play an important role. There are now many examples, from these and other growth systems, which show characteristic Auger curves associated with layer, island and SK growth. It is only by detailed quantitative analysis, with for example known imfp values, that one can make such definitive statements from Auger curves on their own.

Some recent examples are discussed by Li *et al.* (1995), Venables *et al.* (1996), Venables & Persaud (1997) and Persaud *et al.* (1998), where the layer growth analysis was used to investigate interdiffusion in Fe/Ag/Fe(110) and surface segregation in Si/Ge/Si(001) hetero-structures. The purpose of a detailed layer growth analysis is often to show that the system does *not* follow the layer growth mode, as discussed for these cases in sections 5.5.3 and 7.3.2. Many authors have used *deposition time* as the dose variable, and then state that the break point corresponds to 1 ML coverage; this is unfortunately bad logic, since the coverage is the *independent* variable and the Auger intensity is the *dependent* variable. Used in the way described here, the Auger curve needs to be calibrated independently, e.g. via a quartz oscillator or Rutherford back-scattering spectroscopy (RBS), as discussed, for example, by Feldman & Mayer (1986).

3.5 Microscopy-spectroscopy: SEM, SAM, SPM, etc.

3.5.1 *Scanning electron and Auger microscopy*

Scanning electron microscopy (SEM) is now a routine tool which has been extensively commercialized; many experimental scientists need access to a high performance, but non-UHV, SEM. For example, in nanotechnology, the sizes of the structures are so small that they cannot be seen with optical microscopes; SEM is often the easiest choice to obtain better resolution. Electron beam lithography is an important fabrication technique, particularly for mask manufacture, and SEM is used routinely in quality control of the devices produced in this (and other) ways. Via my son's profession in biology/biochemistry, I have seen that SEM pictures of cells and cell organization have taken a central place in even the introductory text books on these subjects. Thus SEM is in danger of being taken for granted, due to the life-like, apparently 3D images which are so readily produced. At the same time, contrast mechanisms are typically not understood in detail, due to the relatively complex nature of secondary electron production in solids and the ill-defined collector geometry.

In a UHV, clean surface environment, there are several SEM-based techniques which are surface sensitive at the ML and sub-ML level. The main SEM signal is based on collecting secondary electrons, which typically form a large proportion of the emitted electrons (see figure 3.7). In several papers with collaborators, we have shown that the low energy secondary electron signal is very sensitive to work-function and other surface-related changes; by biasing the sample negative to a bias voltage V_b between -10 and $-500\,$V, we can obtain biased secondary electron images (b-SEI), and ML and multi-layer deposits can be readily visualized and distinguished (Futamoto *et al.* 1985). UHV-SEM is particularly useful when combined with AES and RHEED (Venables *et al.* 1986, Ichikawa & Doi 1988). Progress in understanding secondary electron contrast mechanisms, including in a clean-surface context, has been reviewed by Howie (1995).

Scanning Auger microscopy (SAM) is the child of AES and SEM. A fine primary beam is used, scanned sequentially across the sample at positions (x, y) as in SEM, and the emitted electrons are energy analyzed as in AES. We can now (attempt to) perform various types of analysis, such as a *spot mode analysis* (scan the analysis energy E at fixed (x_0, y_0)), an *energy-selected line scan* (scan x at fixed y_0 and analysis energy E_A), or an *energy-selected map or image* (scan x and y at analysis energy E_A). These attempts are subject to having a long enough data collection time and high enough beam current to achieve a satisfactory SNR. Some examples are shown in figures 3.21 and 3.22. In particular, you can notice that the SNR of the b-SE linescans is very good, and the b-SE images are reasonably clear; next good are the energy-selected line scans. The most difficult/time-consuming are energy-selected images, especially if difference images are

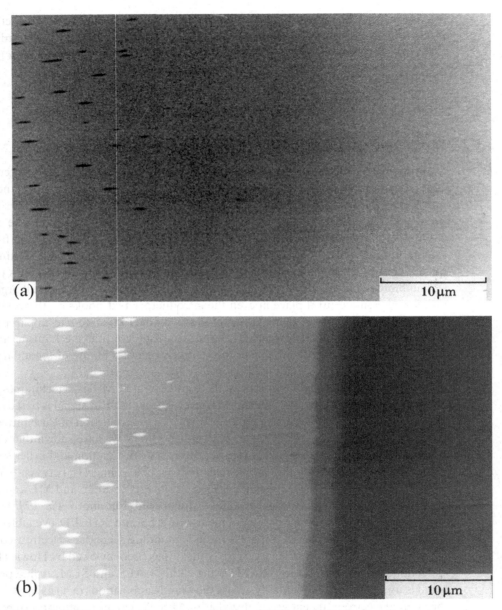

Figure 3.21. Biased secondary electron images of 5 ML Ag/W(110), deposited past a mask edge at $T = 673$ K and deposition rate $R = 0.5$ ML/min, observed with a 30 keV probe beam at $\theta_0 = 70°$; the tilt is around the horizontal axis, so the vertical scale is compressed by $\operatorname{cosec}\theta_0$. (a) Zero bias, with Ag islands showing dark, and the layers essentially invisible; (b) bias voltage $V_b = -200$ V, showing the islands bright, and the 1 and 2 ML steps clearly visible (after Jones & Venables 1985, reproduced with permission).

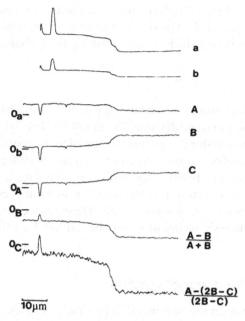

Figure 3.22. Biased secondary and Auger line scans of the edge shown in figure 3.21. Secondary electron scans (a) and (b) with and without zero suppression respectively. Energy selected line scans at A 346, B 385 and C 424 eV electron energy images, plus Auger line scans based on the algorithms shown (after Jones & Venables 1985, reproduced with permission).

needed. Monte Carlo modeling of electron scattering in solids has been developed to study contrast mechanisms, including high spatial resolution studies of analytical techniques based on SEM and SAM (El Gomati *et al.* 1979, Shimizu & Ding 1992).

Following on the discussion in section 3.4, we can think about the extraction of Auger data from energy-selected line scans and images, and the quantification of such information. We need to use ratio techniques for several reasons. First, typical samples are not flat, and may be extremely rough, or can involve changes in backscattering factor. This leads to variations in $(r\sec\theta_0 \cdot T)$; such changes in the Auger peak channel (A) can often go in the opposite direction from what is expected, for example in figure 3.22. In this case, the back-scattering from the Ag islands is less than that from the W substrate. Thus the signal at channel A *reduces* as the scan crosses the islands; but at the background channel B, it reduces more.

Second, one needs to have Auger information which (if possible) is independent of these changes in the background spectrum, and of beam current fluctuations. The first goal is not entirely possible, but one can make a good attempt. By taking line scans or images at one or more energies in the background above the peak (B and maybe C), then difference techniques can be constructed to extract better Auger information. The ratio most commonly used is a quasi-logarithmic measure of Auger intensity based on two channels only,

$$I(\text{difference})/I(\text{sum}) = (A - B)/(A + B), \tag{3.17}$$

which is a first approximation to $(2EN(E))^{-1} \cdot d(EN(E))/dE$ in the fixed retard ratio mode (Janssen *et al.* 1977, Prutton *et al.* 1983). The simplest linear measure is based on an extrapolation of the background from C to B to A. Assuming these channels are equally spaced, then the

Peak to background ratio $(P/B) = (A - 2B + C)/(2B - C)$.   (3.18)

We can see from figure 3.22 that this measure is usually noisier than the ratio based on the two channels A and B, as discussed in some detail by Frank (1991). This is because one is in effect measuring the background slope at each pixel, as well as the peak height. If you are unhappy about the noise level in images, you can always trade SNR for image resolution by digital image processing. The result is not always very pleasing, nor even necessary, since the eye does this for you anyway. It is amazing how well our eyes/brain are able to extract feature information from very noisy data. These issues, which are common to many imaging and analysis techniques, can be explored further via problem 3.3.

3.5.2   Auger and image analysis of 'real world' samples

A particular problem faced by analysts in the 'real world' is that their samples contain many different elements; they may have rough surfaces, and this may interfere with quantitative analysis. However, they may not be so concerned about quantitative information at every point in the image; association of specific types of qualitative information with each point may be more informative. This type of ratio technique has been developed by several groups, and an illustration is given in figure 3.23 (Browning 1984, 1985). Prutton's group at York has furthered these techniques, originally developed for satellite imaging by NASA/JPL, as described in more recent papers (Walker *et al.* 1988, Prutton *et al.* 1995).

In such an approach an SEM picture is taken of the whole field of view and a survey spectrum is taken from this area. This shows many peaks, some of them very small (figure 3.23(a)). The spectrum is used to identify energies (channels) at which information will be recorded at each point on the image, typically a peak and a background channel at higher energy, for each of the elements of interest. This information is collected and stored digitally. Before images are made with this information, scatter diagrams, such as figures 3.23(b), (c) and (d) are constructed. These show that the ratio data cluster in well defined regions of the scatter diagram, and it is easy for the analyst to identify the clusters as particular phases, at least tentatively.

At this stage, one can put a software 'mask' over the data, as shown schematically by the rectangles in figure 3.23(c) and (d), and use all the data which fall within this mask to form an image. An unknown phase was identified which contained Ti, Si, some S and also P. By setting limits on the various ratios, an image can now be produced from the stored data set, which shows the spatial distribution of this particular range of compositions. In this case it was shown that the 'phase' was formed in the reaction zone between the SiC fiber and the Ti alloy which made up the composite material.

A simpler two component system, such as an evaporated tungsten pad on silicon can

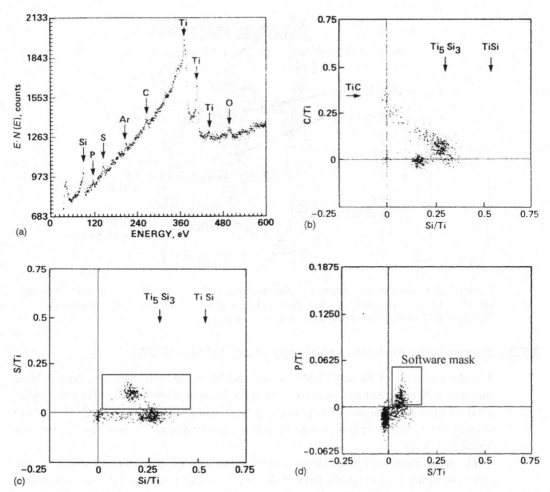

Figure 3.23. Use of scatter diagrams and ratio techniques for an early example of image analysis, adapted from Browning (1984) and reproduced with permission: (a) a survey $E \cdot N(E)$ spectrum from an anomalous region in the reaction zone between a Si fiber and the Ti–6A1–4V alloy matrix; (b)–(d) ratioed scatter diagrams for (b) C/Ti, Si/Ti, (c) S/Ti, Si/Ti and (d) P/Ti, S/Ti. A software mask is indicated on (d), setting limits on the ratios P/Ti and S/Ti, with a less restrictive mask for Si/Ti on (c). See text for discussion.

be used to explain the principles clearly (Kenny *et al.* 1994). There is no real limitation to 2D data; experiments with 3D data sets have been reported in this same paper. These 'associative' or pattern recognition techniques are very powerful, but they do require that a lot of effort be expended on a particular small area. This concentration on one small area may well result in radiation or other forms of damage, and it is always possible that you could have got the answer you really needed faster by another technique. At high spatial resolution one needs to beware of various artifacts associated with sharp edges, essentially because part of the information comes from backscattered electrons. Some of these effects were studied by El-Gomati *et al.* (1988), and are discussed by Smith (1994), Kenny *et al.* (1994) and Prutton *et al.* (1995).

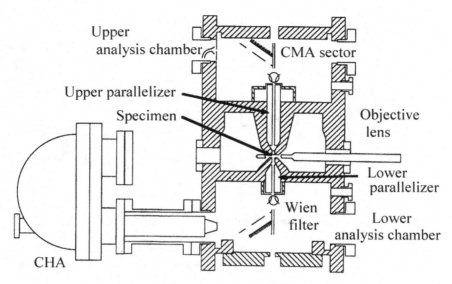

Figure 3.24. Cross sectional diagram of the specimen region of the MIDAS column, showing the relationship of parallelizer coils, objective lens, extraction optics and the sample (after Hembree & Venables 1992, reproduced with permission).

3.5.3 Towards the highest spatial resolution: (a) SEM/STEM

The development of SEM/STEM and AES/SAM at the highest resolution has been pursued at Arizona State University in what has become known as the MIDAS project, a Microscope for Imaging, Diffraction and Analysis of Surfaces. Figures 3.24–3.27 are shown here to illustrate this project, which is described in more detail by Hembree & Venables (1992).

The innovation as regards electron optics is to use the spiraling of the low energy electrons in the high magnetic field of the objective lens to contain the secondary and Auger electrons close to the microscope axis. These electrons are further controlled by auxiliary magnetic fields (parallelizers on figure 3.24) in the bores of the lens, and by biasing the sample negatively. A special combination Wien filter/deflector is then used to deflect the low energy electrons off axis through a right angle, while keeping the 100 keV beam electrons on axis. The low energy electrons then enter a commercial CHA. Because they spiral in the high **B** field and their angle to the axis decreases as the field weakens, quite a large proportion of the emitted electrons can be collected. This higher collection angle compensates for the lower yield at higher beam energy, and the smaller current available in the fine probe.

The quality of the spectra obtained is relatively high, both with respect to energy resolution (figure 3.25(a)) and to sensitivity (figure 3.25(b)). Auger mapping is obtained by taking images A and B and using ratios $(A - B)/(A + B)$ as explained above. Figure 3.26 shows the comparison of the b-SE image, with good SNR, and the Auger image, with relatively poor SNR, even after smoothing. For imaging, we have to be clear about distinctions between 'image' and 'analytical' resolution. This is because of the non-local

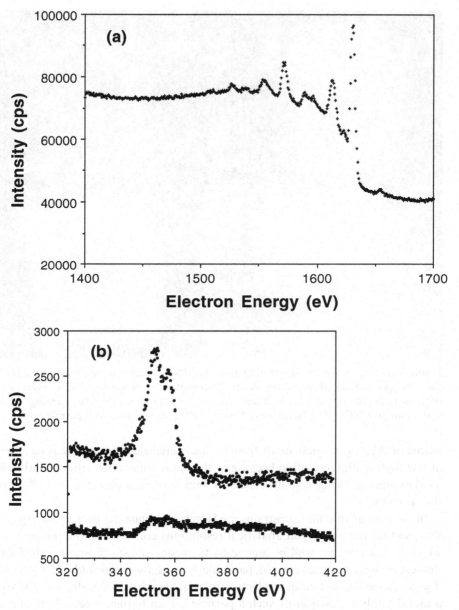

Figure 3.25. Pulse counted Auger electron spectra obtained with the 100 keV probe in MIDAS (a) Si KLL from clean Si (10 nA probe current, 10 min acquisition time for 512 point spectrum); (b) Ag MNN from 100 nm wide island on Si(001) and from <0.5 ML Ag layer between the islands (1.6 nA probe current, in 10 min (upper data) and 20 min (lower data) for the two cases respectively) (after Hembree & Venables 1992, reproduced with permission).

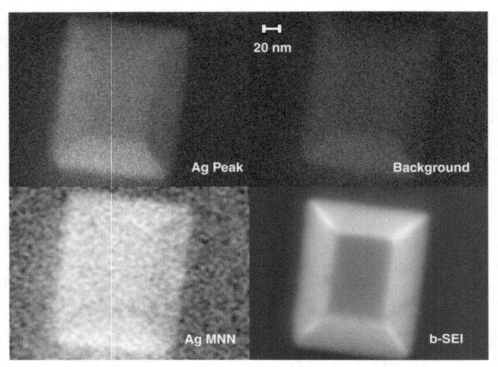

Figure 3.26. Energy selected electron images, Ag MNN Auger intensity map derived from those images and biased secondary electron image of three-dimensional silver island on Si (001), with probe current 1.5 nA, 20 min acquisition time for energy selected images; 0.3 nA and 1 min for b-SEI (after Hembree & Venables 1992, reproduced with permission).

nature of the Auger signal, firstly from backscattered electrons as discussed already, but also at high spatial resolution because of the finite Auger attenuation length, and non-local excitation. Image resolutions below 3nm have been obtained on bulk samples in this instrument.

In the case of thin film substrates, we largely eliminate the backscattering contribution, so that the image and analytical resolutions converge on the image resolution, which in practice may well be limited by the probe size. Such high resolutions are of interest in small particle research, particularly in catalysis. The old joke used to be that if you can see the particle in an electron microscope, then it was already too large to be a useful catalyst. Analysis of such a particle is even harder, especially if one is interested in minority elements. Work on such samples has been pursued by Liu *et al.* (1993) as illustrated in figure 3.27, which shows energy selected images of small Ag particles on a thin carbon substrate.

Here it is not so clear what the quantification routine ought to be, and in practice Auger information has been portrayed using the raw A and B and simple difference (A − B) images for various elements. Even for small Ag particles, backscattering effects can be seen in the intensity of carbon Auger peak images (figure 3.27(c) and (d)), but

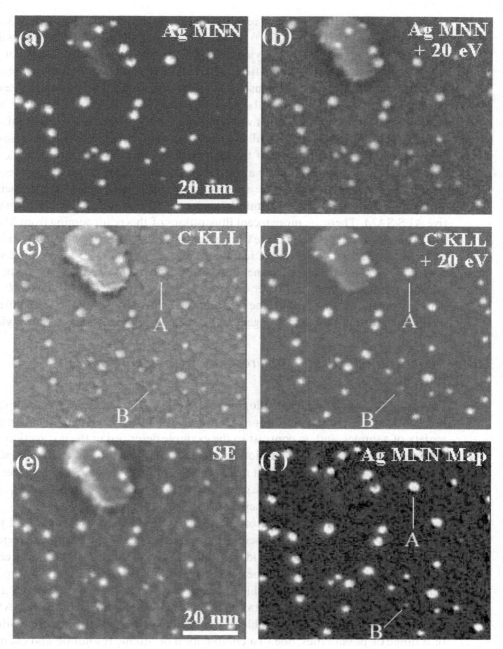

Figure 3.27. Energy selected electron images of the same area containing small Ag particles on a thin amorphous carbon substrate obtained using different signals in MIDAS: (a) Ag MNN; (b) Ag MNN + 20 eV; (c) C KLL; (d) C KLL + 20 eV, (e) low energy SE, and (f) $P_{Ag} - B_{Ag}$ (after Liu *et al.* 1993, reproduced with permission). The larger (A) and smaller (B) particles indicated in panels (c), (d) and(f) lie just off the upper end of the corresponding drawn lines. See text for discussion.

the carbon background is largely removed in the Ag difference image (figure 3.27(f)). Secondary electrons (figure 3.27(e)) show a very similar information to energy selected images, via the higher secondary yield of Ag. Even more interesting is that, for particles of size at or below the Auger attenuation length, the number of atoms in the cluster is measured by the integrated intensity of the particle, rather than the image size of the particle, and that such images can be internally calibrated, using large particles such as A in figure 3.27(f). On this basis, it was concluded that particles such as B in this panel (clearly visible in the original, as all microscopists say) contained < 10 Ag atoms.

We should note that, because of the high yield for Ag MNN Auger electrons, this is a favorable case; we are still quite a way from detecting arbitrary minority species on such small particles. Moreover, we are much more likely to be able to detect them first with a high SNR, qualitative, technique, such as b-SEI, than with low SNR, quantitative AES/SAM. There are more recent illustrations of this point coming from MIDAS. For example, oxygen KLL at 505 eV has a relatively low Auger yield. Small oxide particles on copper can be seen very readily in high resolution b-SE images. Indeed the presence of oxide can be seen in the shape of the (secondary electron) spectrum background, whereas wide beam Auger declares the surface to be clean (Heim *et al.* 1993). This discrepancy is due both to the fact that the oxide particles cover a small fraction of the surface, and that oxides in general have a very high secondary electron yield.

3.5.4   *Towards the highest spatial resolution: (b) scanned probe microscopy-spectroscopy*

Following the revolutionary development of STM by Binnig, Rohrer and co-workers in 1982–83, it is now almost routine that atomic resolution can be obtained on a wide variety of samples, and, in contrast to the example described in the last section, many groups have achieved such resolution, even under UHV conditions. Indeed, these techniques are now so widespread that recent reviews of UHV-based STM have been specialized to particular materials, e.g. metals (Besenbacher 1996) or semiconductors (Kubby & Boland 1996, Neddermeyer 1996).

In my lecture courses, the use of spectroscopy in STM (or other scanned probe) instruments has typically been discussed in a student talk. In principle, such spectroscopic information allows one to identify surface atomic species in favorable cases, if not in general. This is because the STM/STS techniques (Feenstra 1994) probe the valence and conduction bands, which may be sensitive to atomic species, but are not chemical specific in the same sense as AES/SAM. This is not unlike the SEM/SAM distinction; STM/STS may well be able to perform 'chemical' identification possible out of a range of possibilities, due to a combination of atomic resolution and changes of contrast due to electronic effects, and in particular due to a high SNR.

One of the many amazing positive features of STM/STS is that the probing current is also the signal, which may be between 1 nA and 1 pA. In AES/SAM used on a microscopic scale, the probing current may be between 100 nA and 10 pA, but the collected current is down to maybe 100 000 times smaller than the probe current, which does not do good things for the SNR. Thus one typically has to think very carefully about what information is wanted and is practicable to obtain. Some of the examples given in this section are close to the current technical limits. Rather than laboring the virtues of STM, STS, AFM, etc., in this and other respects, specific results are used to illustrate points being made as they arise in the text. To get started in this area, one can consult the references given in section 3.1.3 and the web-based resources listed in Appendix D.

Further reading for chapter 3

Briggs, D. & M.P. Seah (1990) *Practical Surface Analysis, vols. I and II* (John Wiley).

Buseck, P., J.M. Cowley & L. Eyring (Eds.) (1988) *High Resolution Transmission Electron Microscopy and Associated Techniques* (Oxford University Press) especially chapter 13: *Surfaces* by K. Yagi.

Chen, C.J. (1993) *Introduction to Scanning Tunneling Microscopy* (Oxford University Press), especially chapter 1 and the photographic plates which precede this chapter.

Clarke, L.J. (1985) *Surface Crystallography: an Introduction to Low Energy Electron Diffraction* (John Wiley).

Feldman, L.C. & J.W. Mayer (1986) *Fundamentals of Surface and Thin Film Analysis* (North-Holland).

Lüth, H. (1993/5) *Surfaces and Interfaces of Solid Materials* (2nd/3rd Edns, Springer), panels 2, 3 9 and 11, and chapter 6.

Moore, J.H., C.C. Davis & M.A. Coplan (1989) *Building Scientific Apparatus* (2nd Edn, Addison-Wesley) chapter 5.

Prutton, M. (1994) *Introduction to Surface Physics* (Oxford University Press), chapters 2 and 3.

Rivière, J.C. (1990) *Surface Analytical Techniques* (Oxford University Press).

Stroscio, J. & E. Kaiser (Eds.) (1993) *Scanning Tunneling Microscopy* (Methods of Experimental Physics, Academic), volume 27.

Smith, G.C. (1994) *Surface Analysis by Electron Spectroscopy* (Plenum).

Walls, J.M. (Ed.) (1990) *Methods of Surface Analysis* (Cambridge University Press).

Wiesendanger, R. (1994) *Scanning Probe Microscopy and Spectroscopy* (Cambridge University Press) especially chapters 4 and 5.

Woodruff, D.P. & T.A. Delchar (1986, 1994) *Modern Techniques of Surface Science* (Cambridge University Press) especially chapters 2 and 3.

Problems, talks and projects for chapter 3

These problems, talks and projects are to test and explore ideas about surface techniques and surface electronics.

Problem 3.1. Some questions on surface techniques

Give a short description of the following points in relation to surface techniques, including some examples.

(a) Explain why we say that we have conservation of $\mathbf{k}_{//}$, but not of $\mathbf{k}_\perp$, in surface scattering experiments.

(b) Explain why surface X-ray diffraction can be understood quantitatively in terms of 'kinematic' scattering, whereas the various forms of electron diffraction require a 'dynamic' theory.

(c) Explain why the lineshape in UPS is said to reflect the 'valence band density of states' whereas the AES lineshape may depend on a 'self-convolution of the VB DOS'.

(d) Explain the experimental setup, and energy resolution, needed to observe surface phonons. Comment on the relative energy resolution required for inelastic photon (Raman), electron (HREELS) and helium atom scattering.

Problem 3.2. The role of inelastic scattering in LEED

A quasi-kinematic model of LEED is possible based on the following assumptions. The inner potential of the crystal, is $V_0 \sim 10\,\mathrm{V}$, which increases the wavevector in the crystal over that in free space and refracts the beam at the surface. The attenuation of the incident beam amplitude (and the back-diffracted beams) is exponential with a short mean free path λ, which is inversely proportional to the imaginary potential $V_{0i} \sim 3\text{--}5\,\mathrm{V}$. A single backscattering event happens at a given atom at depth z, and has scattering factor f (or equivalently t, the t-matrix) which is a function of the beam energy E and the scattering angle θ.

Assuming that the surface plane is (001), do the following.

(a) Draw the LEED geometry and Ewald sphere, with a plane wave input beam not necessarily perpendicular to the surface.

(b) Write down an expression for the scattered amplitude from a crystal into the (hk) reciprocal lattice rod, where the spacings between layers parallel to the surface are not necessarily equal to the bulk spacing.

(c) Work out the scattered intensity distribution $I(V)$ along the (hk) rod for the case of normal incidence, where the spacings *are* equal to the bulk spacing, and draw the intensity profile.

(d) Show that the peak positions and spacings can be used to calculate the c-plane spacing, and V_0 if f is real. Show that the width of the peaks is inversely related to λ, and hence directly to V_{0i}.

Problem 3.3. The importance of a high SNR in AES

One of the main problems in Auger electron spectroscopy is that the signal rides on a non-negligible background, and that the signal to noise ratio (SNR) and the peak/background ratio (P/B or PBR) can be small. This leads to long data collection times and/or noisy signals, which are especially troublesome for imaging. The schemes discussed in section 3.5 are attempts to approximate the desired ratio signal with a simple algorithm which can be implemented using digital data collection and processing. Assuming that the energy channels A, B and C are equally spaced, with A over the peak, B just above the peak and C an equal distance to higher energy:

(a) Show that $(A - 2B + C)/(2B - C)$ is the simplest linear measure of the P/B ratio, and that this reduces to $(A - B)/B$ if the background spectrum has zero slope.
(b) Assuming that the measured counts are limited by electron shot noise, find the SNR of the peak height $(A - 2B + C)$ and of the peak to background ratio, explaining your reasons carefully.
(c) Compare the SNR of the logarithmic measure $(A - B)/(A + B)$ with that of the linear measure, and convince yourself that it is typically higher for comparable values of the numbers of counts.

Student talks related to chapter 3 have included the following

In each case a page of suggestions for narrowing the topic, and suggested references have been given out. The aim is to give the main features of the techniques clearly, with adequate visual aids, in about 20 minutes, taking questions from the class.

1. A comparison of XPS (X-ray photoelectron spectroscopy) and AES.
2. Angular resolved AES and/or X-ray Photoelectron diffraction (XPD).
3. ARUPS and inverse photoemission.
4. Electron stimulated desorption (ESD) and the angular distribution of ions (ESDIAD).
5. High energy ion, or Rutherford back scattering (HEIS-RBS) or medium energy (MEIS).
6. Low energy ion scattering and ICISS.
7. Scanning tuneling spectroscopy (STS) and microscopy (STM) in UHV.
8. STM in solution.
9. Field ion microscopy (FIM) studies of atomic mobility on surfaces.
10. SIMS and SNMS (sputtered neutral mass spectroscopy).
11. Secondary electron spectroscopy and microscopy in UHV.
12. Low energy and photo-electron microscopy (LEEM/PEEM).
13. RHEED and REM.
14. Optical techniques for monitoring semiconductor crystal growth.
15. Surface magneto-optic Kerr effect (SMOKE).
16. SEM with polarization analysis (SEMPA).
17. Film thickness measurements.
18. Nanoindentation.
19. Reactive ion etching.

4 Surface processes in adsorption

4.1 Chemi- and physisorption

A qualitative distinction is usually made between chemisorption and physisorption, in terms of the relative binding strengths and mechanisms. In chemisorption, a strong 'chemical bond' is formed between the adsorbate atom or molecule and the substrate. In this case, the adsorption energy, E_a, of the adatom is likely to be a good fraction of the sublimation energy of the substrate, and it could be more. For example, in chapter 1, problem 1.2(a), we found that in a nearest neighbor pair bond model, $E_a = 2$ eV for an adatom on an f.c.c. (100) surface when the sublimation energy $L_0 = 3$ eV. In that case the atoms of the substrate and the 'adsorbate' were the same, but the calculation of the adsorption stay time, τ_a, would have been valid if they had been different. Energies of 1–10 eV/atom are typical of chemisorption.

Physisorption is weaker, and no chemical interaction in the usual sense is present. But if there were no attractive interaction, then the atom would not stay on the surface for any measurable time – it would simply bounce back into the vapor. In physisorption, the energy of interaction is largely due to the (physical) van der Waals force. This force arises from fluctuating dipole (and higher order) moments on the interacting adsorbate and substrate, and is present between closed-shell systems. Typical systems are rare gases or small molecules on layer compounds or metals, with experiments performed below room temperature. Physisorption energies are ~50–500 meV/atom; as they are small, they can be expressed in kelvin per atom, via 1 eV $\equiv$ 11 604 K, omitting Boltzmann's constant in the corresponding equations. These energies are comparable to the sublimation energies of rare gas solids, as given in section 1.3, table 1.1.

Adsorption of reactive molecules may proceed in two stages, acting either in series or as alternatives. A first, precursor, stage has all the characteristics of physisorption, but the resulting state is metastable. In this state the molecule may reevaporate, or it may stay on the surface long enough to transform irreversibly into a chemisorbed state. This second stage is rather dramatic, usually resulting in splitting the molecule and adsorbing the individual atoms: dissociative chemisorption. The adsorption energies for the precursor phase are similar to physisorption of rare gases, but may contain additional contributions from the dipole, quadrupole, and higher moments, and from the anisotropic shape and polarizability of the molecules. The dissociation stage can be explosive – literally. The heat of adsorption is given up suddenly, and can be imparted to the resulting adatoms. Examples are O_2/Al(111) and O_2/Pt(111), which will be discussed briefly in section 4.5. O_2 and N_2 can be condensed at low temperatures as (long-lived) physisorbed molecules on many substrates. Bulk solid F_2 is, however, quite dangerous, and has an alarming tendency to blow up by reacting dissociatively with its container.

This chapter starts by considering adsorption at low coverage, where the statistical mechanics of adsorption can be worked out precisely in terms of simple models; two of these limiting models are considered in some detail in section 4.2. The next section 4.3 discusses the application of thermodynamic reasoning to the adsorbed state of matter, including how to describe phases and phase transitions. The final two sections 4.4 and 4.5 discuss the application of thermodynamic and statistical models to first physisorption and then chemisorption, with experimental examples and literature references.

4.2 Statistical physics of adsorption at low coverage

4.2.1 General points

We have already discussed, in section 1.3.1, the sublimation of a pure solid at equilibrium, given by the condition $\mu_v = \mu_s$, with

$$\mu_v = \mu_0 + kT \ln (p), \text{ and the standard free energy } \mu_0 = -kT \ln (kT/\lambda^3). \quad (4.1)$$

Now we wish to consider adsorbed layers in more detail, with a corresponding chemical potential μ_a. Thus we have two possible conditions: $\mu_a = \mu_v$ for equilibrium with the vapor, and $\mu_a = \mu_s$ for equilibrium with the solid. The first case is discussed in the following sections 4.2.2 and 4.2.3. The second case was the subject of problem 1.2(b) in chapter 1.

4.2.2 Localized adsorption: the Langmuir adsorption isotherm

In the Langmuir picture, each adatom is adsorbed at a well-defined adsorption site on the surface. The canonical partition function for the adsorbed atoms is $Z_a = \Sigma_i \exp (-E_i/kT)$, and in general the Helmholtz free energy $F = -kT\ln(Z)$, where E_i represents the energies of all the quantized states of the system. For N_a adsorbed atoms distributed over N_0 sites, each of which have the same adsorption energy E_a, $Z_a = Q(N_a, N_0)\exp(N_a E_a/kT)$, where Q represents the configurational (and vibrational) degeneracy. The new element is the configurational entropy, since there are many ways, at low coverage, to arrange the adatoms on the available adsorption sites. This Q is given by (e.g. Hill 1960, chapter 7.1) as

$$Q = N_0!/(N_a!)((N_0 - N_a)!), \quad (4.2)$$

multiplied by a factor q^{N_a} if vibrational effects are included, as discussed below. The expression for $\ln(Q)$ is evaluated using Stirling's approximation for $\ln(N!) = N \ln(N) - N$, valid for large N, to give

$$\mu_a = F/N_a = -(kT/N_a) \ln (Z_a) = kT \ln (\theta/(1-\theta)) - E_a - kT \ln(q), \quad (4.3)$$

where $\theta = N_a/N_0$.

The first term is the configurational contribution in terms of the adatom coverage θ, the second the adsorption energy (measured positive with the vacuum level zero), and the last term is the (optional) vibrational contribution. We can now see that if $\mu_a = \mu_s$, the density of adatoms in ML units is determined, in the high temperature Einstein model, by

$$\mu_s = 3kT \ln(h\nu/kT) - L_0 = \mu_a. \tag{4.4a}$$

Using the form of μ_a in (4.3), and rearranging to find θ gives, at low coverage,

$$\theta = C \exp\{(-L_0 + E_a)/kT\}, \tag{4.4b}$$

where the pre-exponential function C depends on vibrations in both the solid and the adsorbed layer, and the important exponential term depends on the difference between the sublimation and the adsorption energy.

The Langmuir adsorption isotherm results from putting $\mu_a = \mu_v$, using this to calculate the vapor pressure p in equilibrium with the adsorbed layer. We now have

$$p = C_1 \theta/(1 - \theta) \exp(-E_a/kT), \text{ or} \tag{4.5a}$$
$$\theta = \chi(T)p/(1 + \chi(T)p), \text{ with } \chi(T) = C_1^{-1}\exp(E_a/kT); \tag{4.5b}$$

the constant C_1 can be shown by direct substitution to be $kT/q\lambda^3$. The form of this isotherm is shown in figure 4.1(a), using parameters appropriate to xenon adsorbed on graphite. The coverage starts out linearly proportional to p, but goes to 1 as $p \to \infty$.

The internal partition function q is the product of vibrational functions for the three dimensions, i.e. $q = q_x q_y q_z$. If the Einstein model is chosen, then we can think of the z-direction, perpendicular to the surface, having a vibrational frequency ν_a; this is the frequency appropriate to desorption, and in the high temperature limit $q_z = (kT/h\nu_a)$. The other two (x,y) frequencies, in the plane of the surface, will be the same on the square (or triangular, hexagonal) lattice, and correspond to diffusion frequencies ν_d. Thus q is inversely proportional to an 'effective' value ν_e^3, for the adsorbed state, namely $\nu_a \nu_d^2$. As we will see in section 4.4.4, this model is very good for the z-vibrations, but is certainly not exact for vibrations in the surface plane.

The pre-exponential constant in (4.5) is

$$C_1 = kT/q\lambda^3 = (2\pi m\nu_e^2)^{3/2} (kT)^{-1/2}. \tag{4.6}$$

It is instructive to note that this is in exactly the same form as that for the vapor pressure in section 1.3.1. Moreover, the value of E_a includes the zero-point motion, analogously to the sublimation energy L_0. Inserting reasonably realistic values for the vibration frequencies in (4.6) gives the full curve of figure 4.1(a), to be compared with the dashed curve in which vibrational effects are neglected.

4.2.3 The two-dimensional adsorbed gas: Henry law adsorption

If the entropy due to vibrations in the adsorbed layer becomes even more important, the adsorbate can eventually translate freely in two dimensions. This case is appropriate to a very smooth substrate, with shallow potential wells, and/or at high temperatures. Thus,

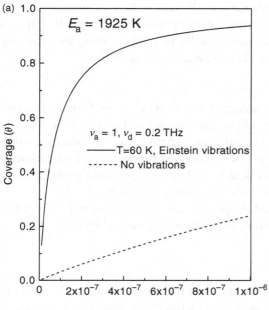

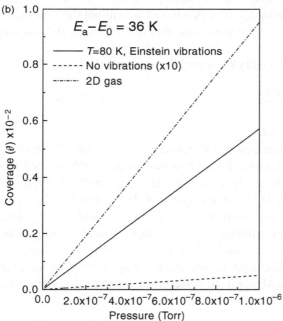

Figure 4.1. Vapor pressure isotherms of a monolayer using parameters approximating to xenon on graphite, with $E_a = 1925$ K/atom (166 meV/atom), but ignoring lateral interactions. (a) Langmuir isotherms for $T = 60$ K: full line, Einstein model for vibration frequencies $\nu_a = 1, \nu_d = 0.2$ THz; dashed line, without vibrational effects so that $q = 1$. (b) Comparison of Langmuir with 2D gas isotherm at $T = 80$ K: full and dashed lines as (a), dot-dash line, 2D gas with average adsorption energy $E_0 = 1889$ K/atom. Note the lower coverage scale ($\div 100$ with respect to (a)) and the extra factor of 10 for the dashed curve without vibrational effects.

in contrast to the previous section, the other limit of isolated adatom behavior is the 2D gas. The mobile adatoms see the *average* adsorption energy E_0, rather than the *maximum* energy E_a at the bottom of the potential wells. In compensation, they gain additional entropy from the gaseous motion. The chemical potential is now

$$\mu_a = -E_0 + kT \ln(N_a \lambda^2 / A q_z), \tag{4.7}$$

where this expression is valid at sufficiently low density for the distinction between classical, Bose–Einstein and Fermi–Dirac statistics not to be important. The derivation involves evaluating the partition function by summing over 2D momenta, analogously to a 3D gas, while retaining the z-motion partition function q_z. The difference between 2D and 3D accounts for λ^2 rather than λ^3, and the N_a/A, the number of adsorbed atoms per unit area, is the 2D version of the 3D density N/V, as in $pV = NkT$.

In fact, there is a 2D version of the perfect gas law of the form $\Phi A = N_a kT$, where Φ is known as the spreading pressure. This means that we could write

$$\mu_a = -E_0 + kT \ln(\Phi \lambda^2 / kT q_z), \tag{4.8a}$$
$$= -E_0 + \mu_2 + kT \ln(\Phi), \tag{4.8b}$$

where $\mu_2 = -kT \ln (kT q_z / \lambda^2)$ is the standard free energy of a 2D gas. This makes the correspondence between 3D gases and 2D adsorption clear: $p \leftrightarrow \Phi$, $\mu_0 \leftrightarrow \mu_2$, and the energy is lower in the 2D case by E_0. Note that it is easy to forget the q_z term, as is often done, since the various qs are dimensionless: this doesn't make them unimportant numerically.

By equating $\mu_a = \mu_v$ we get Henry's law for 2D gas adsorption:

$$p = C_2(N_a/A)\exp(-E_0/kT), \text{ or} \tag{4.9a}$$
$$(N_a/A) = \chi'(T)p, \tag{4.9b}$$

with $\chi'(T) = C_2^{-1}\exp(E_0/kT)$, and $C_2 = kT/q_z\lambda$.

You may feel that detailed discussion of these constants is rather laboring the point, but it is instructive if we stick with it for a while. Note that the 2D gas form has (N_a/A) proportional to p, whereas the localized form has the coverage $\theta = N_a/N_0$ proportional to p. These can be reconciled if we write $(N_a/A) = \theta (N_0/A)$. Here we have defined the monolayer coverage (N_0/A), and then defined (N_a/A), the areal density of adsorbed atoms in terms of this, rather artificial, constant. (Both N_0 and N_a are numbers here, not areal densities, though we can think of them as densities by choosing $A = 1$). This is the identical problem we discussed in section 2.1.4, emphasizing the need for consistency in the definition of the ML unit. If, however, we do make this definition, we can recast the 2D gas equation as

$$p = (kTN_0/Aq_z\lambda)\theta \exp(-E_0/kT), \tag{4.10}$$

which can be compared directly with the corresponding equation for localized adsorption.

This comparison shows that there is a transition from localized to 2D gas-like behavior as the temperature is raised, because $E_a > E_0$, whereas the pre-exponential (entropic) term is larger for the 2D gas. The ratio of coverages at a given p for the two states is

$$(\theta_{gas}/\theta_{loc}) = (2\pi mkT/h^2 q_x q_y)\,(A/N_0)\,\exp\{(E_0 - E_a)/kT\} \qquad (4.11a)$$
$$= (2\pi ma^2 \nu_d^2/kT)\,\exp\{(E_0 - E_a)/kT\}, \qquad (4.11b)$$

where the length a is an atomic dimension ($a^2 = A/N_0$). The comparison of Langmuir and 2D gas isotherms is illustrated in figure 4.1(b) for Xe/graphite parameters at $T = 80\,K$, using the reasonably realistic value of the well depth $(E_0 - E_a) = 36$ K (Kariotis *et al.* 1988). Note that in both models the coverage varies linearly with pressure at low coverage. However, as shown here, the 2D gas model is most appropriate, but if the well depth were much larger, the localized model with vibrations would be a better description. The model without vibrations is numerically quite poor in all such situations.

The second equality (4.11b) is only true for the Einstein model at high temperature. In this limit, where equipartition of energy holds (no term in h), the following argument can be made. Localized atoms vibrate with amplitude x, and $4\pi^2 mx^2\nu_d^2$ is the energy associated with this 2D oscillation, which is equal to $2kT$ at high temperature, if a harmonic approximation is good enough (a big *if*). Thus, the pre-exponential is just a ratio of free areas ($a^2/\pi x^2$), the numerator associated with the 2D gas, and the denominator with the potential well in which the adatom vibrates. Clearly the vibrational model starts to fail as x increases towards $\pi x^2 = a^2$.

4.2.4 Interactions and vibrations in higher density adsorbates

To consider the statistical mechanics of higher density adsorbates, we need both the interaction potentials *and* suitable models of the atomic vibrations. In analogy with the 3D case, moderate densities in a 2D fluid phase can be described by virial expansions (Hill 1960, chapter 15, Bruch *et al.* 1997, section 4.2.2). The spreading pressure is given by

$$\Phi/kT = (N_a/A) + B_2(T)(N_a/A)^2 + B_3(T)(N_a/A)^3 + \ldots, \qquad (4.12a)$$

in which the first term in an expansion in powers of the 2D density (N_a/A) is the second virial coefficient, $B_2(T)$, given by

$$B_2(T) = -1/2 \int_A [\exp(-U(\mathbf{r})/kT) - 1]\,d\mathbf{r}, \qquad (4.12b)$$

where the interaction potential $U(\mathbf{r})$ is between two atoms; the 2D integral is performed over the substrate area A, where for cylindrical symmetry $d\mathbf{r} \equiv 2\pi r dr$. In a relatively low-density gas at high T, this integral is small due to the fact that the atoms spend most of their time outside the range of influence of $U(\mathbf{r})$.

There is a continuous line of reasoning between the argument leading to (4.11), (4.12) and the cell model of lattice vibrations. This model was originally introduced by Lennard-Jones and Devonshire (1937, 1938) as an approximation of the 3D liquid state, and described e.g. by Hill (1960, chapter 16) and Bruch *et al.* (1997, chapter 5).

The free area, $A_f = \pi x^2$ in the discussion following (4.11), is defined by integrating the Boltzmann factor over the 'cell' in which the atom vibrates, namely

$$A_f = \int_A \exp(-U(\mathbf{r})/kT)\,d\mathbf{r}; \qquad (4.13)$$

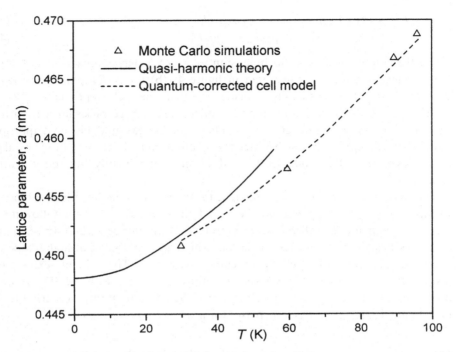

Figure 4.2. Thermal expansion of a 2D Lennard-Jones triangular monolayer solid on a smooth substrate, computed for Xe interaction potential parameters (Bruch *et al.* 1997, after Phillips *et al.* 1981, replotted with permission). Results for the quasi-harmonic theory (QHT) are good at low temperature, with the (quantum-corrected) cell model agreeing closely with classical Monte Carlo calculations at high T. The triple point of 2D Xe (on graphite) is 99 K.

the corresponding quantity in 3D is a free volume, V_f. In a high-density adsorbate with large $U(\mathbf{r})$ at moderate T, the integral is negligible except at positions $\mathbf{r}$ close to the equilibrium spacing. This approximate, classical theory, is very effective in computations for solids at high temperature, since it includes thermal expansion – the response to the spreading pressure exerted by the anharmonic vibrations – which many, apparently more sophisticated models, ignore. For these reasons at least, it deserves to be better known and more widely used as a teaching aid.

One anharmonic model which aims to have to have the correct low temperature limit, and to be useful at higher T, is the quantum (or quantum-corrected) cell model (Holian 1980, Barker *et al.* 1981, Bruch *et al.* 1997, chapter 5). A comparison of lattice dynamical models for a 2D solid monolayer interacting via the approximate Lennard-Jones potential, with parameters appropriate to xenon, is shown in figure 4.2.

4.3   Phase diagrams and phase transitions

One of the intriguing aspects of both physi- and chemisorption is the large number of phases that can exist at the surface, and the transitions that occur between these phases.

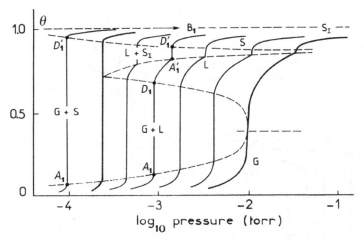

Figure 4.3. Sub-ML isotherms for Xe/graphite between 97 and 117 K. The isotherms, from left to right, are at 97.4, 100.1, 102.4, 105.4, 108.3, 112.6 and 117.0 K. Between 110.1 and 117 K, the layer undergoes two first order phase transitions, showing 2D gas (G), liquid (L) and solid (S) phases, whereas at 97.4 K only the G to S transition occurs; the 2D triple point is at 99 K (after Thomy *et al.* 1981, reproduced with permission).

There is a comparable richness of structure to that displayed in high pressure physics, where there is both a density ρ, and a corresponding structure, at a given p and T. The relation $\rho = f(p, T)$ is called the Equation of State (EOS) in the (3D) physics of bulk matter, or the (p, V, T) relation.

4.3.1 Adsorption in equilibrium with the gas phase

The corresponding equilibrium equation for (2D) adsorbed layers is $\theta = f(p, T)$, and since we have already used $\mu_v = \mu_0 + kT \ln (p)$, and $\mu_v = \mu_a$, we can think of $\theta = f(\mu, T)$ equivalently. For θ, read N_a if we do not define the ML unit in the standard way. So, as we compress a 2D gas or localized adlayer by increasing the (gas) pressure p, the adatoms will come within range of their mutual attractive or repulsive forces, and phase transitions may result, first within the ML, and subsequently from ML to multilayer.

If the substrate and adsorbate are well ordered, the condensation may proceed in well defined steps, as shown in figures 4.3 and 4.4 for physisorbed Xe and Kr/graphite respectively at ML and sub-ML coverages. As studied by several French groups especially (Thomy *et al.* 1981, Thomy & Duval 1994, Suzanne & Gay 1996), these volumetric studies, using high quality exfoliated graphite, established the existence of 2D solid, liquid and gaseous layers. The (p, T) positions of the phase transitions (including multilayer transitions) and fixed points such as triple points and critical points in these layers were also accurately measured.

More recent experimental thermodynamic work has automated the measurement process, and has achieved very high accuracy for quantities which depend on the slope

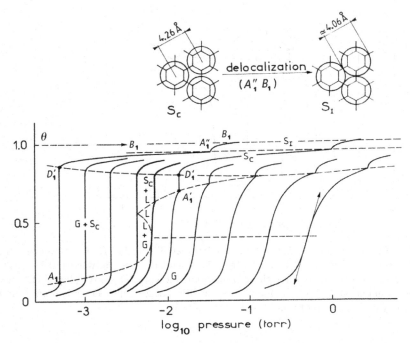

Figure 4.4. Sub-ML isotherms for Kr/graphite between 77 and 110 K. The isotherms are at 77.3, 79.8, 82.3, 84.8, 86.0, 88.0, 91.8, 96.6, 102.6 and 109.5 K. These show 2D gas, liquid (maybe) and *two* solid phases, with a presumed solid–solid phase transition at point A_1'', which is not first order (after Thomy *et al.* 1981, reproduced with permission).

of the isotherm, such as the isothermal compressibility. Examples for Kr in the multi-layer region are given by Gangwar & Suter (1990) and for Xe near ML melting by Gangwar *et al.* (1989) and Jin *et al.* (1989). An example showing the much improved precision of these data is given in figure 4.5 which can be directly compared with figure 4.3.

Before we examine these results, we can note the different forms of graphs and phase diagrams that can be plotted. The problem arises because we have three variables, T, $\ln(p)$ or μ, and θ, but we typically want to output onto paper, so that one of these three is not plotted; the corresponding (third piece of) information may either not be known, or may be discarded, or it may be given as a parameter.

An *isotherm* is a graph of θ against $\ln(p)$ or μ, with T as the parameter, as used in the previous section. A phase diagram using $\log(p)$ and $1/T$ as axes is very convenient for (physisorption) experimentalists, because the pressure can be varied over many orders of magnitude, and this plot results in straight lines for phase transitions (e.g. gas–solid or monolayer–bilayer transitions) which show an Arrhenius behavior – the slope of these lines give the corresponding energies. But typically the coverage information is lost. Theorists are fond of phase diagrams as a function of T and μ: this gives them the chance to investigate the adsorbed phase, and ignore the 3D gas phase, which provides the value of μ, typically $\Delta\mu$ with respect to the bulk 3D phase, when the comparison with experiment is made later.

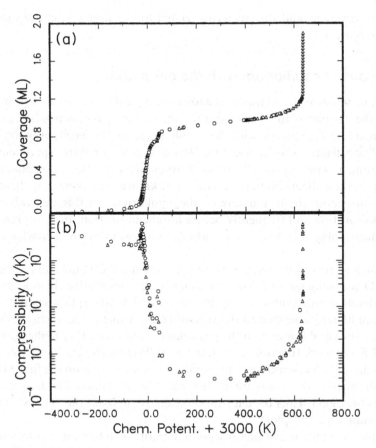

Figure 4.5. (a) Isotherm and (b) compressibility on a logarithmic scale in the monolayer and bilayer region of Xe on graphite at $T = 105.4$ K. Note the large range and high measured accuracy of compressibility, and the reproducibility of two runs taken at $T = 105.456$ and 105.435 ± 0.001 K, with absolute values ± 0.1 K (from Gangwar et al. 1989, reproduced with permission).

An *isobar* is a graph/cut/contour at constant pressure, giving a plot of $\theta(T)$, with $\ln(p)$ or μ as the parameter. The meaning is the same as used on weather charts, but the context is a little different (is the weather an equilibrium phenomenon?). In many single surface experiments, a more or less directed beam is aimed at the substrate to establish a steady state concentration which is almost a true equilibrium, but not quite. In particular, the temperature of the beam T_b is typically not the same as that of the adsorbate T_a; the question of whether or not to correct the pressure for this thermo-molecular effect, of order $(T_a/T_b)^{1/2}$, recurs in the experimental literature.

An *isostere* is a contour on a $p(T)$ plot at constant coverage. Typically $\log(p)$ varies as $1/T$, and the energy associated with such an Arrhenius plot is called the isosteric heat of adsorption. This is the energy associated with the adsorbed phase at that coverage, and it comprises the adsorption energy and lateral binding energies, their derivatives with respect to coverage, and various terms related to the atomic vibrations (as you would by now expect).

4.3.2 Adsorption out of equilibrium with the gas phase

The examples of physisorption, discussed above, are typically, but not always, in equilibrium with the gas phase. In these cases the state of the system depends on T, and also on p. But at low T, exchange with the gas phase can be extremely, even infinitesimally, slow. Phase diagrams which use θ and T as axes are favored by experimentalists in chemisorption, and more generally at low T, where the pressure goes exponentially to zero. Often in these diagrams the pressure is not known, and there may thereby be some uncertainty about the true nature of the equilibrium. In this case, which can occur for physisorption and frequently occurs for chemisorption, the gas pressure is not only immeasurably low, but is irrelevant for discussion of the behavior of the system.

Typically such systems are treated as closed 2D systems, the equilibrium (or lack of it) with the 3D gas being ignored. This is reasonable for dissociative chemisorption at low and moderate temperatures, owing to the very high adsorption energy of the atoms: they are literally confined to the surface layer. A metal–metal chemisorption example where the equilibrium with the gas is taken into account at higher temperature is the AES and work function ($\Delta\phi$) data for Au/W(110) (Kolaczkiewicz & Bauer 1984). In this data, AES is sensitive to the total Au coverage θ within the first ML, but $\Delta\phi$ depends on whether the atoms are in the form of large islands (ϕ higher) or as isolated adatoms (ϕ lower). Thus the data are sufficient to map out the 2D gas–2D solid phase equilibrium on a θ–T plot.

Two examples from the recent physisorption literature will be sufficient to illustrate these various points. There have been several sets of experiments where sub-ML amounts of Xe have been condensed onto metal surfaces. One of these involved STM experiments at liquid helium temperatures (4 K), where the STM tip was used to move the Xe atoms over the surfaces and construct the impressive if somewhat predictable *IBM* (Eigler & Schweizer 1990). Xe/Ni(110) is a typical physisorption system, yet at 4 K the atoms stay where they are pushed/put for hours, and never leave the surface during the duration of the experiment, unless one engages in (again non-equilibrium) experiments to pick them up and transport them with the STM tip.

A second example is the detailed T-dependent study of Xe/Pt(111) (Horch *et al.* 1995). Good STM pictures could be produced below about 30 K, where nuclei of solid ML Xe were shown to grow; above this temperature, however, STM pictures were blurred, due to the motion of Xe atoms over the surface. This temperature is well below that needed for Xe to desorb from the surface – only then is the full equilibrium state obtainable. Note that observations of the average structure are then quite possible with diffraction techniques, but that observation of the local structure by STM is impossible. At low T, what we are observing is really the first stage of Xe crystal growth, rather than equilibrium adsorption. Another way to look at this is to note that we can have a local equilibrium within the 2D system at lower temperature than that needed

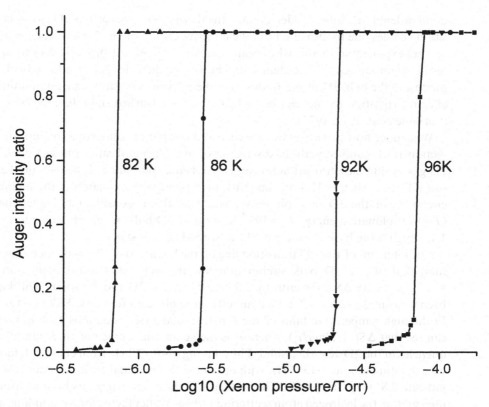

Figure 4.6. AES amplitude of Xe/graphite as a function of log(p), showing a first order gas–solid phase transitions at p and T values indicated (after Suzanne *et al.* 1973, 1974, 1975, reproduced with permission).

for full 3D equilibrium: this point recurs when considering models of (epitaxial) crystal growth in chapter 5 and section 7.3. Chemisorption examples are discussed in section 4.5.

4.4 Physisorption: interatomic forces and lattice dynamical models

4.4.1 *Thermodynamic information from single surface techniques*

Once one applies 'single surface' techniques to adsorbed layers with sub-ML sensitivity, several types of phase and phase transition can be observed on many materials; the following examples are highly selective towards rare gases on graphite. Figure 4.6 shows the AES amplitude for Xe/graphite as a function of log(p). These curves are adsorption isotherms, taken as the pressure is varied through the gas–solid transition. The first order character of the transition is seen very clearly. At the same point that the AES amplitude jumps, spots appear in LEED (or other diffraction technique) characteristic of an ordered ML solid. Understanding the thermodynamics of this 2D gas–solid transition enables one to measure both the cohesive energy of the 2D adsorbed solid, and the pre-exponential factor, which can be related to the entropy of adsorption. This results in an estimate of the change in vibration frequencies between the adsorbed 2D phase and the bulk 3D phase. In this case, the entropy is negative, corresponding to the effective vibration frequencies being higher in the adsorbed state than the bulk phase (Suzanne *et al.* 1974, 1975).

We can see how this arises by reference to the vapor pressure equation introduced in section 1.3.1, coupled with the discussion of monolayer vibrations in section 4.2. The 2D gas–solid phase transition line on an Arrhenius plot has a slope measured experimentally as 2780 ± 50 K/atom, and this corresponds approximately to the sublimation energy L_2 of the 2D solid phase. We note that this is considerably higher than the $(T=0)$ sublimation energy $L_0 = 1937$ K/atom of 3D bulk Xe given in chapter 1, table 1.1, which is the basic reason why the adsorbed layer is stable.

The intercept of this 2D transition line on the $\log(p)$ axis at $T^{-1} = 0$ is actually *higher* than that of the 3D bulk sublimation line intercept, and the difference in $\ln(p)$ $= -\Delta S/k$, where ΔS is the entropy difference between 2D and 3D solids. This has also been measured as $\Delta S = -2 \pm 1$ cal/mole/K, or in more useful units, $\Delta S/k = -1.0 \pm 0.5$. In the high temperature limit of the Einstein model (see equations 1.14 and 4.6), we can see that $\Delta S/k = 3\ln(\nu/\nu_e)$, where ν_e is the geometric average of three vibration frequencies in the 2D adsorbed solid. Thus, taking $\nu = 0.73$ THz from chapter 1, table 1.1, we can estimate ν_e as ~ 1.0 THz, with of course a substantial error bar; the error limits indicate $0.86 < \nu_e < 1.20$ THz. The vertical vibration frequency has been measured by interpreting the hydrogen atom scattering Debye–Waller factor for Xe/graphite, as 0.90 THz (Ellis *et al.* 1985). Consequently, the thermodynamic $\Delta S/k$ estimate implies that the lateral vibrations in the completed solid ML are also higher than bulk values. Although this is consistent with the compression of ML solid Xe at low temperatures, such a result is not inevitable. For example, a non-compressed sub-ML solid may well have a larger ΔS than the 3D counterpart at low temperatures, due to low-lying vibrational and translational modes.

The thermodynamics of these models are given in several places (e.g. Cerny 1983 or Price & Venables 1976); many have referred to earlier work by Lahrer (1970). This last reference is a relatively rare example of a Ph.D. thesis which was widely circulated, but which never appeared in the same form in the open literature. Because they are not generally available, some of the more useful thermodynamic relationships are reproduced here in Appendix E.2.

4.4.2   The crystallography of monolayer solids

The crystallography of the 2D solid phase of Xe/graphite was observed by diffraction techniques (LEED and THEED). The THEED work had high enough precision to detect that this solid was incommensurate (I), having a lattice parameter some 6–7% larger than graphite under the conditions of figure 4.6. At lower T and p, these experiments showed that the layer was compressed into a commensurate (C) phase, i.e. an

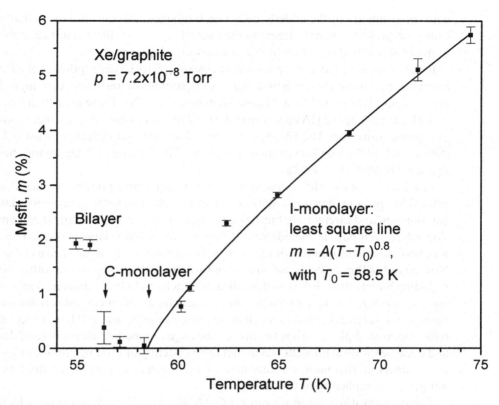

Figure 4.7. I–C and monolayer–bilayer phase transitions for Xe/graphite as measured by THEED, where the misfit is the difference in lattice parameter of the adsorbed layer relative to the commensurate $\sqrt{3}$ structure. As the temperature is lowered at constant pressure ($p = 7.2 \times 10^{-8}$ Torr) within the ML regime, the misfit decreases to zero at $T \sim 58$ K. About 2 K lower, the bilayer condenses with a misfit $\sim 2\%$ (after Kariotis et al. 1987, Hamichi et al. 1989; reproduced with permission).

I–C phase transition was observed, as shown in figure 4.7 (Schabes-Retchkiman & Venables, 1981, Kariotis et al. 1987, Hamichi et al. 1989, 1991). The opposite situation happens for Kr/graphite: Kr first condenses into the C-phase, and then compresses into the I-phase, where the Kr lattice parameter is a bit smaller than the corresponding graphite spacing. This C–I transition was observed by both THEED and LEED.

The I-phase has a modulated lattice parameter; this gains energy from having more of the adsorbate in the potential wells of the substrate, but costs energy in the alternate compression and rarefaction of the adsorbate. For example, if we consider the substrate to provide a template on which the adsorbed monolayer sits, then the interaction potential varies periodically, and can be expressed as a Fourier series:

$$V(\mathbf{r}) = V_0 + \Sigma_\mathbf{g} \, V_\mathbf{g} \exp(i\mathbf{g} \cdot \mathbf{r}), \tag{4.14}$$

where the sum is over as many 2D reciprocal lattice vectors $\mathbf{g}$ as are needed to describe the corrugation of the potential adequately. Typically, only one term in $V_\mathbf{g}$, consistent with the symmetry of the underlying lattice, is retained; this does not mean that higher order components are not present in the potential, just that there is not enough detail in the model to find out more by comparison with experiment.

If the geometry of all these phases is not clear, a pictorial description of the I-phase, and its representation in terms of domain walls, solitons or misfit dislocations, is shown here in figure 1.17 (Venables & Schabes-Retchkiman 1978). There are in fact two types of I-phase: the aligned (IA) phase and the rotated (IR) phase, with another possibility of a phase transition. The IR phase was first discovered for Ar/graphite using LEED (Shaw *et al.* 1978), and is even more pronounced in the case of Ne/graphite shown in figure 4.8 (Calisti *et al.* 1982).

In a rotated phase, the diffraction spots are split, corresponding to two domains rotated in opposite directions. Why does this happen? The misfit is accommodated by compression and rarefaction; but typically shear waves cost less energy than compression waves, so it pays to include a bit of shear if the misfit is large enough. This effect was first described quantitatively as a static distortion, or mass density, wave[1] by Novaco & McTague (1977), and has been further developed by several other workers including Shiba (1979, 1980), as described by Bruch *et al.* (1997, chapters 3 and 5). The energies/atom gained by rotation for the various rare gases on graphite are indicated in figure 4.9. It is remarkable how small these energies can be, and still be sufficient to stabilize the rotated phase; this is because of the large numbers of atoms in each domain, and because the domain walls cannot act independently of their neighbors, unless they are far apart. In that limit, we enter new regimes, such as a domain wall fluid; but let's not get too complicated at this stage.

These observations mean we can get C–IA–IR transitions in sequence, which have been observed for both Kr and Xe/graphite. In the case of Xe/graphite, a large body of THEED data has been obtained at relatively low pressures, close to the C–IA and IA–IR transitions; one data set is shown in figure 4.7. These C, IA and IR phases have also been observed for Xe/Pt(111) using helium atom scattering (Kern *et al.* 1986). We can also get 1D incommensurate, or 'striped' phases, where the misfit is zero in one direction, and non-zero in the other. Then the symmetry is reduced, for example from hexagonal to rectangular as observed for Xe/Pt(111) at low misfit. The reasons why such striped phases occur (or not) depend on details of the domain wall interactions, as discussed by Kern & Comsa (1988) and by Bruch *et al.* (1997, chapter 5).

Near to the C–IA transition, the lattice dynamics can be split into two components, involving low-lying vibrational modes of the domain walls, and faster vibrations of the atoms within their local cells. This is seen both in computer simulations (Koch *et al.* 1984, Schöbinger & Abraham 1985) and in various analytical models (Kariotis *et al.* 1987, 1988, Shrimpton & Joós 1989). For Xe/graphite, it was possible to use the position of the transition in the (T, p) plane, shown in figure 4.7, to determine the depth of the potential well $\Delta V = -61 \pm 3$ K/atom; similar analyses have been attempted for other adsorption systems. Here, ΔV is the difference in energy between the atoms in the

---

[1] Note that the acronym SDW is sometimes used for *static distortion wave*, but that SDW more usually means *spin density wave* in relation to magnetic materials.

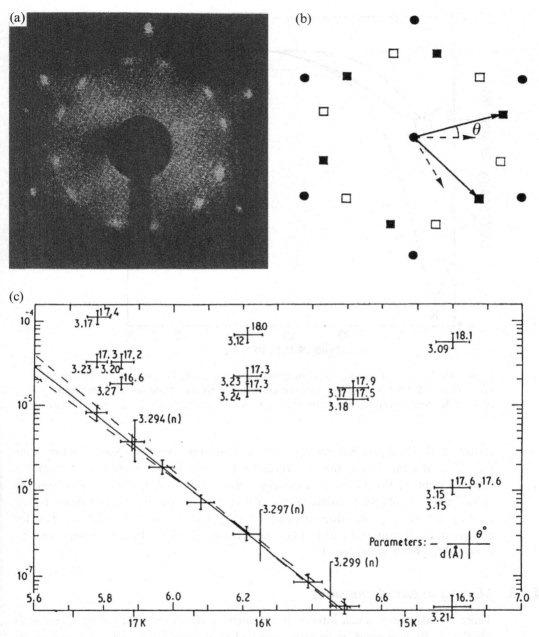

Figure 4.8. Rotated monolayer phases: the LEED pattern of (a) adsorbed Ne/graphite; (b) schematic diagram of the diffraction spots from the two domains; (c) measurements of lattice parameter and rotation (after Calisti *et al.* 1982, reproduced with permission).

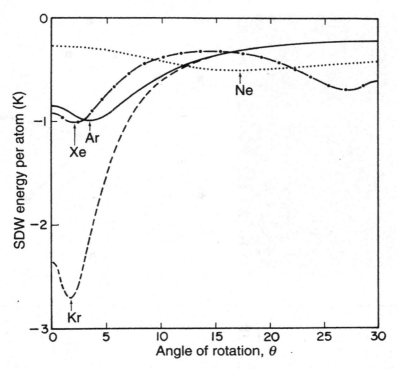

Figure 4.9. Calculated energies as a function of rotation angle for the rare gases on graphite, after Novaco & McTague (1977), reproduced with permission. Note the small energies (0.5–2.5 K/atom) involved, which are nevertheless sufficient to stabilize the rotated phases.

center of the hexagon and on top of a carbon atom, which is some factor times $(E_a - E_0)$, introduced in section 4.2. By following through the algebra in problem 4.2, one can evaluate this factor, and thereby show that $(E_a - E_0) = 36$ K is a reasonable value for the calculation presented in figure 4.1b. We can note in passing that the factor relating ΔV to $(E_a - E_0)$ does depend on the assumption that only one Fourier coefficient V_g is important in (4.14). The error involved in this assumption can be explored further via problem 4.2.

4.4.3 Melting in two dimensions

There has also been much interest in the melting transition. The major question is whether melting proceeds in two stages, via a hexatic phase, in which long range positional order is lost, but some 'bond orientational order' is preserved. Interest in this topic is primarily due to the importance of liquid crystals, where the hexatic phase in free-standing films is well documented (Brock et al. 1989, Strandburg 1992).

Additional interest arises from regarding physisorbed monolayers as model 2D systems, in which properties specific to two dimensions can be demonstrated. Melting has been thought to be such a case, following an influential series of papers describing the statistical mechanics of topological defects in ML systems. These papers

(Kosterlitz & Thouless 1973, Halperin & Nelson 1978, Nelson & Halperin 1979, and Young 1979), which together became known as KTHNY, formed a springboard for subsequent work. The central idea was that dislocations that thread the ML are point defects, which are present in thermal equilibrium. At low temperature, such defects exist only in bound pairs, but they can *unbind* above a certain critical temperature.

In the context of the six-nearest neighbor hexagonal crystal structure of rare gas layers, isolated point dislocations destroy translational order exponentially, but orientational order decays more slowly as a power law. A high density of isolated, or free, dislocations was identified with the hexatic phase. However, the cores of dislocations themselves can unbind to produce *disclinations*, where the local orientational order is alternately five- and seven-fold; if we have many of these, we have a true 2D liquid. The questions arising were then whether the hexatic phase in general existed, and if so, what was the order of the phase transitions. The role of computer simulation in this argument was interesting. These studies essentially always gave first order transitions, but theorists could always invoke 'finite size effects' to claim that the sample size in the simulation was not large enough.

Experimental melting studies on Xe/graphite include thermodynamic studies (Gangwar *et al.* 1989, Jin *et al.* 1989) and many sets of diffraction data. Our own THEED studies of melting (Zerrouk *et al.* 1994) do indeed observe the hexatic phase in a narrow temperature region. But it seems that the thermodynamic experiments are most sensitive to the expansion involved in the hexatic–liquid transition, whereas diffraction is most sensitive to orientation changes across the solid–hexatic transition. Models are complicated by the fact the solid ML Xe melts from the aligned (IA) phase. This means that the orientational order provided by the substrate, via the term V_g in (4.14), must be a significant term in the free energy balance. Detailed discussions of such systems have also emphasized the importance of the organization of dislocations into grain boundaries (Venables & Schabes-Retchkiman 1978, Chui 1983), and the role of anharmonicity (Joós & Duesbery 1985, Bruch *et al.* 1997). Both of these effects tend to cut off the second order transitions predicted by KTHNY, producing weakly first order transitions. These arguments tend to back the claims of computer simulations (Abraham 1982), who claimed that disclination unbinding represents the highest T at which the 2D solid can be metastable, rather than the thermodynamic melting temperature; this type of argument can get quite heated!

One can appreciate that the details of all these phases and their transitions involve competing interactions, often quite subtle. Physisorbed layers are thus test-beds for exploring interatomic forces and dynamics at surfaces, and the rich variety of experimental observations produce constraints on feasible models; however, the search for *truly 2D* model systems with which to test the elegant theories is, from an experimental viewpoint, somewhat elusive.

4.4.4 *Construction and understanding of phase diagrams*

The combination of all the information from different types of experiment into a phase diagram is still very much a research project, but one which proceeds in fits and starts, depending on individual enthusiasm, patience and the availability of funds and

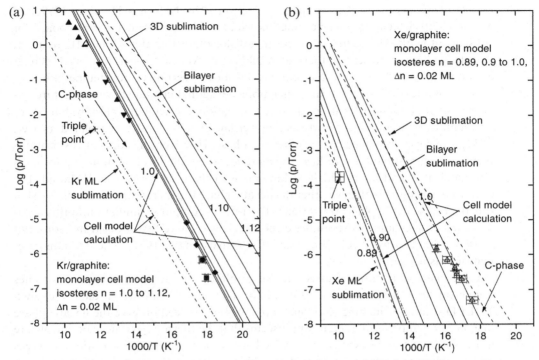

Figure 4.10. Outline phase diagrams of (a) Kr/graphite and (b) Xe/graphite in a multi-parameter potential model using Einstein vibrations for the vertical motion and the cell model for lateral motion. The isosteres plotted have a coverage parameter n, where $n = 1$ corresponds to the C-phase (one atom to every third graphite hexagon). Experimental points from the literature (after Schabes-Retchkiman & Venables, 1981, reproduced with permission).

time. For example, the current Ne/graphite ($\log(p)$, $1/T$) phase diagram dates from Calisti *et al.* (1982). Phase diagrams (T, θ) for Ar, Kr, and Xe/graphite are shown by Zangwill (1988, chapter 11), and many examples of both types of diagram are given in Bruch *et al.* (1997, chapter 6). Note how it is impossible to portray all the information on these 2D cuts of the 3D (T, $\log p$, θ) data. With rotation, as shown for Ne/graphite in figure 4.8(c), we have really four-dimensional information, and this has only been mapped in the barest outline. Quantum cell models of Ne/graphite have been developed by Bruch and by others in order to understand these results, as referenced in Bruch *et al.* (1997).

For the heavier gases, the cell model works remarkably well, and this simple model has been carried through for Kr and Xe/graphite (Schabes-Retchkiman & Venables, 1981) as shown in figure 4.10; parts of this diagram have been elaborated in later papers. These diagrams show the expansion of the layers with increasing T at constant p, and the compression of the layers with increasing p at constant T. The lines are isosteres, lines of constant coverage, or density as determined by the lattice spacing, where $n = 1$ corresponds to the C-phase. As can be seen they are slightly steeper than the phase transition lines. These diagrams are, however, by no means complete; there are now a

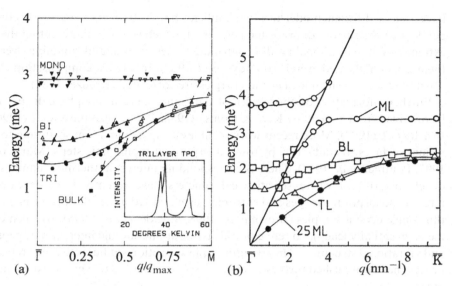

Fig. 4.11. Vibrational energies of phonons as determined by helium atom scattering for: (a) Kr/Ag(111), with insert showing the thermal desorption spectrum of the trilayer, and (b) Xe/Pt(111), after Gibson & Sibener (1985) and Kern & Comsa (1988), both reproduced with permission. See text for discussion.

lot more data in the literature, which are awaiting the time and energy to analyze and present them as a coherent story.

A particularly satisfying set of helium atom scattering (HAS) experiments has been performed on rare gases adsorbed on close packed metal surfaces, Kr/Ag(111) (Gibson & Sibener, 1985) and Xe/Pt(111) (Kern et al. 1986, Kern & Comsa 1988, 1989). By analyzing the detected beam to determine both energy and momentum, HAS can determine surface phonon energies in a similar manner that neutron scattering yields phonon energies in bulk crystals. These results, for both Kr/Ag(111) mono- and multi-layers shown in figure 4.11(a), indicate that the Einstein model is a good model for vertical vibrations within the first ML, with a constant energy $E = h\nu = 2.9 \pm 0.1$ meV, or $\nu = 0.70 \pm 0.03$ THz. The corresponding value for Xe/Pt(111) at the zone center, Γ, is 0.92 ± 0.04 THz, and more interestingly around 0.80 THz at the zone boundary, K, where the effect of the substrate Rayleigh wave is less; both the Kr and Xe values are similar to the bulk values given previously in Table 1.1. The insert in figure 4.11a shows the thermal desorption spectrum (TDS) of trilayer Kr/Ag(111); this relatively simple technique is very powerful, here distinguishing the sublimation energy of all three layers.

Progressively, as the layer thickness is increased, the vibrational energy spectrum goes over to that of the bulk crystal. In physisorption, adsorbate modes are typically lower in energy than substrate modes, except near Γ. However, in the case of Xe/Pt(111) the coupling between the substrate modes and the adlayer modes is seen for both monolayer and bilayer modes in figure 4.11(b). This coupling is stronger when the masses (and binding energies) of the adsorbate and substrate atoms are similar. We have used the Einstein model here for simplicity, but the frequency ν is not strictly independent of the surroundings or temperature, the Debye model representing the other extreme. Many vibrational modes contribute to desorption, and summing over all of them is a complicated exercise (Goldys *et al.* 1982). In TDS, a compensation effect is often observed in which log (ν) and energies are correlated (Kreuzer 1982).

Further information on this wealth of data, and the resulting phase diagrams, can be obtained from reviews by Kern & Comsa (1988, 1989), Suzanne & Gay (1996) and Bruch *et al.* (1997). More recent work includes using helium atom scattering to study vicinal surfaces to study 'row by row' adsorption at monatomic steps (Marsico *et al.* 1997, Pouthier *et al.* 1997). There is corresponding interest in thermodynamic studies of chemisorption, where sensitive calorimeters have recently been developed to measure adsorption energies and entropies, using pulsed molecular beams incident on thin single crystal samples (Brown *et al.* 1998, Stuckless *et al.* 1998). All such experiments eventually lead us to refine our ideas about interatomic interactions, geometric and vibrational structures. The strength of physisorption studies is that this refinement process has been pushed furthest; thus we are forewarned as to what to expect in other areas.

4.5   Chemisorption: quantum mechanical models and chemical practice

Chemisorption in practice is strongly linked to the study of catalytic reactions, and the onset of irreversible reactions such as oxidation. There are many fascinating reactions, some of which are described by Zangwill (1988) and Hudson (1992); more are described in the chemical physics literature, including in King and Woodruff's series *The Chemical Physics of Solid Surfaces and Heterogeneous Catalysis*; several chapters are cited here. Henrich & Cox (1996) and several review articles survey the experimental literature on particular topics such as oxides. Masel (1996) gives a general introduction, containing many details and worked examples. There are also several useful tutorial reviews presented in the series *Chemistry and Physics of Solid Surfaces*, resulting from summer schools organized and edited by Vanselow & Howe. The present section draws on some of these sources; those aspects are described that can be used as the basis of simple models, making contact with the latest research in a few exemplary case studies. We compare these studies with the physisorption results of the last section.

4.5.1   Phases and phase transitions of the lattice gas

The discussion of phases in chemisorption systems relies fairly heavily on the concept of a lattice gas, although this is not the only use of lattice gas models. Such models consider that the entities, atoms in this case, are fixed to particular sites i,j which can either be occupied or not. The starting point (ansatz) is isomorphous with magnetic Hamiltonians, such as the Ising model, which was solved exactly for the 2D square lattice by Onsager in 1944 (Brush 1967, Stanley 1971, Temperly 1972, Roelofs 1982,

1996). As with all these models, their beauty is that they provide an explicit solution to a well-posed question. The simplest magnetic Hamiltonian is

$$\mathcal{H} = -H\Sigma S_i - J_{ij}\Sigma S_i S_j + \ldots, \tag{4.15}$$

where the magnetic field H biases the occupation of a particular site, equivalent in adsorption language to $-(E_a + \mu)$. The exchange interaction J_{ij} is the interaction between neighbors i and j, and is equivalent to the lateral interaction, E_b, in a nearest neighbor model. The spin S_i can have two values in the original magnetic problem, $\pm\frac{1}{2}$; in the Ising model the convention is to use $S_i = +1$ for a full site and $S_i = -1$ for an empty site. Thus the model is symmetric in the spin variables with average spin $\langle S \rangle = 2\theta - 1$ (below 1 ML coverage); $\langle S \rangle$ is analogous to the magnetization $\mathbf{M}$. This correspondence is well described by Stanley (1971), Schick (1981) and Roelofs (1996).

Connection with thermodynamics is made via the general relation

$$F = -kT \ln (\text{Tr} \exp(-\mathcal{H}/kT)), \tag{4.16}$$

where the trace (Tr) is the sum of the diagonal elements of the Hamiltonian matrix $\mathcal{H}$, as in the previous relation used in section 4.2, $Z = \Sigma_i \exp(-E_i/kT)$. A major effort has been made to construct and solve such models, both with analytic solutions and with Monte Carlo simulations. These models are most reasonable when we have a very site-specific bond, and when lateral interactions are considerably smaller. The phase diagrams constructed via (4.15) exhibit particle–hole symmetry; assuming the saturation coverage in the Langmuir isotherm is defined as $\theta = 1$, then coverages of θ and $(1 - \theta)$ are equivalent, and the (T, θ) phase diagram is symmetric about $\theta = 1/2$. Extra terms can be added via trio, or three-body interactions of the form

$$\mathcal{H}_1 = \Sigma J_{ijk} S_i S_j S_k; \tag{4.17}$$

such higher order interactions allow lower symmetry phase diagrams to emerge.

Much of the theoretical interest rests on critical phenomena, and on the value of 'critical exponents' for various thermodynamic properties either side of the critical point. This is where departure from mean field results is most marked, and where theoretical techniques such as the renormalization group have made their name. Several models can be solved in 2D but not in 3D. Discussion and tabulation of these exponents for 2D systems are given by Schick (1981), Wu (1982) and Roelofs (1996), where terms such as the 3- and 4-state Potts, and XY models (with or without various forms of anisotropy) are introduced. Chemisorption studies have relied on and developed earlier work in magnetism, discussed here in section 6.3.2. A full account of higher-order interactions in such models is given by Einstein (1996).

An example of the agreement of such a model, with interactions up to fifth-neighbor interactions included in a five-parameter fit, for the much studied case of O/W(110) (Lagally et al. 1980, Rikvold et al. 1984, Rikvold 1985) is shown in figure 4.12. Other examples include the phase diagram of Se/Ni(001) as studied by Bak et al. (1985) using the Askin–Teller model, and disordering of the 3×1 reconstruction on Si(113) by Yang et al. (1990) using the chiral three-state Potts model. The level of agreement with experiment reached using these models is interesting but not the last word. They

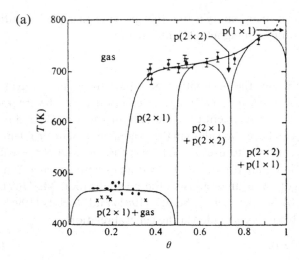

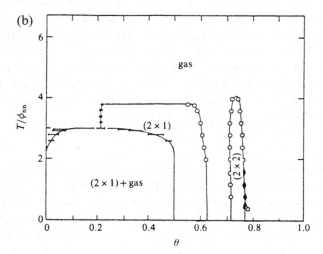

Figure 4.12. Phase diagram for sub-ML O/W(110) on a (T, θ) plot. (a) Experimental phase boundaries determined by LEED (after Lagally *et al.* 1980); (b) theoretical lattice gas calculation using a five parameter fit (after Rikvold 1985, both diagrams reproduced with permission).

describe well the configurational entropy associated with lattice occupation; but the entropy due to vibrations are not included directly, only via effective interaction parameters; most interest has been in functional forms rather than absolute values.

4.5.2 The Newns–Anderson model and beyond

A detailed model of chemisorption has to start with the energy levels and density of states of the adsorbate atom or molecule, and of the substrate. However, in the words of

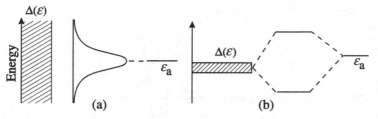

Figure 4.13. Energy levels and local density of states of the substrate to the left, and the adsorbate to the right in the Newns–Anderson model. There are two limiting cases: (a) a broad band (e.g. s-p band metal) substrate creates a resonance on the adatom, with a Lorentzian LDOS; (b) a narrow band (e.g. an insulator, semiconductor or d-band metal) substrate creates bonding and antibonding states on the adatom (after Nørskov 1990, redrawn with permission).

Hammer and Nørskov (1997), 'adsorption and reactions at surfaces are intriguing processes that are not simply described in the usual vocabulary of chemistry or that of solid state physics'. Einstein (1996) observes that this is why Desjonquères and Spanjaard (1996), in their detailed treatment of quantum mechanical models as applied to surfaces, leave chemisorption to the last chapter! Nonetheless, there is a strong history of models at work here too, and it is a reasonable question to ask where a newcomer should start.

A good candidate is what has become known usually as the Newns–Anderson, or more fully as the Anderson–Grimley–Newns model (Grimley 1967, 1983, Newns 1969; see Einstein 1996, Desjonquères & Spanjaard 1996, or Hammer & Nørskov 1997). The basic features of this model in two opposing limits are illustrated in figure 4.13. The one-electron states $\mathbf{k}$ of the combined system, with energies $\varepsilon_{\mathbf{k}}$, result from the competition between the unperturbed adsorbate energy levels ε_a, and the perturbation caused by the presence of the substrate, which can be included via a matrix element V_{ak}. The Hamiltonian analogous to (4.15) is

$$\mathcal{H} = \Sigma \varepsilon_{\mathbf{k}} n_{\mathbf{k}} + \varepsilon_a n_a + \Sigma V_{ak} (c_a^+ c_{\mathbf{k}} + c_a c_{\mathbf{k}}^+), \tag{4.18}$$

where the second-quantized form of the Hamiltonian uses the creation and annihilation operators c^+ and c as a shorthand notation in the last term, and the density of states n in the first two terms are given by $n = c^+ c$. Additionally a Hubbard $U n_{a\sigma} n_{a-\sigma}$ term may be added to the second term if spin variables σ are explicitly included (Desjonquères & Spanjaard 1996, section 6.4).

The matrix solution of (4.18) gives essentially a two-parameter fit to the changes induced on the adsorbate by the presence of the metal, in the form of an expression for the local projected density of states (LDOS) on the adsorbate

$$n_a(\varepsilon) = \pi^{-1} \Delta(\varepsilon) / [(\varepsilon - \varepsilon_a - \Lambda(\varepsilon))^2 + \Delta(\varepsilon)^2]. \tag{4.19a}$$

The important parameter $\Delta(\varepsilon)$ is the local projection of metal states at the adsorbate, given by

$$\Delta(\varepsilon) = \pi \Sigma_{\mathbf{k}} |V_{ak}|^2 \delta(\varepsilon - \varepsilon_{\mathbf{k}}). \tag{4.19b}$$

The limiting cases shown in figure 4.13 correspond to the following. (a) The case when $\Delta(\varepsilon)$ is independent of energy. This gives rise to a Lorentzian distribution $n_a(\varepsilon)$ of an

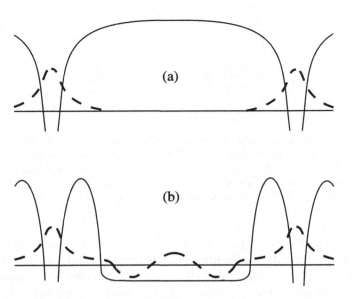

Figure 4.14. Interaction between two chemisorbed atoms interacting via the substrate: (a) potential (full lines) and wavefunctions (dashed lines) when the atoms are in vacuum and separated so that there is no overlap, and no direct interaction beyond the van der Waals interaction; (b) the same atoms chemisorbed on a metal surface showing the indirect interaction via the substrate (after Grimley 1967, and Einstein 1996, redrawn with permission).

unshifted atomic energy level, sometimes called *weak* chemisorption. In detail, the energy shift $\Lambda(\varepsilon)$ goes to zero as the bandwidth W of the metal gets larger, varying as $|V_{ak}|^2\varepsilon/W^2$. (b) The case when $\Delta(\varepsilon)$ is essentially a delta function in energy. In this latter case, *strong* chemisorption, the interaction V_{ak} gives rise to essentially discrete bonding and antibonding states. As in a diatomic molecule AB with different energy levels ε_A and ε_B, the tight binding scheme gives a separation of the bonding and anti-bonding states. There is a bias, indicated here by $\Lambda(\varepsilon)$, which is non-zero if ε_A and ε_B (or in the surface case ε_a and the mean value of ε_k) are different. This discussion therefore starts from a very similar point to section 7.1 where binding in semiconductors is outlined.

The next point to realize is that the strong bonding to the surface creates disturbances in the substrate; if the substrate is a metal, such disturbances will be strongly screened via Friedel oscillations, which are discussed more fully in section 6.1. The schematic picture of figure 4.14, first introduced by Grimley (1967) in the context of the origin of indirect lateral interactions, is dramatically illustrated by the free electron calculations for large clusters shown in figure 6.2(b), and by the experimental quantum corrals of figure 6.4. The asymptotic form of these interaction energies for adatoms on jellium a distance **R** apart is

$$E(R) \sim R^{-5} \cos(2k_F R + \phi), \tag{4.20}$$

k_F being the Fermi wave vector, and ϕ a phase factor having the same meaning as in chapter 6. But (4.20) only applies when the interactions are isotropic, and the form does not remain the same at short distances, where the R^{-5} term will diverge. Although this asymptotic result is correct to all orders of perturbation, it is only strictly accurate in the case when the answer is too small to matter in real life, according to several authors.

Einstein (1996) has discussed in detail the form of lateral interactions in the region where they are more substantial, making careful distinctions between tight binding and other schemes. In a tutorial spirit, he has introduced a model of two 'chemisorbed' atoms placed at different positions on a closed loop of 50 'substrate' atoms: this yields a 52×52 matrix to solve this '1D' quantum mechanics problem exactly. This is now practical as a student exercise, using a computer package such as *Mathematica*™. However, the main problem, as in chapters 6 and 7, is how to make sense of, or 'understand', the results, since each electron interacts with all the others.

Many of the schemes that yield insight are semi-empirical but computationally fast, enabling them to illustrate trends in experimental data. Of these various schemes, embedded atom methods (EAM) and effective medium theory (EMT) have been widely applied, and are relatively transparent. Computationally, they are now fast enough that the progress of an adsorption reaction can be followed in real time on the pico- to (almost) nanosecond time scale. (for EMT see Nørskov 1990, 1993, 1994, Hammer & Nørskov 1997). This model, and other versions of density functional theory (DFT) which have their starting point in chemisorption, are beginning to be applied to study surface processes at metal surfaces (e.g. Ruggerone *et al.* 1997). Some of this work is discussed in chapters 5 and 6. The more ambitious claim of molecular dynamics, to do a full *ab initio* quantum mechanical calculation in real time still consumes amazing amounts of computer time to study relatively small systems. To follow a reaction for a nanosecond is beyond most calculations, and the typical timescale is picoseconds. A particularly important but demanding project is to understand the reaction dynamics and trajectories of (diatomic) molecules as they arrive at the surface, react and split up, as discussed by Darling & Holloway (1995). However, some codes have produced results that can be shared in the form of web-based animations. The web addresses of some active groups can be found via Appendix D.

4.5.3 Chemisorption: the first stages of oxidation

A reasonable question to ask next is simply: why we do want to know all this? What is at stake? The first answer is that chemisorption is the first major exothermic process in the range of processes which occur in a chemical reaction, whose end product is a stable compound such as an oxide. Given the widely different starting and end points (e.g. Si and SiO_2, Al and Al_2O_3, or iron and rust in all its forms), it is not surprising that very different models are used depending on whether one is interested in the first stages of chemisorption, the overall rate of the reaction, or the stability of devices based on these materials.

An example is provided by $O_2/Al(111)$ (Brune *et al.* 1992, 1993). Here, STM was used at sub-ML coverage to investigate the nature of dissociation of O_2 into chemisorbed O. The precursor oxygen molecule is highly mobile at room temperature, but the final state of the O is completely immobile. By observing that the positions of these oxygen

Table 4.1 *Nearest neighbor metal–oxygen bond lengths, d, and vertical vibration frequencies, ν in oxygen chemisorption on metals*

System	Phase	θ (ML)	d (nm) expt	d (M–O) calc	ν (THz) expt	ν (THz) calc
O–Cu(100)	$2\sqrt{2} \times \sqrt{2}$	0.5	0.191	0.190	8.7	7.3
O–Cu(110)	2×1	0.5	0.181	0.188	11.7	11.6
O–Ni(100)	p2×2	0.25	0.193	0.192	9.2	7.5
O–Ni(110)	2×1	0.5	0.177	0.181	11.6	18.4

Source: After Besenbacher & Nørskov, 1993.

atoms were largely uncorrelated, they deduced that pairs of O-atoms were up to 8 nm away from their point of dissociation. An alternative realized subsequently by molecular dynamics calculations (Engdahl & Wahnström, 1994, Wahnström *et al.* 1996), was that only one of the O-atoms may remain on the surface. In either case we can visualize this transition as both irreversible, and essentially explosive. The energy liberated during the chemisorption 'event' (estimated to be of order 10 eV/ molecule, i.e. large) is transferred in part to the motion of the O-atoms, which then skid to a halt some distance away, or desorb. This process is just the first of a long series of reactions, whose end point is the formation of the stable phase, alumina, Al_2O_3.

Oxygen chemisorption alone is a huge topic, with STM having contributed greatly in recent years, the combination with EMT calculation being particularly effective for understanding the variety of structures found on noble metals (Besenbacher & Nørskov 1993) as well as on aluminum (Jacobsen *et al.* 1995). Although many models, such as the Newns–Anderson model of section 4.5.2, do not discuss vibrations in the adsorbed state, these can be accurately measured using infrared, HREELS, or helium atom probes. Table 4.1 shows that EMT models the metal–oxide bond lengths and vibrational frequencies, mostly with reasonable accuracy. We can note that the vertical vibrational frequencies (given here in THz rather than in meV in the original reference, see Appendix C for conversion factors) are an order of magnitude larger than those encountered in physisorption in sections 4.2–4.4.

A reaction between a known single atom and a well-defined (single crystal) substrate can initially be described in the terms outlined here. However, it becomes a much more complex, possibly out of control, process in which the substrate is an active partner and in the later stages will be consumed. In these later stages, electron, ionic and heat transport, and microstructural evolution are dominant, and may reach a kinetic limit due to such factors at relatively small oxide thickness. Examples include the passivation of Al and Si by their oxides (at a thickness of a few nanometers), without which we would not be able to use these common elements. Iron oxide 'scale' is unstable over time in a damp atmosphere. Our low-tech remedy of applying a new coat of paint will keep surface and materials chemists, as well as the painters, fully employed for many years yet: very expensive, but so much a part of everyday life that most of us don't give it any thought.

4.5.4   Chemisorption and catalysis: macroeconomics, macromolecules and microscopy

At the other end of the same scale, but also driven by the need to understand and improve industrial processes, are the catalytic industry and the emerging sensor market. This sector provides the second type of answer to the question posed in the previous section. Here we typically are interested in relatively weak chemisorption, since although we want the atoms or molecules to stick on the surface long enough to react at moderate temperature, we also want the reaction products to desorb, and leave the catalyst surface free for the next molecules to arrive. If this doesn't happen the catalyst is said to be poisoned.

As mentioned already in section 2.4.4, there are three major types of catalyst that are the subject of intense study: these are (single crystal) *metal* and *oxide* catalysts, and *supported metal* catalysts, where small metal particles (SMP) are suspended, typically on oxide surfaces. Examples of SMP catalysts are Pt, Pd and/or Rh dispersed on polycrystalline alumina, zirconia and/or ceria; a selection of these form the principal components of the catalytic converters in car exhaust pipes, converting partially burnt hydrocarbons, CO and NO_x (nitrous oxides) into CO_2, N_2 and H_2O. The role of the catalyst is traditionally defined as promoting reactions, while not itself being changed in the process. But the present view is that SMP catalysis is a highly dynamic process, in which the particles move, change shape and eventually coalesce, at the same time as enabling the reactions between the adsorbed species and subsequent desorption to take place. In other words, the whole system may behave like a giant molecule with almost biological properties. This behavior is reminiscent of the changes which take place in hemoglobin during breathing in (uptake of O_2) and out (giving off CO_2); even the sizes of the two types of structure are similar, around 2–5 nm diameter for SMPs and 5.5 nm for hemoglobin.

This picture of the *interactive* substrate is essentially the opposite of the *inert* substrate invoked in section 4.3 for physisorption, and is one of the reasons why catalysis is considered a difficult topic scientifically.[2] As in the case of breathing, we should not let a minor difficulty of understanding get in the way of continuing the practice. Catalyst-based industry is worth billions of pounds/dollars annually, and is central to the production of all petroleum and pharmaceutical products. And in addition, diffraction and imaging tools (and a lot of determination and patience) have been instrumental in finding out what we know about SMPs as well as hemoglobin. It took Perutz 23 years before he drew blood on the famous molecule (Perutz 1964, Stryer 1995). We probably need a similarly patient attitude to catalysis.

The literature on SMPs in the context of catalysis is extensive, and there have been some successes. Campbell (1997) gives a review with a 'surface processes' viewpoint. A

---

[2] Of course, the inert substrate is not strictly true for physisorption either. Measurement of the stress caused by adsorbing Xe and other gases on thin graphite shows that at low coverage, the substrate tends to wrap around the adsorbate, and at higher coverage the adsorbate bends the substrate in the other direction (Beaume *et al.* 1980).

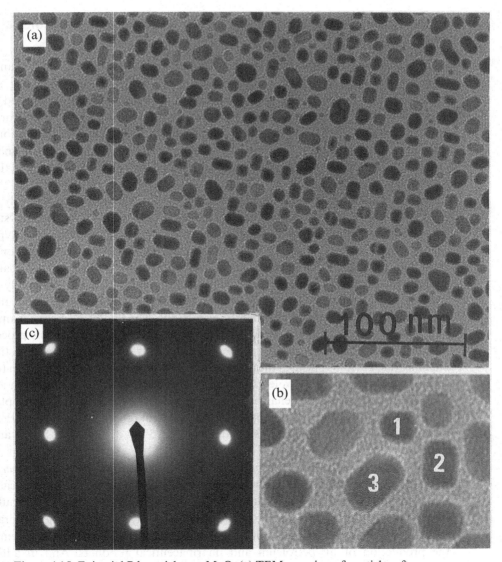

Figure 4.15. Epitaxial Pd particles on MgO: (a) TEM overview of particles after some coalescence has occurred; (b) higher magnification view of particles with different shapes numbered 1–3; (c) transmission diffraction pattern, giving epitaxial orientation of all such islands (after Henry *et al.* 1991, 1992, reproduced with permission).

combination of microscopy and diffraction to characterize the particles, and mass spectrometry to measure desorption products has been usefully employed by the group of Claude Henry in Marseille (not the *other* (William) Henry, who worked during the first third of the nineteenth century). For the case of Pd/MgO(001), they characterized the particle density, sizes and shapes and epitaxial orientation by TEM and THEED (Henry *et al.* 1991,1992), as shown in figure 4.15. In parallel, they used a chopped molecular beam to deliver CO to the sample at a given temperature, and a mass

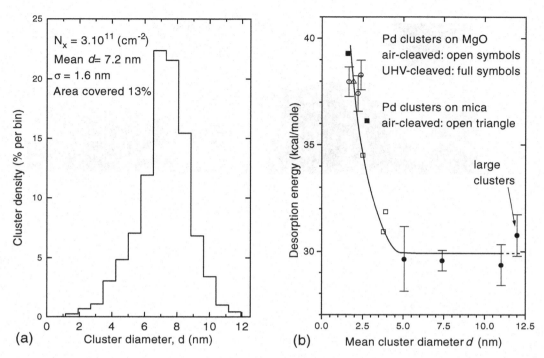

Figure 4.16. Epitaxial Pd particles on MgO: (a) size distribution histogram, nucleation density and other quantities derived from figure 4.15; (b) variation of the initial desorption energy of CO as a function of mean particle size for CO adsorbed on size-selected Pd particles on MgO(001) and mica (after Henry *et al.* 1992, replotted with permission).

spectrometer with phase sensitive detection to detect CO desorption. In this way they were able to determine the residence time (in the millisecond–second range) of CO as a function of T, and hence to deduce the effective activation energy and prefactors for desorption from the composite sample. Figure 4.16 shows a typical particle size distribution, and the resulting energy as a function of particle size, which is constant down to 5nm, but rises dramatically below 2.5 nm. Reviews of this work are given by Henry *et al.* (1997) and Henry (1998).

SMPs may additionally have a non-crystalline structure, with pentagonal symmetry, distorted, multiply twinned particles (MTPs) being observed in many systems (Ogawa and Ino 1971, 1972). In addition, these particles change shape frequently, on the second time scale, under observation by high resolution electron microscopy. While there is some discussion as to whether such effects are induced by the electron beam, they are certainly happening rapidly at relevant catalytic temperatures (Poppa 1983, 1984). The idea of the surface which changes its morphology *in response to* the reaction took a while to take hold, but some of the evidence has been in the literature for a long time.

An example from the oxidation of much larger, $\sim 5\,\mu$m diameter, Pb crystals on graphite at 250°C is shown in figure 4.17 (Métois *et al.* 1982). This *in situ* UHV-SEM picture is of the same type of crystal used to establish the equilibrium form, as

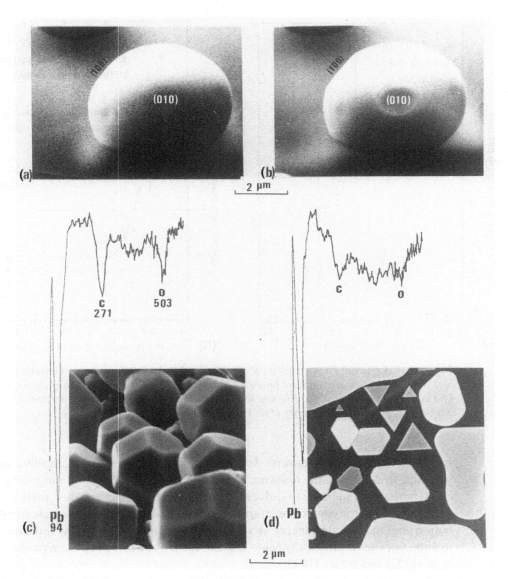

Figure 4.17. SEM pictures of the change in form of Pb crystals, following adsorption of oxygen at 250 °C: (a) equilibrium form, showing small {100} facets; (b) 100 L O_2 showing increased size of {100}; (c) further increase after 10^4 L exposure, and corresponding Auger spectrum; (d) insensitivity of tabular {111} crystals to the same O_2 exposure (after Métois *et al.* 1982, reproduced with permission).

described in chapter 1, figure 1.7. The major facets in the equilibrium shape are {111}, followed by {100} and {110}. However, exposure[3] to ~100 L O_2 in 100 s is sufficient to increase the size of the {100} at the expense of the {111} facets, and by 10^4 L the crystal is bounded by greatly enlarged {100} faces. AES shows that we are dealing with ML quantities of oxygen on the surface, not more. This exposure, however, has little effect on the tabular {111} crystals shown in figure 4.17(d).

This type of surface movement is typically mediated by mass transfer surface diffusion, where adatoms and/or vacancies have to be both created at steps *and* move to the next one, under the driving force of surface energy reduction. In this case the distance moved r in a given time scales as $(Dt)^{1/4}$ (Mullins 1957, Nichols & Mullins 1965, Bermond & Venables 1983). Since we are seeing effects at the ~1 μm scale in 100 s in the example shown, the same effects on the 10 nm scale would take place in an estimated 1 μs. However, nothing happens to the {111} tabular Pb crystals of a similar size. This indicates both how face-specific these arguments can be, and also that there may be severe nucleation barriers before the reactions can take place. In this example, {111} crystallites exhibit a nucleation barrier to melting (Spiller 1982, Métois *et al.* 1982). Similarly, there can be substantial barriers to incorporation of diffusing adatoms on perfect crystals, which is the reason why such tabular crystals are formed during vapor deposition and can co-exist with the equilibrium forms (Bermond & Venables, 1983).

A recent case of weak chemisorption which has been studied using low temperature STM is O_2/Pt(111) (Winterlin *et al.* 1996, Zambelli *et al.* 1997). The initial chemisorbed O_2 appears as pairs of atoms, some two–three atom spacings apart. It was shown that the presence of already adsorbed atoms catalyzed the breakup of O_2 arriving later, leading to the formation of linear chains and then networks. This system shows interesting nonlinearity, which are characteristic of many such reactions, and also anisotropy, even though the O atoms are adsorbed in symmetric three-fold hollow sites. This may be due to stresses, both caused and relieved by adsorption, and the possibility that adsorption can change the reconstruction of the substrate. The input of calculations to the discussion of what is going on is at an interesting stage (Feibelman, 1997).

One of the most fascinating phenomena is the occurrence of space- and time-dependent reactions which have been observed in real time by photo-electron emission microscopy (PEEM), as shown in figure 4.18. The reactions can be periodic or chaotic in time, and spatial patterns evolve on the surface, often resembling spiral waves. The original work by the Ertl–Rotermund group in Berlin (Rotermund *et al.* 1990, Jakubith *et al.* 1990, Nettesheim *et al.* 1993, Ertl 1994) showed that the reaction between CO and O_2 to produce CO_2, on a Pt(110) substrate, proceeds at the boundary between two adsorbed phases, one primarily CO and the other primarily O; this reaction was followed by TV observation in real time with a typical length scale of 10–50 μm, at CO pressure up to a few 10^{-4} mbar.

There are many reasons why one would want to follow such reactions at higher pressures, in order to simulate the conditions of real catalysts. Optical observation is

---

[3] One langmuir (L), the unit of exposure to a gas, is equal to 10^{-6} Torr·s; do not confuse 1 ML = 1 Torr·s with the symbol for a monolayer (ML).

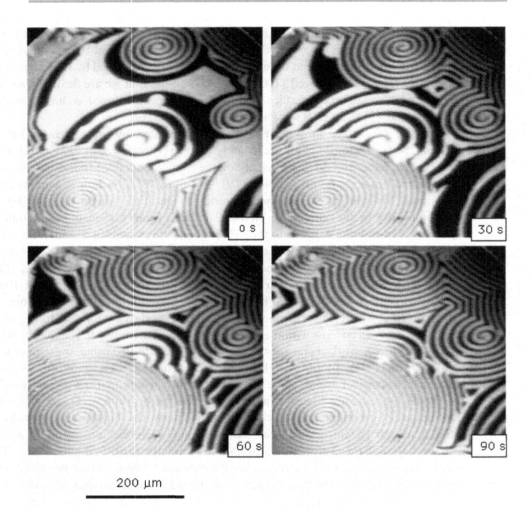

200 µm

Figure 4.18. PEEM pictures of the spatio-temporal reaction $CO + O_2$ to produce CO_2 on a Pt(110) substrate, at $T = 448$ K and partial pressures $\sim 4 \cdot 10^{-4}$ mbar. The darker areas show adsorbed O (work function change $\Delta\phi = 0.5$ V) and the lighter areas adsorbed CO ($\Delta\phi = 0.3$ V relative to Pt), with the reaction proceeding at the moving boundary between the phases (Nettesheim *et al.* 1993, reproduced with permission).

advantageous, even if the spatial resolution is limited. A development of ellipsometry from the same group (Rotermund *et al.* 1995, Rotermund 1997) has observed the same reactions at CO pressures $> 5 \cdot 10^{-2}$ mbar, and at higher $T \sim 550$ K. The reactions have been identified as being associated with the following features. These are: (a) oxygen needs two adjacent Pt sites to chemisorb, which suppresses O-adsorption at high CO coverage; and (b) CO lifts the 2×1 reconstruction which is present, both on the clean and O-covered surfaces (Eiswirth *et al.* 1995). The coupling of these reactions has been modeled with three non-linear coupled rate-diffusion equations, for the local concentrations of CO, O covered and 1×1 uncovered structures (the areas of 2×1

then make up the missing fraction). Such models contain several parameters, but the argument is made that the structure of how the reactions proceed is not unduly influenced by the detailed choice of such parameters (Eiswirth *et al.* 1990, Krischer *et al.* 1991).

The use of rate and diffusion equations is a powerful tool for modeling chemical kinetic experiments, and other related topics such as population biology. There are many features in common to all these fields that would be wonderful to study, if only one weren't limited by *time*! As a result, many of the developments have proceeded in parallel in the different groups, without interaction. Here we use these techniques to discuss the (arguably simpler) case of epitaxial crystal growth in chapter 5.

Further reading for chapter 4

Bruch, L.W., M. W. Cole & E. Zaremba (1997) *Physical Adsorption: Forces and Phenomena* (Oxford University Press).

Desjonquères, M.C. & D. Spanjaard (1996) *Concepts in Surface Physics* (Springer) chapter 6.

Henrich, V.E. & P.A. Cox (1994, 1996) *The Surface Science of Metal Oxides* (Cambridge University Press) chapter 6.

Hill, T.L. (1960) *An Introduction to Statistical Thermodynamics* (Addison-Wesley, reprinted by Dover 1986) especially chapters 7–9, 15 and 16.

Hudson, J.B. (1992) *Surface Science: an Introduction* (Butterworth-Heinemann) chapters 12,13.

Masel, R.I. (1996) *Principles of Adsorption and Reaction on Solid Surfaces* (John Wiley) chapter 3.

Stanley, H.E. (1971) *Introduction to Phase Transitions and Critical Phenomena* (Oxford University Press).

Zangwill, A. (1988) *Physics at Surfaces* (Cambridge University Press) chapters 8, 9, 11 and 14.

Problems and projects for chapter 4

These problems and projects explore ideas about surface crystallography, adsorption potentials and the question of localization in adsorbed layers.

Problem 4.1. Monolayer structures on a honeycomb lattice

Consider rare gas adatoms on graphite, consulting figure 1.16 for the incommensurate aligned (IA) phase of Xe/graphite, and figure 4.8 for the incommensurate rotated (IR) phase of Ne/graphite, as needed. Diffraction from a square or rectangular lattice does not give any conceptual problems, but the graphite, or honeycomb lattice is more difficult because the lattice vectors **g** are not perpendicular to each other, so we have to keep in mind what **g·r** means.

(a) Convince yourself that the 2D unit cells of the *real* lattice of graphite (lattice parameter a_c) and of the commensurate (C) phase of adsorbed rare gases (lattice parameter a) are as shaded in the bottom right hand corner of figure 1.16 in chapter 1.

(b) Construct the lowest order region of the *reciprocal* lattice $\mathbf{g} = h\mathbf{a}^* + k\mathbf{b}^* + 0\mathbf{c}^*$ of both the graphite and the Xe ML C-phase, remembering that reciprocal lattice vectors are perpendicular to real lattice vectors and inversely proportional to the corresponding (hk0) plane spacing. Show that the lowest order (10.0) and (01.0) C-phase diffraction spots are in the center of the triangles formed from the lowest order graphite spots. (The dot in (hk.0), representing $-(h+k)$, is the four- axis notation for hexagonal crystals, see e.g. Kelly and Groves 1970.)

(c) Explain which features of the Ne/graphite diffraction pattern (figure 4.8) indicate that the IR phase has a smaller unit cell than the C-phase, rotated $\sim \pm 18°$ from the IA orientation.

Problem 4.2. Adatom energies on graphite

Equation (4.14) gives the formal expression for an expansion of the substrate potential $V(\mathbf{r})$ seen by a single adatom in terms of the average potential V_0 and the Fourier coefficients V_g.

(a) Construct the potential $V(\mathbf{r})$ for the graphite surface if only the six lowest order **gs** are important, which have a common value of V_g. Show that the adsorption energy for an adatom in the middle of a graphite hexagon is $(V_0 + 6V_g)$, whereas at the bridge position between two carbon atoms is $(V_0 - 4V_g)$, and at the ontop site, over one carbon atom it has the value $(V_0 - 3\sqrt{3}V_g)$. Given that the lowest energy position is calculated to be in the center of the hexagon, show that V_g is negative, and that the diffusion path is via the bridge site with a diffusion energy $E_d = -10V_g$.

(b) If a calculation gives the adsorption energies in these positions (hexagon center E_h, bridge E_b, on top of carbon E_c), show that the description of $V(\mathbf{r})$ in terms of the Fourier series (4.14) requires three parameters V_0, V_{g1} and V_{g2}. Make a sensible choice for $\mathbf{g}_2$, and set up a matrix to solve for the Fourier coefficients. If a particular calculation gives $E_h = -1427$, $E_b = -1392$ and $E_c = -1388$ K per atom, calculate V_0, V_{g1} and V_{g2} in the same units. Compare this calculation with literature values for Kr/graphite (e.g. Price 1974, Bruch 1991, Bruch *et al.* 1997, chapter 2), and note that the calculation does *not* depend on the height of the adatom above the surface being the same at the three positions.

Problem 4.3. Localized or 2D gas adsorption?

(a) Differentiate the same Fourier series (4.14) (with two or three sets of terms) twice, to derive the diffusion frequency ν_d when the Kr atom is placed at the hexagon center position on graphite. Using the energy parameters given above, find the value of ν_d in THz units given that Kr has atomic mass 83.8.

(b) Use the arguments of section 4.2 to discuss whether a low density gas of Kr adatoms should be considered to be localized, vibrating about a particular lattice site, or in a 2D gas. Estimate the transition temperature between these two states.

Project 4.4. Chemisorption on d-band metals

This project should not be attempted until after reading chapter 6 in addition to this chapter.

The Anderson–Grimley–Newns model described in section 4.5.2 is useful for correlating a large range of data, where trends can be analyzed. One such correlation is the effect of strain in the substrate on reactivity. As shown by several authors (Hammer and Nørskov 1997, Ruggerone *et al.* 1997, Mavrikakis *et al.* 1998) moderate strains are calculated to produce substantial changes in reactivity, with expanded surfaces having higher reactivity. Given the importance of d-band metals as catalysts, explore how this comes about, and how adsorption and dissociation energies can be correlated with movement of the center of the d-band relative to the Fermi level.

5 Surface processes in epitaxial growth

This chapter discusses models of nucleation and growth on surfaces, in the context of producing epitaxial thin films. Section 5.1 gives some of the reasons for studying this topic and introduces some ideas needed as background. In section 5.2, we discuss differential equation formulations used to describe nucleation experiments quantitatively, and show how experiment–model comparisons can yield energies for characteristic surface processes. Sections 5.3 and 5.4 describe such comparisons in the case of metals growing on insulators and on metals. Section 5.5 explores steps, ripening and interdiffusion on insulator and metal surfaces. Although many of the same experiments and models are used in studying the growth of semiconductors, the role of surface reconstruction is much more important, so this topic is deferred to chapter 7.

5.1 Introduction: growth modes and nucleation barriers

5.1.1 Why are we studying epitaxial growth?

Epitaxial growth is a subject with considerable practical application, most obviously in relation to the production of semiconductor devices, but also to a whole range of other items. For example, magnetic devices such as recording heads have been produced by using metallic multilayers, in which alternating layers of magnetic and nonmagnetic materials produces high sensitivity to magnetic fields; other magnetic examples are bistable switches, where the alignment of the magnetic moments can be parallel or antiparallel in the neighboring layers. Many such films are required to be single crystals with low defect density, and are produced via epitaxial growth processes. The term epitaxy has come to mean the growth of one layer in a particular crystallographic orientation relationship to the underlying, or substrate layer (Schneider & Ruth 1971, Bauer & Poppa 1972, Matthews 1975, Kern *et al.* 1979).

In most of these applications, the end-point interest is almost always electrical, magnetic or optical, and there may also be an interest in the mechanical properties; some of these features are explored in chapters 6–8. However, it is not enough to be interested just in the end-point, since we need to know how to get there, and what influences the final properties. It is here that the science behind the atomic and molecular processes in epitaxial growth can find a good part of its (societal) justification.

However, in delving into this topic for its own sake, we should realize that the techno-logical ends may be better served in other ways. For example, many multilayer films are produced by sputtering, and are polycrystalline, albeit with a preferred orienta-tion; another current example is that it may be better to produce films by depositing (ionized) clusters rather than single atoms. A chapter which tried to interpret all the different growth methods, including those listed in section 2.5, would look rather different from this one.

5.1.2 Simple models – how far can we go?

In this situation, it seems a good idea to study a relatively simple approach in some depth. This enables one to say clearly what one does, and does not understand. Although it *may* help to offer advice on what is or is not a good recipe for producing better films or devices, this is certainly not straightforward. This dichotomy is an inter-esting example of the relationship between science and technology. It means that one can use the understanding so gained as background to appreciate the next technolog-ical advance, but that trying to advance the science and the technology are usually rather different endeavors. But the role of scientists in producing new instruments such as the SEM, the STM or AFM should not be underestimated; such developments can completely change our perception of what is observable and interesting.

What is attempted here is to see how far one can go with simple models involving adsorption and diffusion of atoms, and the new element, *binding* between atoms on surfaces. Binding introduces cooperative features into the models, which are non-linear in the adatom concentration. This opens the way to a discussion of the kinetics of crystal *nucleation* and growth, as contrasted with the thermodynamics of adsorption studied in chapter 4. For both experiment and models, we can discuss these topics in atomistic terms; indeed the behavior of single atoms or molecules influences the final microstructure of thin films in many cases.

The nucleation and growth patterns observed experimentally reflect directly the different types of bonding in solids. Thus we can discuss what is expected in the growth of metals on metals, metals on ionic crystals or on semiconductors, and semiconduc-tors on semiconductors. This subject has a long history, and a full literature citation is neither possible nor sensible in the present context. *Growth and Properties of Ultrathin Epitaxial Layers* (King & Woodruff 1997), *Surface Diffusion: Atomistic and Collective Processes* (Tringides 1997) and *Thin Films: Heteroepitaxial Systems* (Liu & Santos 1999) are recent research compilations; the main arguments are given here. One advan-tage of discussing this material in textbook form is that we can build on concepts, examples and problems discussed in earlier chapters using consistent notation. The great variety of notation is a major hazard in consulting the research literature directly.

5.1.3 Growth modes and adsorption isotherms

The classification of three growth modes shown in figure 5.1 dates from a much quoted paper in *Zeitschrift für Kristallographie* (Bauer, 1958). The layer-by-layer, or

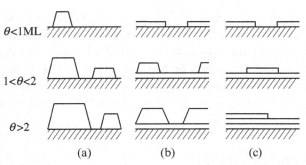

$\theta < 1\text{ML}$

$1 < \theta < 2$

$\theta > 2$

(a)   (b)   (c)

Figure 5.1. Schematic representation of the three growth modes, as a function of the coverage θ in ML: (a) island, or Volmer–Weber growth; (b) layer-plus-island, or Stranski–Krastanov growth; (c) layer-by-layer, or Frank–van der Merwe growth (after Bauer 1958, and Venables *et al.* 1984, redrawn with permission).

Frank–van der Merwe, growth mode arises because the atoms of the deposit material are more strongly attracted to the substrate than they are to themselves. In the opposite case, where the deposit atoms are more strongly bound to each other than they are to the substrate, the island, or Volmer–Weber mode results. An intermediate case, the layer-plus-island, or Stranski–Krastanov (SK) growth mode is much more common than one might think. In this case, layers form first, but then for one reason or another the system gets tired of this, and switches to islands.

Bauer was the first to systematize these (epitaxial) growth modes in terms of surface energies, similarly to earlier work on adhesion and contact angles by Young and Dupré. When we deposit material A on B, we get layer growth if $\gamma_A + \gamma^* < \gamma_B$, where γ^* is the interface energy, and vice versa for island growth. The SK mode arises because the interface energy increases as the layer thickness increases; typically this layer is strained to (more or less) fit the substrate. Pseudomorphic growth is the term used when it fits exactly. For each of these growth modes, there is a corresponding adsorption isotherm, shown in figure 5.2. In the island growth mode, the adatom concentration on the surface is small at the equilibrium vapor pressure of the deposit; no deposit would occur at all unless one has a large supersaturation. In layer growth, the equilibrium vapor pressure is approached from below, so that all the kinetic processes occur at undersaturation, as in the discussion of adsorbed monolayers in section 4.3.[1] In the SK mode, a finite number of layers are on the surface in equilibrium. The new element here is the idea of a nucleation barrier, shown as dashed lines on figure 5.2, which in the SK example occurs after the second ML has been formed, as opposed to after the first ML in figure 5.1. The existence of such a barrier means that a finite supersaturation is required to nucleate the 3D deposit. Since these are kinetic phenomea, metastable (supersaturated) layers can also co-exist with the islands.

---

[1] Note that although it may be advisable to read chapter 4 first, this is certainly not necessary. All the material in this chapter can be understood on the basis of chapter 1, especially section 1.3.

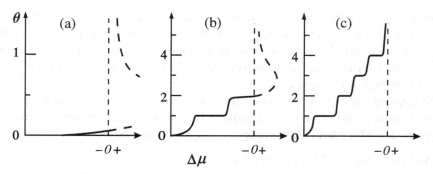

Figure 5.2. Adsorption isotherms corresponding to the three growth modes shown in figure 5.1. $\Delta\mu$ represents the chemical potential of the growing deposit relative to the bulk material, and θ the coverage in ML. In (b) two stable intermediate layers are indicated (after Venables *et al.* 1984, redrawn with permisison).

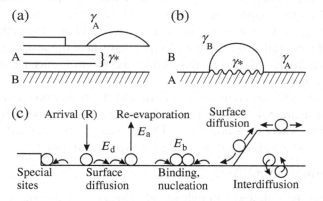

Figure 5.3. (a) Growth of A on B, where $\gamma_A < \gamma_B$; misfit dislocations are introduced, or islands form after a few layers have been deposited; (b) growth of B directly onto A as islands. The interfacial energy γ^* represents the excess energy over bulk A and B integrated through the interface region; (c) surface processes and characteristic energies in nucleation and film growth (after Venables 1994, redrawn with permission).

5.1.4 Nucleation barriers in classical and atomistic models

These phenomena look a lot more complex when one tries to compare this surface energy picture with what is going on at the atomic level, as shown schematically in figure 5.3. In figure 5.3(a) we see the deposition of A on B, and can visualize the way that islands form after a few layers due to the increase in γ^* with increasing film thickness. If we grow B on A, as shown in figure 5.3(b), we get islands right away. Thus the growth of a multilayer film A–B–A typically consists of alternate *good* and *bad* interfaces; there is no thermodynamic way to avoid this, though, as we shall see, kinetics plays a very important part in what actually happens.

If one looks at either interface in atomistic terms, as in figure 5.3(c), there are many

kinetic processes which can occur at both surfaces; in general only a few of these processes can be put into quantitative models at the same time. In particular, maybe only
one (combination) of these processes will be rate-limiting, and thus be responsible for
a nucleation barrier; this concept can be explored in both classical (macroscopic
surface energy) or in atomistic terms.

Nucleation theory proceeds classically as follows. If we draw the case where $\gamma_A > \gamma_B$,
so that we have 3D islands, as in figure 5.3(b), then we can construct a free energy
diagram $\Delta G(j)$ for islands containing j atoms, which has the form

$$\Delta G(j) = -j\Delta\mu + j^{2/3}X, \text{ at supersaturation } \Delta\mu, \tag{5.1a}$$

where X is a surface energy term of the form

$$X = \Sigma_k C_k \gamma_k + C_{AB}(\gamma^* - \gamma_B), \tag{5.1b}$$

where the first term represents the surface energy of the various faces of the island A,
and the second represents the interfacial energy between A and B; the geometrical constants C_k, C_{AB} depend on the shape of the islands. Note that the reference state for
which $\Delta G(j) = 0$ is both a cluster containing no atoms ($j=0$), and also the bulk solid
A in equilibrium with its own adsorbed layer and vapor, but where the surface energy
is neglected, so that both $\Delta\mu$ and X are zero; this second case was introduced in section
1.3 to define $\Delta\mu = kT\ln(p/p_e)$.

The form of such curves for different values of $\Delta\mu$ and X (in arbitrary units, but
think kT) is shown in figure 5.4. The nucleation barrier results because there is a
maximum in these curves, where the slope is zero. Differentiating, we can see that this
maximum occurs at $j=i$, where i is called the *critical* cluster size, and that

$$i = (2X/3\Delta\mu)^3 \text{ and } \Delta G(i) = 4X^3/(27\Delta\mu^2). \tag{5.2}$$

The same argument can be followed through for 2D clusters, i.e. monolayer thick
islands. In this case, the relevant supersaturation is expressed in relation to the corresponding step in the adsorption isotherm, i.e. $\Delta\mu' = kT\ln(p/p_1)$ for nucleation of the
first monolayer. The form is now

$$\Delta G(j) = -j\Delta\mu' + j^{1/2}X, \tag{5.3}$$

where the square root expression results from the extra *edge energy* X. Finding the
maximum in the same way leads to

$$i = (X/2\Delta\mu')^2 \text{ and } \Delta G(i) = X^2/(4\Delta\mu'), \tag{5.4}$$

where $\Delta\mu' = \Delta\mu - \Delta\mu_c$ and $\Delta\mu_c = (\gamma_A + \gamma^* - \gamma_B)\Omega^{2/3}$, with Ω as the atomic volume of the
deposit. Thus, in this formulation, a measurement of the pressure of the steps in the
adsorption isotherm directly determines the surface energy difference $(\gamma_A + \gamma^* - \gamma_B)$.

This way of looking at the problem is less than 100% realistic, perhaps not surprisingly. It is rather artificial to think about surface energies of monolayers and very small
clusters in terms of macroscopic concepts like surface energy. Numerically, the critical
nucleus size, i, can be quite small, sometimes even one atom; this is the justification for
developing an atomistic model, as discussed in the next section. The form of the free

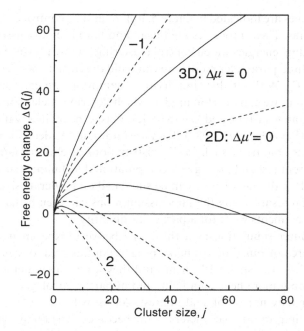

Figure 5.4. Free energy of nucleation $\Delta G(j)$ for 3D (full line) and 2D (dashed line) clusters. These curves are to scale for the surface free energy term $X=4$, and for $\Delta\mu$ or $\Delta\mu'=-1$, 0, 1 and 2. $\Delta G(j)$, X and $\Delta\mu$ are all in the same (arbitrary) units. However, if the unit is taken as kT, then the scale of free energy is the same as used by Weeks & Gilmer (1979) as shown in figure 1.12 and discussed in section 1.3 (replotted after Venables *et al.* 1984).

energy in an atomistic calculation is very similar to figure 5.4, but has a discrete character which can show secondary minima at particularly stable sizes, which are sometimes referred to as magic clusters. An early example of such a calculation is given by Frankl & Venables (1970). However, an atomistic model should be consistent with the macroscopic thermodynamic viewpoint in the large-i limit. To ensure this is not trivial, and most models don't even try; if I rather emphasize this, it is because I am attempting to sort this out in my research papers (Venables 1987). In other words, there are (at least) two traditions in the literature; it would be nice to unify them convincingly.

5.2 Atomistic models and rate equations

5.2.1 *Rate equations, controlling energies, and simulations*

We have considered simple rate equations for adatom concentrations in section 1.3, and in problem 1.2, adding a diffusion gradient in problem 1.3. Now we need to add nonlinear terms to describe clustering and nucleation of 2D or 3D islands. These equations are governed primarily by energies, which appear in exponentials, and also by frequency and entropic preexponential factors.

The most important energies are indicated in figure 5.3(c): E_a and E_d control desorption and diffusion of adatoms. These processes are linear, and have been discussed in chapter 1; E_j and E_i are binding energies which control clustering – these processes are non-linear. In the simplest three-parameter model, we can build the cluster energies out of pair bonds of strength E_b. Without this simplifying assumption, we can't make explicit predictions; but with it, we can develop models which describe nucleation and growth process over a large range of time and length scales. This is the main advantage of such 'mean field' models (Venables 1973, 1987). They are known not to describe fluctuations very well, so various quantities, such as size distributions of clusters, are not described accurately. In current research, using fast computational techniques such as 'Kinetic Monte Carlo' (KMC), the early stages can be simulated on moderate size lattices. These KMC simulation results using the same assumptions can then be used to check whether mean field treatments work for a particular quantity.

The emergence of computer simulation as a third way between experiment and theory is clearly a growth area of our time. To make progress in this area, one has to start with the simplest models, and stick with them until they are really understood. You need to beware generating more heat than light, and in particular of generating special cases which may or may not be of real interest. As we will see later in this chapter, the number of important parameters can become alarmingly large. Simulations can however be very illuminating, and may suggest inputs for simple models that one hadn't thought of. Animations are immediately appealing, and if Spielberg can do it, why shouldn't we? The problem lies only in the subsequent claims for correspondence with reality; then a measure of self-discipline is needed, both from the author and the reader. An extensive list of methods and suitable warnings are provided by Stoltze (1997).

5.2.2 Elements of rate equation models

To develop an atomistic model, we consider rate equations for the various sized clusters and then try to simplify them. If only isolated adatoms are mobile on the surface, we have

$$dn_1/dt = R \text{ (or } F) - n_1/\tau_a - 2U_1 - \Sigma U_j, \tag{5.5}$$

and for larger clusters

$$dn_j/dt = U_{j-1} - U_j \qquad (j \geq 2), \tag{5.6}$$

where U_j is the net rate of capture of adatoms by j-clusters. This is not very useful yet, since we need expressions for the U_j, and we need the simplification introduced by the idea of a critical nucleus size. In its simplest form, this means that (a) we can consider all clusters of size $> i$ to be 'stable', in that another adatom usually arrives before the clusters (on average) decay; the reverse is true for clusters of below critical size; and (b) these subcritical clusters are in local equilibrium with the adatom population.

The first consideration leads to defining the stable cluster density, or nucleation density n_x, via the nucleation rate

$$dn_x/dt = \Sigma_{j \geq i} (U_j - U_{j+1}) = U_i - \ldots, \tag{5.7}$$

since all the other terms cancel out in pairs. The ... means that we can add other terms, such as the loss of clusters due to coalescence, in various approximations. The second consideration leads to arguments about detailed balance, and the Walton relation, named after a paper where local equilibrium was first discussed in this context (Walton 1962). These detailed balance arguments lead to all the U_j, for $j < i$, being zero separately, and hence $dn_j/dt = 0$. Note that this is not the same as a steady state argument where $dn_j/dt = 0$ because $U_{j-1} = U_j$; it is more stringent.

A typical form of the U_j contains both growth and decay, and in local equilibrium these two terms are numerically equal; the growth term due to adding single adatoms by diffusion to $j - 1$ clusters is of the form $\sigma_j Dn_1 n_{j-1}$, where σ is known as a capture number. The decay term has the form $-\nu_d n_j \exp\{-(E_d + \Delta E)/kT\}$, where ΔE is the binding energy diference between j and $j - 1$ clusters. The key point is that if there is local equilibrium, then the ratio

$$n_j/n_{j-1} = n_1 C \exp(\Delta E/kT), \tag{5.8}$$

where C is a statistical weight, which is a constant for a particular size (and configuration) cluster. Note that this equilbrium must not depend on D, which is only concerned with kinetics. This argument can then be cascaded down through the subcritical clusters, yielding the Walton relation

$$n_j = (n_1)^j \Sigma_m C_j(m) \exp (E_j(m)/kT), \tag{5.9}$$

where (m) denotes the mth configuration of the j-sized cluster. This formula gives essentially the equilibrium constant, in the physical chemistry sense, of the polymerization reaction j adatoms $\Leftrightarrow 1$ j-mer. It can thus be derived using classical statistical mechanics on a lattice, with N_0 sites.

In the above formulae, ML units have been used for n_j for simplicity, but sometimes the N_0 is put in explicitly. In that case the n_j are areal densities, and we need n_j/N_0 and n_1/N_0 in the above equation. You may note that at low temperature, we would only need to consider the most strongly bound configuration, because of the dominant role of the exponential in (5.9). However, the critical cluster size is large typically at high temperature, so we need to be on our guard. If $i = 1$, at low temperature, the above discussion is not required anyway, since pairs of adatoms already form a stable cluster, and so are part of n_x.

At this point, we do have something useful, because we can simplify the rate equations down to just two coupled equations, namely

$$dn_1/dt = R - n_1/\tau_a - (2U_1 + \Sigma_{j<i} U_j) - \sigma_x Dn_1 n_x, \tag{5.10}$$

where the term in brackets is almost always numerically unimportant, and the last term describes the capture of adatoms by stable clusters, and can be written as n_1/τ_c, and

$$dn_x/dt = \sigma_i Dn_1 n_i - U_{cl}. \tag{5.11}$$

In (5.11) the assumption of local equilibrium for n_i, which is only a first approximation, will make the first term explicit, and proportional to the $(i + 1)$th power of the adatom concentration: this is very non-linear if the critical nucleus size is large! The second term in (5.11) is typically due to coalescence of islands; this U_{cl} is proportional to $n_x dZ/dt$, where Z is the coverage of the substrate by the stable islands. Thus dZ/dt is related to the (2D or 3D) shape of the islands, and how they grow (Venables 1973, 1987, Venables & Price 1975, Venables *et al.* 1984); all these details are not repeated here, but in the simplest 2D island case the last two loss terms in (5.10) equal $N_a dZ/dt$, where N_a is the 2D density of atoms in the deposit. Note that Z is measured in ML, and is therefore dimensionless.

The capture numbers are related to the size, stability and spatial distribution of islands. The simplest mean field model, which was worked on long ago (Venables 1973, Lewis & Anderson 1978, Stoyanov & Kaschiev 1981), and which others have worked on more recently (Bales & Chrzan 1994, Brune *et al.* 1999), is referred to as the uniform depletion approximation; it considers a typical cluster of size k immersed in the average density of islands of all sizes. Then one can set up an ancilliary diffusion equation for the adatom concentration in the vicinity of the k-cluster (size specific), or x-cluster (the average size cluster), which has a Bessel Function solution. This model gives exactly in the incomplete condensation (re-evaporation dominant) limit

$$\sigma_k = 2\pi X_k K_1(X_k)/K_0(X_k) \text{ and} \tag{5.12a}$$

$$\sigma_x = 2\pi X K_1(X)/K_0(X), \tag{5.12b}$$

where $X_k^2 = r_k^2/D\tau_a$ and $X^2 = r_x^2/D\tau_a$, r_k and r_x being the corresponding island radii, and K_0 and K_1 the Bessel functions. For complete condensation the mean field expressions are the same, but the arguments of the Bessel functions contain τ_c instead of τ_a; in general we should use τ as defined in the next section. In complete condensation, these capture numbers are just functions of the coverage of the substrate by islands, Z.

The details of these capture numbers are the subject of problem 5.2, but for the moment we need to remember that they are simply numbers, with σ_i in the range 2–4 and σ_x often in the range 5–10. Using these expressions one can compute the evolution of the nucleation density with time, or more readily with Z, as the independent variable, as first proposed by Stowell *et al.* in the early 1970s. There are also other approximations for the various σs, such as the lattice approximation. However, the appropriateness of any of these depends on the spatial correlations between islands which develop as nucleation proceeds. The key point of Bales & Chrzan's (1994) paper was to show, for the particular case of $i = 1$ in complete condensation, that (5.12) with τ_c plus the other small terms as the argument, *is* the correct expression in the absence of spatial correlations, in agreement with their KMC simulations.

5.2.3 Regimes of condensation

The above reasoning needs a bit of explaining to make it clear; if you are interested in the details it may be worth doing problems 5.1 and 5.2 at some point. Let us start by focusing our attention on the rate equation for the adatoms, where we can write as

$$dn_1/dt = R - n_1/\tau, \text{ where } \tau^{-1} = \tau_a^{-1} + \tau_n^{-1} + \tau_c^{-1} + \dots \tag{5.13a}$$

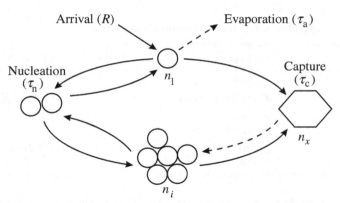

Figure 5.5. Schematic illustration of the interaction between nucleation and growth stages. The adatom density n_1 determines the critical cluster density n_i; however, n_1 is itself determined by the arrival rate R in conjunction with the various loss processes which have characteristic times as described in the text (Venables 1987).

We can define the various time constants by comparison with (5.10), to obtain

$$\tau_n^{-1} = (2U_1 + \Sigma_{j<i} U_j)/n_1 \text{ and } \tau_c^{-1} = \sigma_x Dn_x. \tag{5.13b}$$

It is reasonable that the nucleation term (in brackets) is almost always numerically unimportant, as we have already convinced ourselves that U_j is close to zero for sub-critical clusters. The ratio $r = \tau_a/\tau_c = \sigma_x n_x D\tau_a$ then determines whether we are in the complete ($\gg 1$) or incomplete ($\ll 1$) condensation regime. So at high temperatures, $\tau \Rightarrow \tau_a$ and at low temperatures (and/or high deposition rates), $\tau \Rightarrow \tau_c$. This is set out pictorially in figure 5.5, where the different reaction channels are illustrated. It is useful to think of competitive capture; equations (5.13a,b) describe processes in which all the adatoms end up somewhere, and the different competing processes (or channels) add as resistances in parallel.

Often, the condensation starts out incomplete, and becomes complete by the end of the deposition. This is the *initially incomplete* regime. If the diffusion distance on the surface is so short that only atoms which impinge directly on the islands condense, then we have the *extreme incomplete* regime. In each of these limiting cases, the two coupled equations ((5.10) for n_1 and and (5.11) for n_x) can be evaluated explicitly, and give the nucleation density of the form $n_x \sim R^p \exp(E/kT)$, with p and E dependent on the regime considered. The formulae are given in table 5.1, and can be explored further via problem 5.1. Perhaps the most important regime for what follows is *complete* condensation. Re-evaporation is absent in this regime; one can notice from table 5.1 that the adsorption energy E_a does not appear in the expression for the cluster density.

In the complete condensation regime the (non-linear) interplay between nucleation and growth is most marked. In general, it is clear that if one includes different processes, then one would expect to get different power laws and energies. For example, Markov (1996) and Kandel (1997) explore different models to study the role of surfactants in promoting layer growth during complete condensation, and obtain different power laws which could be tested by experiments. There have been many discussions

Table 5.1. *Parameter dependencies of the maximum cluster density*

Regime	3D islands	2D islands
Extreme incomplete	$p = 2i/3$ $E = (2/3)[E_i + (i+1)E_a - E_d]$	i $[E_i + (i+1)E_a - E_d]$
Initially incomplete	$p = 2i/5$ $E = (2/5)[E_i + iE_a]$	$i/2$ $[E_i + iE_a]$
Complete	$p = i/(i+2.5)$ $E = (E_i + iE_d)/(i+2.5)$	$i/(i+2)$ $(E_i + iE_d)/(i+2)$

about how surfactants might work, especially in relation to semiconductor growth; at this stage a surfactant can be thought of as any foreign species which remains at the surface as growth proceeds.

5.2.4 *General equations for the maximum cluster density*

The final question following this type of reasoning is to ask whether there is a general equation for the maximum cluster density, which yields these three regimes as limiting cases. The answer is *yes*, but the resulting equation is not especially simple. We can see that (5.11) will lead to a maximum in the stable cluster density at the point where the (positive) nucleation term is balanced by the (negative) coalescence term. At this point, $dn_x/dt = 0$ and the coverage of the substrate by islands $Z = Z_0$. If we make substitutions for all the various terms, then we will obtain an explicit expression for $n_x(Z_0)$. As a practical point, we can calculate the Z-dependence of n_x, *within* each of the condensation regimes, and obtain pre-exponentials $\eta(Z, i)$ for each of these regimes, as illustrated for 3D and 2D islands in complete condensation in figure 5.6. The following arguments are aided by the fact that these pre-exponentials, which multiply the parameter dependencies of table 5.1, are only weakly dependent on both Z and the critical size i.

Although the coverage Z_0 depends on the formula chosen for the coalescence term U_{cl}, which in turn depends on the spatial correlations that develop during growth, none of this influences the exponential terms in the equations. Here we use the coalescence expression due to Vincent (1971), where $U_{cl} = 2n_x dZ/dt$. The rest is algebra, starting from (5.11). Inserting the Walton equation (5.9) for n_i, the steady state equation (5.13) for n_1 (neglecting the nucleation term $R\tau_n$), and specializing to 2D islands, n_x is then given, after considerable rearrangement, by

$$n_x (g+r)^i (Z_0 + r) = f (R/D)^i \{\exp (E_i/kT)\}(\sigma_x D\tau_a)^{i+1}. \tag{5.14}$$

The slowly varying numerical functions f and g involve the capture numbers σ_i and σ_x. For 3D islands, n_x is changed to $n_x^{3/2}$ on the left hand side of (5.14), and the constants change a little (Venables 1987).

A point that may have been misunderstood is the following: the arguments of this section have been carried through *on the assumption that* the critical nucleus size is i. The *actual* critical nucleus size is that which produces the *lowest* nucleation rate and density; it is only then that the critical nucleus is consistent with the *highest* free energy

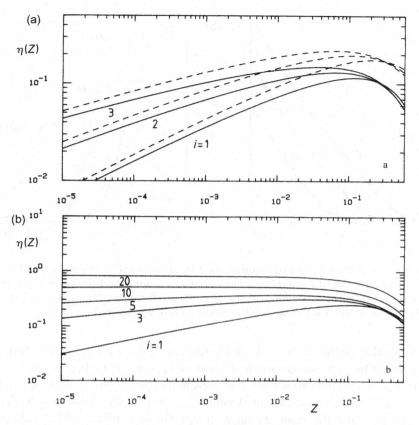

Figure 5.6. Pre-exponential factors $\eta(Z,i)$ in the complete condensation regime: (a) 3D islands for $i=1$, 2 and 3 with σ_x evaluated in the lattice (full line) and uniform depletion (dashed line) approximations; (b) 2D islands for $i=1$–20, with σ_x approximated by $4\pi/(-\ln Z)$ which is very close to the uniform depletion approximation (Venables *et al.* 1984, reproduced with permission).

$\Delta G(j)$, for $j=i$, as discussed in section 5.1.3. The critical nucleus size is thus determined self consistently as an output, not an input, of an iterative calculation for all feasible assumed critical sizes. In complete condensation the ratio $r \equiv \tau_a/\tau_c$ is much greater than both g and Z_0, corresponding to adatom capture being much more probable than re-evaporation. In the extreme incomplete regime both r and $g \ll Z_0$. In between we have $Z_0 < r < g$, where most cluster growth occurs by diffusive capture, at least initially. It is a straightforward exercise to check that these conditions on (5.14) lead to the parameter dependencies given in table 5.1; keeping track of all the pre-exponential terms requires patience, and cross checks with the original literature.

5.2.5 Comments on individual treatments

The argument of the last section indicates that nucleation equations are only soluble if we know the E_i to enter into (5.11) or (5.14); however, we have also noted that the value of i is itself determined implicitly. This means that the predictions are only explicit if

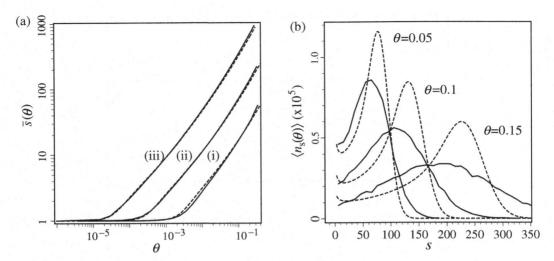

Figure 5.7. Comparison of rate equation (dashed lines) and KMC calculations (solid lines) of (a) the average number of atoms ($\bar{s}$) in a monolayer island, for the ratio $(D/R) = $ (i) 10^5, (ii) 10^7 and (iii) 10^9; (b) the size distribution of islands $\langle n_s(\theta) \rangle$ as a function of coverage θ, for $\theta = 0.05$, 0.1 and 0.15 (after Bales & Chrzan 1994, reproduced with permission).

we can calculate, within the model, the binding energy E_j for all sizes j. This is the principal reason why a *pair binding* model is invoked: life is too complicated otherwise. For 2D clusters, this simplification allows us to estimate $E_j = b_j E_b$, where b_j is the number of *lateral* bonds in the cluster, each of strength E_b. However, by retaining E_a as the *vertical* bonding to the substrate, we have enough freedom to model large differences between vertical and lateral bonding, within a three-parameter fit to nucleation and growth data. This feature of the pair binding model is important in giving us enough latitude to mirror, in the simplest way, the different types of bonding which actually occur in the growth of one material on another.

In developing the above model (Venables 1987), vibrations were included in a self-consistent way within the mean field framework outlined above. It is very easy to construct a model which is inconsistent with the equilibrium vapor pressure of the deposit unless the vibrations are treated reasonably carefully. This paper builds on the Einstein model calculations which we have done as problems in chapter 1, and is the basis for subsequent model calculations described here.

In the past few years there have been many related treatments by several groups, mostly in response to the new UHV STM-based experimental results, which have studied nucleation and growth down to the sub-monolayer level. Recent papers include a detailed comparison of rate equations and KMC simulations for $i = 1$ in the complete condensation limit (Bales & Chrzan 1994). The KMC work is important for checking that the rate equation treatment works well for average quantities, such as the nucleation density, n_x, or the average number of atoms in an island, w_x, as shown in figure 5.7(a); but it also shows that the treatment does not do a good job on quantities such as size distributions, shown in figure 5.7(b), which are dependent on the local environment, i.e. on spatial correlations which develop during nucleation and growth. Many authors (e.g. Myers-Beaghton & Vvedensky 1991, Bartelt & Evans 1992, 1994, Amar & Family 1995, Mulheran & Blackman 1995, 1996, and Zangwill & Kaxiras 1995) have evaluated size distributions during simulations, and shown that they are characteristic both of the critical cluster size, i, and of the spatial correlations. Brune *et al.* (1999) have made a detailed comparison of the various approximations, including coalescence terms, for the case of 2D sub-monolayer growth.

5.3 Metal nucleation and growth on insulating substrates

Some results of atomistic nucleation and growth models are described in the context of specific experimental examples for the remainder of this chapter. We concentrate on metal deposition, on insulators in this section and on metals in section 5.4. These examples illustrate the kinds of experimental tests to which atomistic models have been subjected over almost 30 years.

5.3.1 *Microscopy of island growth: metals on alkali halides*

An example of the use of non-UHV TEM to study nucleation and growth is shown in figure 5.8 from the Robins group (Donohoe & Robins 1972, Venables & Price 1975). The deposition of Au/NaCl(001) was done in UHV, but the micrographs were obtained by (1) coating the deposit with carbon in UHV; (2) taking the sample out of the vacuum; (3) dissolving the substrate in water; and (4) examining the Au islands on the carbon by TEM. By this means island densities, growth, coalescence and nucleation on defect sites could all be observed. By performing many experiments at different deposition rates R, and temperatures T, as a function of coverage, their group and others have produced quantitative data of island growth and rate equation models have been tested. It is clear that this type of technique is destructive of the sample: it is just as well that NaCl is not too expensive, and that gold/silver, etc. are relatively unreactive, or the technique would not be feasible.

The work on noble metals Ag and Au deposited onto alkali halides constitutes quite a long story which can be summarized roughly as follows. A full review of early work has been given by Venables *et al.* (1984) and by Robins (1988), including extensive tabulation of energy values deduced from experiment using the models described in section 5.2. Typically, these values were deduced by first showing that the initial nucleation rate J at high temperatures varied as R^2, and so corresponded to $i=1$ in (5.11). In this regime where $n_1 = R\tau_a$ we can see that

$$J= \mathrm{d}n_x/\mathrm{d}t = \sigma_i D n_1^2 \text{ which is proportional to } R^2 \exp\{(2E_a - E_d)/kT\}, \qquad (5.15)$$

so the T-dependence of the nucleation rate yields $(2E_a - E_d)$. This information can be combined with the low temperature (complete condensation) nucleation density, which for $i=1$ yields E_d, as can be seen from table 5.1. An alternative piece of data is the island growth rate at high temperature, determined by the width of the BCF diffusion zone

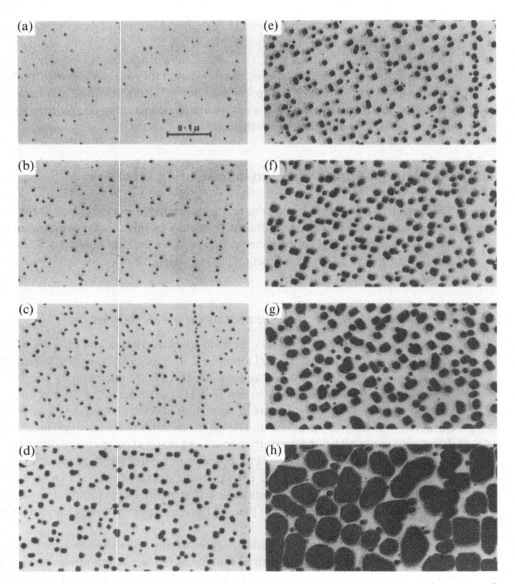

Figure 5.8. TEM micrographs of Au/NaCl (001) formed at $T = 250\,°C$, $R = 1 \times 10^{13}$ atoms cm^{-2} s^{-1} and deposition times of (a) 0.5, (b) 1.5, (c) 4, (d) 8, (e) 10, (f) 15, (g) 30 and (h) 85 min (from Donohoe & Robins 1972, reproduced with permission).

around each island, which leads to (5.12) for the capture numbers. The growth rate thus has a T-dependence given by $(E_a - E_d)$ via an equation which depends on the 2D or 3D shape of the islands.

The energies deduced depended a little on the exact mode of analysis, but were in the region $E_a = 0.65–0.70$ and $E_d = 0.25–0.30$ eV for Au/KCl. The corresponding quantities deduced for Au/NaCl were similar; for Ag/KCl they were around 0.5 and 0.2 eV, and for Ag/NaCl 0.65 and 0.2 eV respectively (Stowell 1972, 1974, Venables 1973, Donohoe & Robins 1976, Venables *et al.* 1984, Table 2). These E_a and E_d values are much lower than the binding energy of pairs of Ag or Au atoms in free space, which are accurately known, having values 1.65 ± 0.06 and 2.29 ± 0.02 eV respectively (Gringerich *et al.* 1985). We can therefore see why we are dealing with island growth, and why the critical nucleus size is nearly always one atom. The Ag or Au adatoms re-evaporate readily above room temperature, but if they meet another adatom they form a stable nucleus which grows by adatom capture. This type of behavior was observed for all metal/alkali halide combinations.

5.3.2 *Metals on insulators: checks and complications*

The combination of experiment and model calculations presented in the previous section is satisfying, but is it *correct*? What do I mean by that? Well, the experiment may not be correct, in that there may be defects on the surface which act as preferred nucleation sites. It is very difficult to tell, simply from looking at TEM pictures such as figure 5.8, whether the nuclei form at random on the terraces, or whether they are nucleated at defect sites; only nucleation along steps is obvious to the eye. In the previous section we described the classic way to distinguish true random nucleation, with $i = 1$. But there are several other ways to get $J \sim R^2$, including the creation of surface defects during deposition. In this case we might have nucleation on defects ($i = 0$), but with the defects produced in proportion to R by electron bombardment. Alkali halides are very sensitive to such effects, which were subsequently shown to have played a role in early experiments (Usher & Robins 1987, Robins 1988, Venables 1997, 1999).

As substrate preparation and other experimental techniques improved, lower nucleation densities which saturated earlier in time were observed (Velfe *et al.* 1982). This has been associated with the reduction in impurities/point defects, and the mobility of small clusters. From detailed observations as a function of R, T and t, some energies for the motion of these clusters have been extracted. Qualitatively, it is easy to see that if all the stable adatom pairs move quickly to join pre-existing larger clusters, then there will be a major suppression of the nucleation rate (Venables 1973, Stowell 1974). This was studied intensively for Au/NaCl(001) by Gates & Robins (1987a), who found that a model involving both defect and cluster mobility parameters were needed to explain the results of Usher & Robins (1987). The revised values of E_a and E_d for this system are given in table 5.2; in particular, $(E_a - E_d)$ has been determined in several independent experiments to be 0.33 ± 0.02 eV (Robins 1988, Venables 1994).

There are several further interesting experiments, including the study of alloy deposits, which has now been performed for three binary alloy pairs, formed from Ag, Au and Pd on NaCl(100) (Schmidt *et al.* 1990, Anton *et al.* 1990). In such experiments the atoms with the higher value of E_a, namely Au in Ag–Au, or Pd in Pd–Ag and Pd–Au, form nuclei preferentially, and the composition of the growing film is initially enriched in the element which is most strongly bound to the substrate. The composition of the films was measured by X-ray fluorescence and energy dispersive X-ray analysis, and only approached that of the sources at long times, or under complete condensation conditions.

Table 5.2. *Values (in eV) of E_a and E_d of Ag, Au and Pd adatoms on NaCl(001)*

Alloy δE	Element	$(E_a - E_d)$	E_a	E_d
Ag–Au: 0.11 ± 0.03	Ag	0.22	0.41	0.19
Au–Pd: 0.12 ± 0.03	Au	0.33 ± 0.02	0.49 ± 0.03	0.16 ± 0.02
Pd–Ag: 0.25 ± 0.05	Pd	0.45	0.78	0.33

Note: Values without errors are derived from data combinations (Robins 1988, Anton *et al.* 1990).

Table 5.3. *Calculated values of E_a and E_d of Ag and Au on alkali halide(001) surfaces*

Parameter (eV)	Ag/NaCl	Au/NaCl	Ag/NaF	Au/NaF
E_a	0.27 [0.27] (0.41)	0.15 [0.69] (0.49)	0.26	0.18 [0.59] (0.63)
E_d	0.15 [0.09] (0.19)	0.07 [0.22] (0.16)	0.24	0.14 [0.08] (0.08)
$(E_a - E_d)$	0.12 [0.18] (0.22)	0.08 [0.47] (0.33 ± 0.02)	0.02	0.04 [0.51]

Source: From Harding *et al.* 1998, with experimental values (round brackets) and previous calculations [square brackets].

These experiments can be analyzed to yield energy differences δE, where $\delta E = \delta E_x - \delta E_y$, and δE_x, $\delta E_y = (E_a - E_d)_{x,y}$ for the two components. Values of δE have been obtained for the pairs, namely Au–Ag, 0.11 ± 0.03; Pd–Au, 0.12 ± 0.03; Pd–Ag, 0.25 ± 0.05 eV. These experiments measure, very accurately, differences in integrated condensation coefficients, $\alpha_{x,y}(t)$, which are determined by the BCF diffusion distances of the corresponding adatoms, as explored in problem 5.1. Coupled with nucleation density measurements, the data give accurate values for E_a and E_d for these three elements on NaCl(100), as given in table 5.2.

More recently, efforts have been made to understand some of these energy values in terms of metal–ionic crystal bonding. Earlier estimates maintained that a considerable part of the binding was of van der Waals type, but recent work has shown that this contribution was almost certainly overestimated. In particular, Harding *et al.* (1998) calculated the pairwise ion–ion interactions within the relativistic Dirac–Fock approximation, the metals Ag and Au being most attracted to the halide ion; the van der Waals energy was reduced, due to overlap and better values of dispersion constants, from 0.5 to around 0.15 eV/atom. Atomic polarization within the shell model was included, and the whole assembly relaxed to equilibrium, with a claimed accuracy of ± 0.1 eV. Some results are given in table 5.3. The results suggest that the calculated values of E_d are within the quoted accuracy, but that E_a seems to be systematically underestimated, possibly because of the neglect of charge transfer in the calculation. A Hartree–Fock cluster calculation (Mejías 1996) also obtains very low values, ~0.1 eV for E_a; so this method presumably also gives very small values of E_d, which must by definition be less than E_a.

What survives from previous work (Chan *et al.* 1977, Gates & Robins 1988) is that the metal–metal binding energies E_b are high, close to free space values, and that the diffusion energies of dimers and small clusters are very low, often as low as E_d itself. This arises because one can fit the first atom of a pair on the surface optimally, but the second one is constrained by being a member of a pair; the resulting energy surface, while quite complex, is less corrugated.

Extension of these calculations to point defects has been attempted (Harding *et al.* 1998), with the result that the noble metal adatoms are generally attracted to surface cation, but not anion, vacancies. This is especially the case if the adatom size is sufficiently small to fit inside the surface vacancy, the attractive energies increasing as the height of the adatom above the surface plane decreases. The role of surface charges, and their effect on the charge state of vacancies, could be very important, as has been found for the case of Ag and other metals, particularly Pd, on MgO(001) (Ferrari & Pacchioni 1996).

5.3.3 Defect-induced nucleation on oxides and fluorides

There are many examples in the literature where defect nucleation seems to be needed (Harsdorff 1982, 1984, Venables 1997, 1999). The transition from $i=0$ to 1 was observed for Au/mica (Elliott 1974). In this case defect sites were used up initially, and nucleation then proceeded on the perfect terraces in the initially incomplete condensation regime. A more recent example is furnished by high resolution UHV-SEM observations of the growth of nanometer-sized Fe and Co particles on various CaF_2 surfaces, typically thin films on Si(111), as indicated in figure 5.9. In this work (Heim *et al.* 1996) the nucleation density, for $Z_0 \cong 0.2$ close to the maximum density, was independent of temperature over the range 20–300 °C. At 400 °C and above the substrate itself is unstable. This behavior is not understandable if nucleation occurs on defect-free terraces, but may be understood if defect trapping is strong enough.

Defects of various types can be incorporated into either analytical treatments or simulations, at the cost of at least two additional parameters, the trap density n_t, and the trap energy E_t, or the binding energy of adatoms to steps E_s, as indicated schematically in figure 5.10. The nucleation density of islands on defective substrates can be derived by considering the origin of the various terms in (5.14). The right hand side of this equation is proportional to the nucleation rate (via the term in $\exp(E_i/kT)$), which is enhanced by a ratio $B_t = 1 + A_t$ with defects present (Heim *et al.* 1996, Venables 1997).

We start by considering the point defect traps shown in figure 5.10(a), constructing a suitable differential equation for the number of adatoms attached to traps, n_{1t},

$$dn_{1t}/dt = \sigma_{1t} D n_1 n_{te} - n_{1t} \nu_d \exp(-(E_t + E_d)/kT), \tag{5.16}$$

where n_{te} is the number of empty traps $= n_t - n_{1t} - n_{xt}$. In steady state, this equation is zero, and inserting (1.15) for D, we deduce

$$n_{1t}/(n_t - n_{xt}) = A/(1 + A), \text{ with } A = n_1 C_t \exp(E_t/kT), \tag{5.17}$$

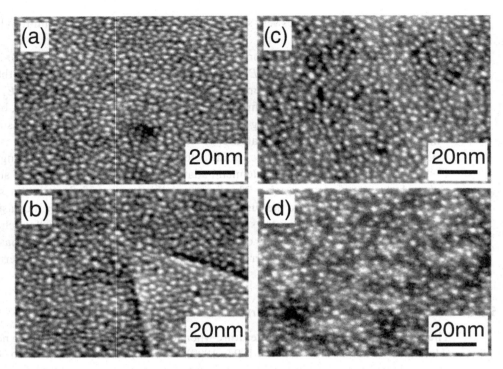

Figure 5.9. Nucleation and growth of small Fe crystals on CaF_2(10 nm)/Si(111) at (a) room temperature, (b) 140 °C, (c) 300 °C and (d) 400 °C, observed by *in situ* high resolution SEM. The average thickness is between 3.1 and 3.6 ML, and the coverage of the substrate, $Z \sim 20\%$ (from Heim *et al.* 1996, reproduced with permission).

where C_t is an entropic constant, which has been put equal to 1 in the illustrative calculations performed to date. Equation (5.17) shows that the traps are full ($n_{1t} = n_t - n_{xt}$) in the strong trapping limit, whereas they depend exponentially on E_t/kT in the weak trapping limit, as expected. This equation is a Langmuir-type isotherm (section 4.2.2) for the occupation of traps; the trapping time constant (τ_t, in analogy to 5.13) to reach this steady state is very short, unless E_t is very large; but if E_t is large, then all the traps are full anyway.

The total nucleation rate is the sum of the nucleation rate on the terraces and at the defects. The nucleation rate equation becomes, without coalescence, analogously to (5.11),

$$dn_x/dt = \sigma_i Dn_1 n_i + \sigma_{it} Dn_1 n_{it}, \tag{5.18}$$

where the second term is the nucleation rate on defects, and n_{it} is the density of critical clusters attached to defects, σ_{it} being the corresponding capture number. In the simplest case where the traps only act on the first atom which joins them, and entropic effects are ignored, we have

$$A_t = n_{1t}/n_1 = (n_t - n_{xt})A/[n_1(1 + A)]. \tag{5.19}$$

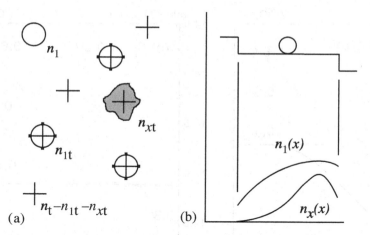

Figure 5.10. (a) Model for nucleation at attractive random point defects, which can be occupied by adatoms, density n_{1t}, clusters, density n_{xt}, or can be empty; (b) schematic diagram of line defects (steps), with adatom density $n_1(x)$ and nucleation density $n_x(x)$ for position x from up-steps (with attractive forces) and down-steps (maybe repulsive forces).

A high value of A gives strong trapping, in which almost all the sites unoccupied by clusters will be occupied by adatoms; in the simplest model we assume that *clusters* cannot leave the traps.

However, even the simplest behavior ensures that the defect processes are not linear. The clusters which form on the defect sites get established early on and thereby deplete the adatom density on the terraces. As a result, the overall nucleation density, which appears in the left hand side of (5.14), grows only as a fractional power [typically $1/(i+2.5)$ for complete condensation, see table 5.1] of the trap density. In this weak trapping limit, the main effect is the reduced diffusion constant D due to the time adatoms spend at traps (Frankl & Venables 1970). Nucleation on terrace sites is strongly suppressed, due to adatom capture by already nucleated clusters. But when $n_x > n_t$, there is little effect on the overall nucleation density. These effects result in the s-shaped curves shown in figure 5.11, illustrated for $n_t = 0.01$ ML, $E_t = 0.5$ and 1.0 eV, and Fe/CaF$_2$(111) parameters. If the trapping is very strong, and the diffusion energy is low, there is a large regime where $n_x = n_t$; conversely, weak trapping will lead only to a point of inflection at, and a change of slope above and below, the trap density.

Comparison with experiments puts bounds on the energies E_a, E_b and E_t, all $\cong 1$ eV, and suggests a low value, 0.1–0.3 eV, for E_d. Note that the reason why a low value of E_d is needed is so that the adatoms can migrate far enough at *low* temperatures to reach the defect sites. The high values of the other energies are needed, so that something else doesn't intervene at *high* temperatures. For example, if E_t is as low as 0.5 eV, the density does not reach n_t over a large enough temperature range; if E_a is too small, condensation becomes incomplete too early. Note also that the transition from $i = 1$ to 2 is observed for $E_d = 0.1$–0.3 eV at the highest temperature; this

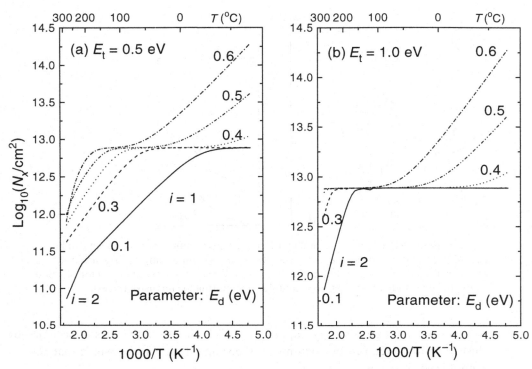

Figure 5.11. Nucleation density on point defects predicted with trap density $n_t = 0.01$ ML, trap energy (a) $E_t = 0.5$ eV, and (b) 1.0 eV, parameter E_d, with $E_a = 1.16$ eV and $E_b = 1.04$ eV (recalculated after Heim *et al.* 1996, and Venables 1999).

means that the limiting process can become breakup of the cluster (on a trap), rather than removal of the adatom from the trap; E_t is not then itself important, provided it is high enough.

This type of model thus contains several sub-cases, depending on the values of the parameters. A remarkable example is Pd/MgO(001), studied with AFM by Haas *et al.* (2000); this data also requires a high trapping energy E_t, in agreement with the calculations of Ferrari & Pacchioni (1996) and Venables & Harding (2000) for trapping of Pd in surface vacancies, and a low value of E_d. Models involving point defects are typically indicative, rather than truly quantitative, because of the possibility of other effects, such as cluster mobility and cluster detaching from defects, and the possibility of a range of defect binding energies. One can see qualitatively that if E_t is moderate and dimer motion is easy, then the traps may become reusable at high enough temperature; this further complication, needing even more parameters, has been thought necessary on occasion (Gates & Robins 1987a, Usher & Robins 1987). This is an ongoing tension between science and technology; both need conditions to be *reproducible*; technology can be successful even if *complicated*, but science needs the models to be relatively *simple*: we cannot sensibly deal with too many parameters at once.

5.4 Metal deposition studied by UHV microscopies

UHV microscope-based techniques (SEM, STEM, FIM and more recently STM, AFM, LEEM) plus diffraction techniques (including X-ray, SPA-LEED and helium scattering) examine the deposit/substrate combination *in situ*, without breaking the vacuum; we have discussed procedures for specific cases in chapters 2 and 3. Compared to *ex situ* TEM described in section 5.3.1 and 5.3.2, a wider set of substrates and deposits have been observed in recent years; some examples for metals on insulators were given in section 5.3.3.

In situ studies of metal deposition on metals are described in this section. These results are illustrative, in order to give a feel for the work; this is *not* a review article. Many review articles start with a statement: *this is not an exhaustive treatment . . .!* Perhaps this is just as well, no-one wants to be exhausted; but it means that one often has to go back to the original literature to be sure what went on. There are many examples where the primary literature represents a progress report, updated and maybe actually negated by subsequent work. Attempting to find out what happened is often hard, but is preferable to repeating the work in ignorance of the original!

5.4.1 In situ *UHV SEM and LEEM of metals on metals*

Ultra-high vacuum SEM and related techniques were developed by our group in Sussex, and used to study Stranski–Krastanov growth systems at elevated temperature. The SEM is good for visualizing strongly 3D objects, such as the (001) oriented Ag islands seen on Mo(001) in figure 5.12. Since adatoms must diffuse across the substrate to form the islands, we can see very directly that diffusion distances can be many micrometers (Hartig *et al.* 1978, Venables *et al.* 1984).

The systems Ag/W(110) and Ag/Fe(110) have been examined in detail with research students and other collaborators over a span of several years (Spiller *et al.* 1983, Jones & Venables 1985, Jones *et al.* 1990, Noro *et al.* 1995, 1996, Persaud *et al.* 1998). In these systems, 2ML of Ag form first, and then flat Ag islands grow in (111) orientation. The experimental nucleation density $N(T)$ is a strong function of substrate temperature, and the results of several Ag/W(110) experiments are shown in figure 5.13, in comparison with a nucleation calculation $n_x(T)$ of the type outlined in section 5.2. In practice this proceeds by solving (5.14) iteratively, using the complete condensation solution as the starting point.

Condensation is complete in this system, except at the highest temperatures studied, and the critical nucleus size is in the range 6–34, increasing with substrate temperature. Energy values were deduced, $E_a = 2.2 \pm 0.1$, and the combination energy $(E_d + 2E_b) = 0.65 \pm 0.03$ eV, within which $E_d = 0.15 \pm 0.1$ and $E_b = 0.25 \mp 0.05$ eV (Jones *et al.* 1990). Note the errors are *anti*-correlated, since the combination energy is quite well determined by the absolute values of the data. What, however, makes these values interesting is that they can be compared with the best available calculations of metallic binding.

Comparison has been made with effective medium theory, and the agreement is

$2\,\mu m$

Figure 5.12. *In situ* UHV SEM pictures of nucleation and growth of Ag on Mo(001), shown after 20 ML was deposited at $T = 550\,^{\circ}$C (from Hartig *et al.* 1978, reproduced with permission).

striking, as shown in table 5.3 with other values later. In particular, the results demonstrate the non-linearity of metallic binding with increasing coordination number. In the simplest nearest neighbor bond model which we explored in chapter 1, the adsorption energy on (111) corresponds to 3 bonds, or half the sublimation energy for a f.c.c. crystal. So for Ag, with $L = 2.95$ eV, such a model would give $E_a = 1.47$ eV, whereas the actual value is much larger. The same effect is at work in the high binding energy of Ag_2 molecules, quoted in section 5.3.1 in connection with island growth experiments. However, the last bonds to form are much weaker, so that in this case E_b is much less than $L/6 = 0.49$ eV. This is a general feature of effective medium or embedded atom calculations on metals, discussed further in section 6.1.

The comparison between Ag/W and Ag/Fe(110) shows that the first two layers are different crystallographically, with two distorted Ag(111)-like layers on W(110) (Bauer *et al.* 1977), compared to a missing row c5×1 structure for the first layer, followed by an Ag(111)-like second layer on Fe(110) (Noro *et al.* 1995). But adatom behavior on top of these two layers is very similar. UHV SEM methods can visualize the first and second MLs, and the islands on top of these MLs in finite deposits, using biased secondary electron imaging described in chapter 3. Diffusion in these systems is discussed in section 5.5.2. Some effects in the first MLs of the related, but rather more reactive systems Cu and Au deposited onto W and Mo(110) have been studied by LEEM

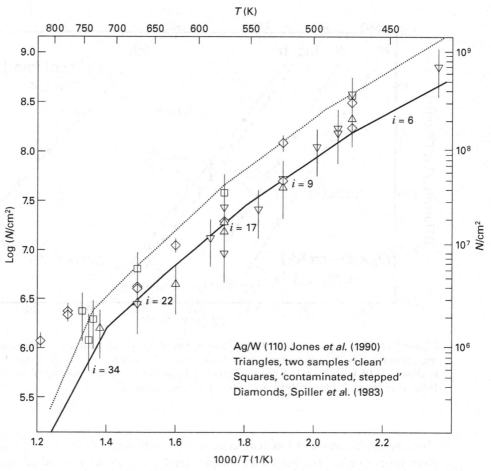

Figure 5.13. Arrhenius plot of nucleation density for silver islands on 2MLAg/W(110) with nucleation model superimposed. Full line: $E_a = 2.1$, $E_b = 0.25$, $E_d = 0.135$ eV, compared to data on the flattest, cleanest samples; dashed line: $E_d = 0.185$ eV, other parameters unchanged, compared to data on stepped and/or slightly contaminated samples. Deposition rate $R = 0.3$ ML/min (from Spiller *et al.* 1983, and Jones *et al.* 1990, reproduced with permission).

(Bauer 1997). Although one might attempt to extract detailed binding parameters from such kinetic observations, this has not been emphasized, possibly because of their complexity, including quite marked crystallographic anisotropy. Several of the adsorption energies are however known from older thermal desorption and other thermodynamic measurements (Bauer 1984).

5.4.2 FIM studies of surface diffusion on metals

At the other end of the length scale, the field ion microscope (FIM) has been used to study individual atomic jumps and the formation and motion of small clusters. These

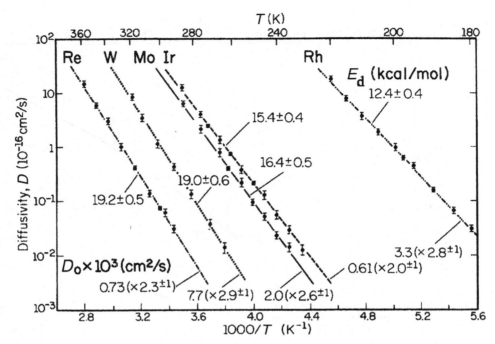

Figure 5.14. FIM measurements of surface diffusion of individual adatoms at a function of temperature on W(211), measured on the same sample. Note that energies E_d are quoted in kcal/mol (23.06 kcal/mol = 1 eV/atom). Errors in D_0 are the factors in brackets (from Ehrlich 1991, after Wang & Ehrlich 1988, reproduced with permission).

observations are made at low temperature, with annealing at higher temperature for given times to effect the jumps; the imaging field is of course switched off during the annealing periods. FIM works best for refractory metals, and high quality information has been obtained on diffusion coefficients of individual atoms and small clusters in systems such as Re, W, Mo, Ir and Rh on tungsten, as shown in figure 5.14 for the W(211) surface (Ehrlich 1991, 1995, 1997). Diffusion on this surface is highly anisotropic, essentially moving along 1D channels parallel to the [01$\bar{1}$] direction, and avoiding jumps between channels. On higher symmetry surfaces, e.g. diffusion on (001) is observed to be isotropic, as it should be.

A particularly elegant application of FIM is to distinguish the *hopping* diffusion (pictured schematically in figure 5.3 and often assumed implicitly as the adatom diffusion mechanism) from *exchange* diffusion. If we draw a (001) surface of an f.c.c. crystal such as Pt, we know from chapter 1, problem 1.2, where an adatom will sit on this surface. So you can convince yourself that hopping diffusion will proceed in the close-packed ⟨110⟩ directions. By contrast, the exchange process consists of displacing a nearest neighbor of the adatom, and exchanging the adatom with it. The substrate atom 'pops out' and the adatom becomes part of the substrate. In this case you can see that the direction of motion during diffusion is along ⟨100⟩, at 45° to hopping diffusion and with $\sqrt{2}$ times the jump distance. By repeated observation of adatom diffusion over a single crystal plane, FIM has been able to map out the sites which the adatoms visit, and thus to distinguish exchange and hopping diffusion. Such measurements taken at different annealing temperature can show the cross-over from one mechanism to the other (Feibelman 1990, Kellogg & Feibelman 1990, Chen & Tsong 1990, Kellogg 1994, 1997, Tsong & Chen 1997). Although the (001) surface presents particularly clear-cut examples of exchange diffusion, it is interesting to remember that the first studies were actually done a decade earlier on f.c.c. (110) surfaces of Pt and Ir (Bassett & Webber 1978, Wrigley & Ehrlich 1980). Diffusion in the (cross-channel) [001] direction was found to proceed by an exchange process; this early work is reviewed by Bassett (1983) and Ehrlich (1994).

Many such interesting results have been obtained by the relatively few groups working in this field. In particular, observations of linear rather than close-packed clusters, cluster diffusion, and adatom incorporation at steps by displacement mechanisms were all surprises when they were first discovered, and warn against us making oversimple assumptions. Another use of FIM is for direct observation of the probability of different spacings of pairs of adatoms within the first ML. Applying Boltzmann statistics to these observations enables the lateral binding energy to be mapped in 2D as a function of spacing and direction. These interactions for Ir on W(110) are found to be in the range 30–100 meV, but can have either sign (Watanabe & Ehrlich 1992, Einstein 1996); thus the model introduced in this chapter, where a single pair binding energy E_b is used to describe lateral interactions, and nearest neighbor binding and directional isotropy are assumed, would be a serious oversimplification if applied uncritically to such systems.

The same statistical methods have been used to identify the proportion of 'long jumps' and/or 'alternative paths' in surface diffusion, both by FIM and more recently by STM. Although these are typically a small proportion of the total, they could be important in particular circumstances, and are an important test of our understanding of rate processes at surfaces (Jacobsen *et al.* 1997, Lorensen *et al.* 1999). The full detail of these FIM and STM results are however very specific to each system; this is a reminder that the amazing complexity of dynamical cluster chemistry is involved in particular surface systems, but that we also need simple models to categorize broad classes of behavior.

5.4.3 Energies from STM and other techniques

Until the advent of the STM, it was very difficult to observe monolayer thick nuclei, except in special cases by REM and TEM, where high atomic number deposits were used (Klaua 1987, Yagi 1988, 1989, 1993). In the past few years UHV STM, with a variable low temperature stage, has become the most powerful technique for quantitative work on nucleation and growth. The sub-ML sensitivity over large fields of view, and the large variations in cluster densities with deposition temperature, have provided detailed checks of the kinetic models described in section 5.2. In particular, STM has enabled the experiments to be done at high density, which occurs at low T, and so typically $i = 1$. In this

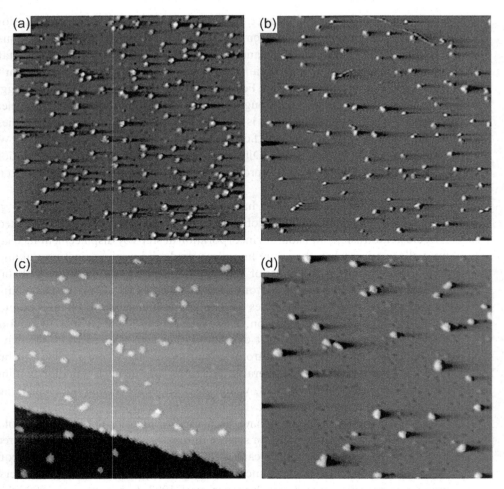

Figure 5.15. STM pictures of a Pt(111) surface after deposition of 0.0042 ML at sample temperatures of (a) 23 K, (b) 115 K, (c) 140 K and (d) 160 K. Each picture is 48 nm wide and was taken at 20 K (after Bott *et al.* 1996, reproduced with permission).

limit the only energy parameter is E_d, which has been measured with high accuracy in several cases.

An example of the data obtained is shown in figure 5.15, from the work of Bott *et al.* (1996) on Pt/Pt(111); it is clear that nucleation densities, size and position distributions can be extracted from such (digitally acquired) images. This study, and several other studies on similar systems, have now made it possible to do detailed comparisons with effective medium and related density functional calculations of metal–metal binding. One illuminating comparison is that of Ag/Ag(111) with Ag on 1 ML Ag on Pt(111), and with Ag/Pt(111). The systematic variations that are found reflect small differences in the lattice parameter (strain) and in strength of binding between these closely similar systems (Brune *et al.* 1994, 1995, Brune & Kern 1997); these features are also reproduced, more or less anyway, by the calculations (Ruggerone *et al.* 1997, Brune 1998). In Ag on 2 ML Ag on Pt(111), an interesting example of pattern formation was

Table 5.4. E_a, E_d and E_b for f.c.c. (111)-like metal substrates

Deposit/Substrate	E_a	E_d	$E_d + 2E_b$	$L - E_a - E_d$ *	Technique
Pt/Pt(111)		0.26 ± 0.02			STM [a]
		0.26 ± 0.003			FIM [b]
Cu/Cu(111)		0.035 ± 0.01			HAS [c]
				0.76 ± 0.04	STM [d]
Ni/Cu(111)		0.08 ± 0.02			HAS [c]
Ag/2MLAg/W(110)	2.20 ± 0.10	0.15 ± 0.10	0.65 ± 0.03		SEM [e]
Ag/1MLAg/Fe(110)				0.86 ± 0.05	SEM [e]
Ag/Ag(111)		0.10 ± 0.01		0.71 ± 0.03	STM [f, h]
	(2.23) [e]	(0.12) [e, g]	(0.68) [e]		Calculation
Ag/Pt(111)		0.16 ± 0.01			STM [f]
	(2.94) [g]	(0.15) [g]			Calculation
Ag/1MLAg/Pt(111)		0.06 ± 0.01			STM [f]
		(0.06) [g]			Calculation

Notes: * For a discussion of values in this column see section 5.5.2
Values in eV; those in brackets are theoretical calculations.
Sources: see table 5.5 on p. 172.

found in which misfit dislocations provided a barrier to diffusing Ag adatoms (Brune *et al.* 1998); this example has much in common with the defect nucleation examples discussed in section 5.3.3.

Several groups are producing results in this field, and as a result a data base is being accumulated as we write, albeit somewhat complicated by slight differences in techniques and analysis methods. Some comparisons between the various experiments are made in the tables which follow, and in a comprehensive review by Brune (1998). We are at an early stage of understanding these values in detail, but it is already clear that (001) f.c.c. metal surfaces are very different from the (111)-like surfaces of table 5.4. The diffusion energies on (001) are quite a bit higher than on (111), and it is possible that several of these values correspond to exchange, rather than hopping diffusion. Some values abstracted from the literature are given in table 5.5.

In the last entry in table 5.5, Cu/Ni(100), Müller *et al.* (1996) observed the transition from $i = 1$ to $i = 3$, which is expected for the (001) surface, and so could deduce E_b in addition to E_d. They also observed both the rate dependence, and the size distributions, showing that this formed a consistent story, as illustrated in figures 5.16 and 5.17. One can see that the temperature dependent region labelled $i = 3$, shows the corresponding rate dependence power law, $p = i/(i+2) = 3/5$, supplementing the lower temperature regimes of $i = 1$ and $i = 0$. The last case arises at the lowest temperature where nucleation happens *after* rather than during deposition, so that the final nucleation density may depend on the amount condensed, but does *not* depend either on how fast it was deposited, or on the difusion coefficient.

At higher temperatures, work on the homoepitaxial system Cu/Cu(001) has failed to

Table 5.5. E_a, E_d and E_b for f.c.c. (001) metal substrates

Deposit/Substrate	E_a (eV)	E_d (eV)	E_b (eV)	Technique
Pt/Pt(001)		0.47 exchange		FIM [b]
Fe/Fe(001)		0.45 ± 0.05		STM [j]
Cu/Cu(001)		0.36 ± 0.03		SPA [k]
Ag/Ag(001)		0.40 ± 0.05		LEIS [l]
Ag/1MLAg/Mo(001)	2.5	0.45 ± 0.05	0.125 ∓ 0.125	SEM [m]
Ag/Pd (001)	(2.67)	0.37 ± 0.03		HAS [n]
Cu/Ni(001)		0.35 ± 0.02	0.46 ∓ 0.19	STM [p]

Note: Values in eV; those in brackets are theoretical calculations.

Sources for tables 5.4 and 5.5.

Techniques: STM = scanning tunneling microscopy; FIM = field ion microscopy; HAS = helium atom scattering; SEM = (UHV) scanning electron microscopy; SPA = SPA-LEED; LEIS = low energy ion scattering.

References: [a] Bott *et al.* (1996); [b] Feibelman *et al.* (1994), Kyuno *et al.* 1998; [c] Wulfhekel *et al.* (1996, 1998), Brune (1998); [d] Giesen & Ibach (1999); [e] Jones *et al.* (1990), Noro *et al.* (1996); [f] Brune *et al.* (1994, 1995); [g] more calculations (and more experiments) can be found via Brune (1998); [h] Morgenstern *et al.* (1998); [j] Stroscio *et al.* (1993), Stroscio & Pierce (1994); [k] Dürr *et al.* (1995), see also Swan *et al.* (1997); [l] Langelaar & Boerma (1996); [m] Venables (1987); [n] Félix *et al.* (1996); [p] Müller *et al.* (1996).

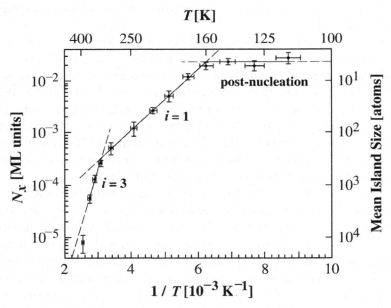

Figure 5.16. Arrhenius plot of the island density of Cu/Ni(001) measured by low temperature STM at coverage 0.1 ML, for a deposition rate 0.00134 ML/s (from Müller *et al.* 1996, reproduced with permisison).

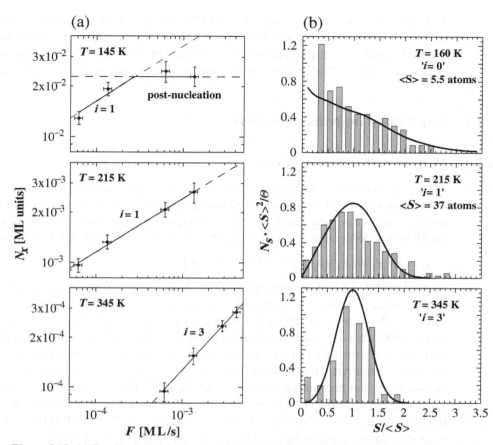

Figure 5.17. (a) Double logarithmic plot of the island density of Cu/Ni(001) versus deposition flux for different temperatures at coverage 0.1 ML; (b) scaled island size distributions deduced from the STM images of Cu/Ni(001) at coverage 0.1 ML, compared with KMC calculations of the corresponding distributions for $i=0$ and 1 (after Amar & Family 1995, Müller *et al.* 1996, and Brune 1998, reproduced with permission).

see the $i=1$–3 transition but instead observed direct transitions to higher i-values (Swan *et al.* 1997). However, no particular sequence of i-values is required by the nucleation model itself; what actually happens is the result of the (lateral and vertical) binding energy of the clusters. On (001) surfaces, the role of second-nearest neighbors is particularly important, since all clusters only have either one or two nearest neighbor bonds. Moreover, it has been suggested that vacancy, in addition to adatom, migration is involved in coarsening (Hannon *et al.* 1997); this is certainly the case at high enough temperatures. Again, one can see that rather careful experimentation and analysis is required to keep the number of parameters in the models at a manageable level.

The cluster size distributions found for $i=0$ as well as other i-values have been calculated, among others by Amar & Family (1995) and Zangwill & Kaxiras (1995). These distributions are compared with the Cu/Ni(001) STM experiments in figure 5.17(b). Note that the case for $i=0$ has a maximum at small sizes. This is also a feature of nucleation in the presence of steps (Bales 1996). The role of steps is introduced in the following section 5.5, along with ripening and interdiffusion; further aspects of nucleation and growth of thin films are deferred to the remaining chapters where they are considered via models of bonding and electronic structure in metals and semiconductors.

5.5   Steps, ripening and interdiffusion

5.5.1   Steps as one-dimensional sinks

Steps often act as one-dimensional sinks, as illustrated schematically in figure 5.10(b). Most interest has historically centered on incomplete condensation in island growth systems. In this case, when both E_a and E_d are typically quite small, the mean diffusion length before desorption is the BCF length x_s, introduced in section 1.3, and given by

$$x_s = (D\tau_a)^{1/2} = (a/2)(\nu_d/\nu_a)^{1/2}\exp\{(E_a - E_d)/2kT\}. \tag{5.20}$$

At moderate deposition temperatures, steps (statistically) capture atoms arriving in a zone of width x_s either side of the step, as was shown in problem 1.4. Alkali halides show the classic example of step decoration (Kern et al. 1979, Mutaftschiev 1980, Keller 1986) where strongly bound pairs of metal atoms nucleate at steps, but nucleation is unlikely on a perfect terrace. In a few cases, step decoration effects have been studied quantitatively. The binding energy, $E_s = 0.23 \pm 0.025$ eV, for Au atoms to steps on NaCl(100) has been measured by comparison with a detailed step nucleation model (Gates & Robins 1982, 1987b), and this value is borne out by a recent calculation (Harding et al. 1998). The case of Cd/NaCl(100) has also been studied by modulated mass spectrometry techniques, resulting in the much higher step binding energy of Cd atoms, 1.1 ± 0.15 eV (Henry et al. 1985).

There is also a large literature devoted to such clusters at steps, including their position relative to the step, which involves long range elastic interactions in addition to other atomic level forces. The 'double decoration' technique is an elegant method of demonstrating such effects, which has been used extensively in TEM experiments over many years (Bassett 1958, Bethge 1962, Kern et al. 1979, Kern & Krohn 1989, Bethge 1990). The decoration technique is best known for demonstrating 2D island and pit nucleation, oscillatory nucleation and growth and dislocation spiral growth (and evaporation) of the alkali halides themselves, as reviewed by Mutaftschiev (1980) and Venables et al. (1984). As seen in figure 5.18, single height steps are rounded, double height steps square, and dislocation spirals are easily recognized.

The effect of adatom capture by steps can be treated using an extension of (5.13) in one of two ways. Either, we can look on a scale between the steps, in which case the steps provide the boundary conditions for a rate–diffusion equation of the form

$$dn_1/dt = R - n_1/\tau + \partial/\partial x(D\partial n_1/\partial x). \tag{5.21}$$

Or, we can average over many steps, in which case adatom capture by steps adds an additional loss term to (5.13) with a characteristic time τ_s. We can show from the

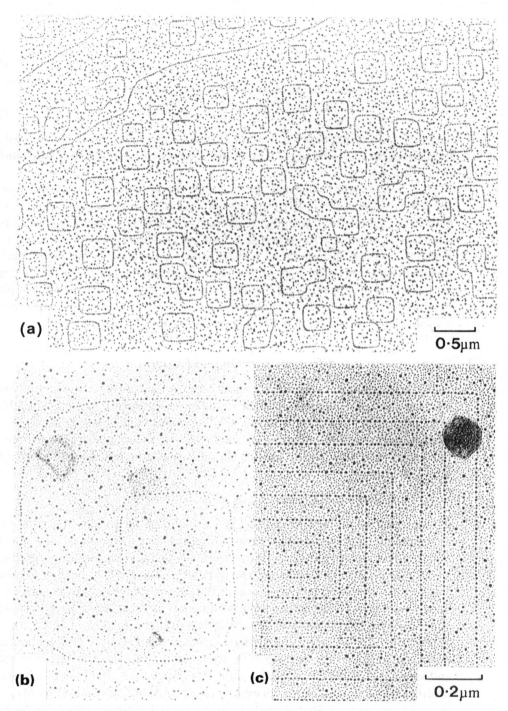

Figure 5.18. Steps on KCl(001) revealed by the decoration technique: (a) single height pits; (b) single height and (c) double height dislocation spirals (from Venables *et al.* 1984, reproduced with permission, after Métois *et al.* 1978 and Meyer & Stein 1980).

equivalence of these two viewpoints in complete condensation that the average adatom density n_1 is

$$n_1 = n_{1e} + R\tau_s ,$$ (5.22)

where n_{1e} is the concentration in equilibrium with the steps, and

$$\tau_s = d^2/12D,$$ (5.23)

where d is the distance between steps; this expression is readily derived on the basis that the adatom concentration $n_1(x)$ is an inverted parabola for complete condensation, as in problem 1.4.

This simple expression may be modified if evaporation, nucleation or the Ehrlich–Schwoebel (ES) barrier are allowed, as can be explored via project 5.3. The name, adopted relatively recently, stems from two papers in which the effect was first discussed (Ehrlich & Hudda 1966, Schwoebel & Shipsey 1966). TEM and FIM methods have demonstrated the ES effect, where diffusing adatoms have difficulty in surmounting a downward step. At low temperatures the barrier is effective, and nuclei are formed on the upper terrace right up to the down-step. An early TEM example was obtained by Klaua (1987), who observed Au ML islands decorating steps, and also nucleating on the terraces of a Ag(111) sample, as shown in figure 5.19.

5.5.2 Steps as sources: diffusion and Ostwald ripening

Steps, in addition to being sinks, can also act as sources of adatoms. Emission of adatoms is part of sublimation, which we studied in section 1.3 via problem 1.2(b), and in the context of adsorption in section 4.2; this happens if $n_{1e} > 0$ in (5.22). In this case, adatoms are created at kink sites and can diffuse over the terrace and become incorporated in other steps. When the adatom concentration is low, this process has an effective diffusion constant

$$D_e = n_1 D, \text{ where } n_1 = K \exp\{-(L - E_a)/kT\},$$ (5.24)

so that the activation energy $Q = (L - E_a + E_d)$, with K an entropic constant. At high temperatures, other mechanisms may be active, including surface vacancies; indeed one of the difficulties of studying surface diffusion is that there are so many mechanisms which may need to be considered (Bonzel 1983, Gomer 1983, 1990, Naumovets & Vedula 1985, Naumovets 1994). However, during annealing at moderate temperatures (5.24), or a variant which allows for small amounts of clustering, is likely to be a good approximation. Such issues can be explored via project 5.4.

Noro et al. (1996) analyzed Ag patches deposited on Fe(110), following the annealing of the 3 ML of deposited Ag as shown in figure 5.20. The broadening of the first and second ML could be followed separately, by biased secondary electron imaging as described in section 3.5; D_e values were deduced for Ag motion on the first ML, giving an activation energy 0.86 ± 0.05 eV. Using (5.24) and assuming that L for the second ML is close to the bulk Ag value of 2.95 ± 0.01 eV, they found $(E_a - E_d) = 2.09 \pm 0.06$ eV. This is to be compared with that deduced for Ag on 2 ML Ag/W(110) from the

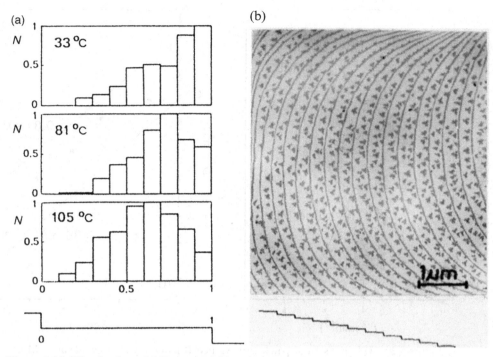

Figure 5.19. Distribution of ML Au nuclei in relation to steps on a Ag(111) surface at the temperatures indicated. The sense of the step train corresponds to the TEM image of the Au islands. Note that the up-steps are decorated with a continuous thin strip of Au, and what is plotted is the island position histogram on the terraces, demonstrating an active ES barrier at the lower temperatures (after Klaua 1987, reproduced with permission).

nucleation experiments described in section 5.4.1: $E_a = 2.2 \pm 0.1$ and $E_d = 0.15 \pm 0.1$eV; the values of $(E_a - E_d)$ are rather similar! Use of the more accurate value for $E_d = 0.10 \pm 0.01$eV from STM experiments (Brune *et al.* 1995, Brune 1998) makes the agreement even closer.

This type of annealing is a particular case of Ostwald ripening, which is a term usually applied to situations where large clusters grow or coarsen, and small clusters shrink or disappear. In the case of ripening controlled via adatom surface diffusion, the effective diffusion coefficient is given by (5.24) with activation energy $Q = (L - E_a + E_d)$. It is noticeable that the high temperature limit of the nucleation model in complete condensation gives this same energy, since $L = E_a + 3E_b$ for large 2D (hexagonal) clusters. The nucleation density expression in table 5.1 has $E_i = 3E_b$ for large critical nucleus sizes i, so the corresponding activation energy is again $(E_d + 3E_b)$; this is thus an internal check on the validity of the nucleation model. Ostwald ripening is generally important in materials science (Martin & Doherty 1976, Voorhees 1985); 2D ripening on surfaces is described more fully by Zinke-Allmang *et al.* (1992) and Zinke-Allmang (1999).

STM experiments have been instrumental in pinning down some of these energies, and the ES barrier has been investigated specifically for Ag(111), where authors have

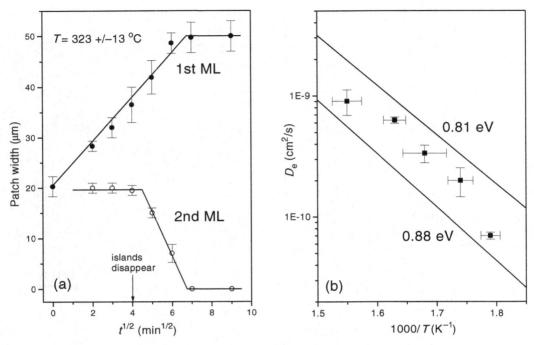

Figure 5.20. (a) Expansion of patches of Ag/Fe(110), initially 20 μm wide and 3ML deposited, during annealing at a typical temperature $T = 323 \pm 13\,°C$, the arrow indicating the time at which islands on the second ML disappear; (b) Effective diffusion coefficient D_e as function of $1000/T$, with activation energies Q indicated for an assumed $K = 4$ (after Noro *et al.* 1996, reproduced with permission). See text for discussion.

found values of $E_S = 0.12 \pm 0.02$ eV (Bromann *et al.* 1995) and 0.13 ± 0.04 eV (Morgenstern *et al.* 1998). This latter experiment is particularly elegant, in that the authors were able to create ML height pits on the surface by ion bombardment, and then place either islands (by deposition) or smaller ML deep pits (by further bombardment) close to the centers of the larger pits. The small pits were found to fill in more slowly than the islands evaporated, in annealing experiments lasting several hours around and below room temperature. This difference in rate is directly due to the presence of the ES barrier, whereas the average rate is governed by the same Q as above in (5.24), with $Q = 0.71 \pm 0.03$ eV from the comparison of the model with experiment.

Comparison with the SEM results on Ag on 1 ML Ag/Fe(110) which gave 0.86 ± 0.05 eV is interesting, in that the SEM experiments (Noro *et al.* 1996) involve diffusion over a large number of steps; in this case the maximum activation energy would correspond to $(Q + E_S)$, i.e. $(0.71 + 0.13) = 0.84$ eV, good agreement! But such arguments are at the limit of current experimental accuracy, and there are uncertainties in frequency factors, and other important effects to consider. One which has been demonstrated is the effect of strain, where Ag adatoms diffusing on the compressed ML phase of Ag/Pt(111) have a lower value of E_d than on Ag(111) (Ruggerone *et al.* 1997, Brune & Kern 1997, Brune 1998), as indicated in table 5.4. These calculations also give a lower value of E_S on compressed Ag(111). Other Ostwald ripening experiments of a similar nature have yielded $Q = 0.76 \pm 0.04$ eV and $E_s = 0.22 \pm 0.01$ eV for adatoms on Cu(111) (Giesen & Ibach 1999). These authors and their co-workers have also observed step fluctuations, and rapid decay processes when multilayer islands coalesce, and have recently invoked electronic mechanisms in explanation. Thus, as we have seen in the previous two sections, research on *atomistic* processes at metal surfaces has encountered the need for *electronic structure* calculations in the search for complete explanation. The background needed for this understanding is given in section 6.1.

5.5.3 Interdiffusion in magnetic multilayers

Magnetic multilayers are typically formed by interspersing a magnetic metal (Fe, Ni, Co, Cr, etc.) with a non-magnetic spacer, often a noble metal (Cu, Ag, Au, Pt etc.). The sequence Co/Cu/Co . . . for example has a giant magnetoresistance whose properties are controlled by the various layer thicknesses and perfection; there are many such systems, whose properties have been extensively reviewed in the last few years, as discussed later in sections 6.3 and 8.3. In this section we concentrate on the growth mode, taking Fe/Ag/Fe(110) as the example.

As described in section 5.4.1, Ag/Fe(110) is a typical SK growth system, with two layers before islands form, the first of which has the c5×1 structure, which has a nominal coverage of 0.8ML, and the second is close to a compact Ag(111) layer. Auger amplitudes from this structure have been measured (Noro *et al.* 1995, Venables *et al.* 1996, Venables & Persaud 1997); there is nothing unusual about the Ag/Fe(110) interface. However, deposition of Fe on thin films of Ag/Fe(110) results in some interdiffusion, the extent of which depends on the Ag film thickness, deposition and annealing conditions. An example is shown in figure 5.21, where the ratio of Ag/Fe AES intensities is plotted against Fe coverage, and is compared with a layer growth calculation (Persaud *et al.* 1998).

The lower curves are calculated assuming no surface segregation, the two curves reflecting some uncertainty in the correct inelastic mean free path for the Auger electrons. For deposits of under 1 ML at room temperature, the data follow this layer growth curve, more or less. But between 1 and 2 ML, there is clearly some segregation, where the calculation assumes that all of the first 0.8 ML Ag has moved to the surface; this is clearly not a bad approximation. But annealing to around 250 °C results in more segregation, and deposition at 250 °C results in almost complete segregation. Results for other Ag layer thicknesses show a similar trend: interdiffusion at the ML level proceeds even at room temperature, and there is long range interdiffusion already at a few hundred degrees Celsius.

From the arguments given in section 5.1.1, we can see that metal deposition systems should follow the island growth mode, if the surface energy of the deposit (Fe ~ 2.9 J/m^2) is greater than that of the substrate (Ag ~ 1.2 J/m^2); surface energy values are discussed and tabulated in section 6.1.4. Thus islands of the strongly bound material, Fe, once formed, could lower their energy by allowing themselves to be coated with a thin skin of Ag substrate material! This corresponds to a curious form of

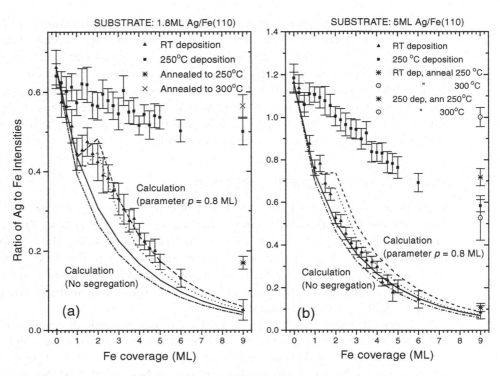

Figure 5.21. AES deposition curves for Fe/1.8 ML Ag/Fe(110), showing (a) rearrangement of the Ag layer between 1 and 2ML, and segregation on annealing, or during deposition at elevated temperature. The Fe/5 ML Ag/Fe(110) curves (b) show rather less complete segregation. The parameter p is the amount interchanged with the surface in the model (Persaud et al. 1998, reproduced with permission).

interdiffusion, in which islands or layers, rather than single atoms, bury themselves in (i.e. burrow into) the substrate. At low temperatures this will not happen, because the substrate atoms will not diffuse. However, STM studies of surface steps on noble metals have shown that steps can move quite rapidly, even at room temperature (Poensgen et al. 1992) and burying of deposited clusters has been observed by in situ TEM at elevated temperatures (Zimmerman et al. 1999). It is now clear that the difficulties various groups have experienced in producing well-defined thin films of magnetic metals on noble metal surfaces are related to effects of this nature. Such magnetic metals generally have higher surface energies than the substrates; they also undergo structural phase changes with increasing thickness; the magnetic features are discussed in sections 6.3 and 8.3.

One case which has been studied by STM is Ni/Ag(111) (Meyer & Behm 1995). Here Ni can both diffuse by hopping over the surface, or, at higher temperature, can exchange with a silver atom. This immobile Ni atom now acts as a nucleus for further growth of Ni clusters. In a fixed temperature deposition, this corresponds to creation of nuclei at a rate proportional to the adatom concentration; if the Ni–Ni bond is strong enough, then $i = 0$. Similar cases are Fe, Co and Ni/Au(111), with a complex reconstruction which orders the Fe, Co or Ni islands (Voigtländer *et al.* 1991, Chambliss *et al.* 1991, Stroscio *et al.* 1992, Meyer *et al.* 1995, Tölkes *et al.* 1997), and Fe or Co/Cu(100), where subsurface and surface ML islands can co-exist (Kief & Egelhoff 1993, Chambliss & Johnson 1994, Healy *et al.* 1994). The size distributions of clusters nucleated on point defects have been studied by KMC; a broad distribution is found, with a high proportion of small islands, as shown earlier in figure 5.17(b) (Amar & Family 1995, Zangwill & Kaxiras 1995, Bales 1996, Brune 1998).

All these cases take us back to figure 5.3, and the difficulty of making high quality multilayers from A/B/A systems: if one interface is 'good', typically an example of SK growth as described in this section, then the other interface is 'bad'. These systems may formally be an example of island growth, but active participation of the substrate makes this classification too naive, in some cases even at room temperature. Nuclei form by exchanging deposit and substrate atoms; clusters of deposited atoms start to form, and then tend to get covered by a substrate 'skin'. Once one realizes what is happening on a microscopic scale, the evidence is already there in the classical surface science results, e.g. from AES as shown in figure 5.21. These data show that segregation of Ag to the surface already happens at the ML level at room temperature; at 250 °C there is widespread interdiffusion.

Further reading for chapter 5

King, D.A. & D.P. Woodruff (Eds.) (1997) *Growth and Properties of Ultrathin Epitaxial Layers* (The Chemical Physics of Solid Surfaces and Heterogeneous Catalysis, Elsevier), **8**.

Liu, W.K. & M.B. Santos (Eds.) (1999) *Thin Films: Heteroepitaxial Systems* (World Scientific).

Matthews, J.W. (Ed.) (1975) *Epitaxial Growth, part B* (Academic).

Tringides, M.C. (Ed.) (1997) *Surface Diffusion: Atomistic and Collective Processes* (Plenum NATO ASI) **B360**.

Problems and projects for chapter 5

Problem 5.1. Growth laws and the condensation coefficient

The rate equation (5.10) is a good approximation when the coverage of the substrate by islands, $Z \ll 1$. When Z is not so small, one might like to correct (5.13a) for *direct impingement*, by writing $R(1 - Z)$ in place of R. The condensation coefficient is the ratio of the amount of material in the film to the amount in the depositing flux, and comes in two forms $\alpha(t)$ and $\beta(t)$.

(a) Identify the terms in the modified (5.10) which lead to the increase in size of clusters, and write down an expression for the cluster growth rate, in atoms per unit area per second.

(b) Assuming 2D islands, express the *instantaneous* condensation coefficient $\beta(t)$ in terms of the rate of atom arrival and departure (per unit area per second) at time t.

(c) Use the two above expressions to derive the form of the *integrated* condensation coefficient $\alpha(t)$, assuming deposition was started at $t=0$.

(d) Using (5.11) in addition, we can now compute $n_x(t)$, $\alpha(t)$ and $Z(t)$. However, as explained in the text, it is preferable to use Z as the independent variable. Compute $n_x(Z)$, $\alpha(Z)$ and $t(Z)$ for parameter values which illustrate the *initially incomplete* condensation regime for a critical nucleus size $i=1$, and $10^{-4} < Z < 0.5$, and identify on the surface processes which dominate at different values of Z.

Problem 5.2. Capture numbers and Bessel functions

The rate equation treatment of capture numbers needs an ancilliary diffusion equation with cylindrical symmetry. Consider the formulation of such an equation in order to understand how the solutions (5.12) arise for incomplete condensation ($\tau = \tau_a$), and how they can be generalized to the more general case when growth also occurs ($\tau^{-1} = \tau_a^{-1} + \tau_c^{-1}$).

(a) Express the adatom diffusion equation $\partial n_1 / \partial t = D \nabla^2 n_1$ in cylindrical polar coordinates, and determine the simplified equation which results when the solution does not depend on the angular variable θ.

(b) Now consider a particular cluster with radius r_k, centered at $\mathbf{r}=0$, and add the source and sink terms from the rate equation (5.13) for the adatom concentration. Explore the relation between the resulting *steady state* equation for $n_1(r)$ outside the cluster and Bessel's equation.

(c) By considering the boundary condition at the edge of the cluster, show that the concentration $n_1(r) = R\tau(1 - K_0(X)/K_0(X_k))$, where the argument of the Bessel function $X = r/\sqrt{(Dt)}$, and X_k is as defined in the text following (5.12).

(d) Given that the derivative of $K_0(X) = -K_1(X)$, derive (5.12) for the capture numbers σ_k and σ_x by considering the diffusion flux $J(r) = -D \nabla n_1(r)$ at the cluster boundary. Show this result is exact for small clusters when $\tau = \tau_a$, and that it is a good approximation when $\tau^{-1} = \tau_a^{-1} + \tau_c^{-1}$. For complete condensation, show that the result for σ_x only depends on the island coverage Z.

Project 5.3. Step capture and diffusion barriers between layers

Adatoms being captured by steps can be formulated as in problem 1.3 by considering an individual terrace and the boundary conditions at the steps at either end. Consider some further 1D step capture problems along the following lines.

(a) The presence of an Ehrlich–Schwoebel barrier at a down step means that an adatom has a temperature-dependent probability to be reflected there. Formulate this problem and apply your equations to the data shown in figure 5.19. What surface processes determine the nucleation density $N(x,T)$ shown, on the assumption of no intermixing between Au and Ag, and what value for the corresponding energies might you deduce from the comparison with experiment.

(b) Problems on a mesoscopic scale require a suitable mixture of atomistic and continuum modeling. One such problem is how to model the effects of step capture on a scale large compared to the distance between the steps. Show that in this limit, step capture contributes and extra characteristic inverse time τ_s to (5.13). Evaluate τ_s in terms of the step spacing d and the adatom diffusion coefficient D to prove the limit given by (5.23), and see if it is valid in general.

(c) Use the results obtained from parts (a) and (b) to discuss the early stages of nucleation and growth of islands on a vicinal surface. Using such a formulation, discuss the occurrence of denuded zones parallel to the steps, and the reduction in nucleation density on vicinal surfaces. What are the effects of step movement and island incorporation into steps at later stages?

Project 5.4. Clustering and intermixing during diffusion

The formulation of adatom diffusion in terms of hopping via (1.16) provides the simplest description of *intrinsic* diffusion, valid at low coverages. Consider some of the forms of the diffusion coefficent which are appropriate to long range diffusion, valid at higher coverage and/or temperatures.

(a) The mass transport or *chemical* diffusion coefficient D^* is expressed here as $n_1 D$ in (5.24). Consider the surface processes involved in Ostwald ripening of clusters, and show that this expression is reasonable at low concentrations when only adatoms are mobile.

(b) At higher adatom concentrations some of the adatoms will spend part of the time in small clusters, size j, which have typically smaller (maybe zero) intrinsic diffusion coefficients, D_j. Show that in this case, the chemical diffusion coefficient is concentration dependent, and is given by $D^* = \Sigma_j j^2 n_j D_j / \{ \Sigma_j j^2 n_j \}$. Hence show, using the Walton relation (5.9) in its simplest form, that at non-zero concentrations D^* depends exponentially on E_b / kT as well as on E_d / kT, and may also depend on other energies due to diffusion of small clusters.

(c) At higher temperatures, surface vacancies are created in addition to adatoms, and at even higher temperatures the surface may become rough over several layers. In alloys we can have exchange diffusion with unlike species involved. Consider how these possibilities affect the interpretation of D^* in terms of individual surface (and near surface) processes.

6 Electronic structure and emission processes at metallic surfaces

This chapter gives, in section 6.1, some generally accessible models of metallic behavior, and tabulates the values of work function and surface energies of selected metals. In section 6.2 we discuss electron emission properties of metals, concentrating on the role of low work function, high surface energy materials as electron sources; we also show that electron emission and secondary electron microscopy can be used to study diffusion of adsorbates. An introduction to magnetism in the context of surfaces and thin films is given in section 6.3.

6.1 The electron gas: work function, surface structure and energy

6.1.1 Free electron models and density functionals

Free electron models of metals have a long history, going back to the Drude model of conductivity which dates from 1900 (Ashcroft & Mermin 1976). The partly true, partly false predictions of this classical model were important precursors to quantum mechanical models based on the Fermi–Dirac energy distribution. If words in the following description don't make sense, now is the time to take a second look at section 1.5. Modern calculations start from a description of the electron density, $\rho^-(\mathbf{r})$ ($\rho^-(z)$ in 1D) in the presence of a uniform density $\rho^+(\mathbf{r}$ or $z)$ of metal ions. This is the jellium model, where the positive charge is smeared out uniformly. At a later stage we can add the effects of the ion cores $\Delta\rho^+(\mathbf{r})$ by pseudopotentials or other approximations. This division into a uniform ρ^+, with a step function to zero at the surface, allows us to consider the electron density ρ^- as the response to this discontinuity. Clearly, a long way inside jellium, $\rho^-=\rho^+$, and there is overall charge neutrality. But at the surface there is a charge imbalance, and the electrostatic potential V varies as a function of z.

To see this response, we draw an energy diagram as in figure 6.1, with $V(-\infty)$ $<V(+\infty)$, with the Fermi energy $E_F=\bar{\mu}$, the chemical potential for the electrons, and note that

$$E_F - V(-\infty) = \bar{\mu}, \tag{6.1}$$

the Fermi level with respect to the bottom of the conduction band, and that the work function, $\phi= V(+\infty) - V(-\infty) - \bar{\mu}$, or equivalently

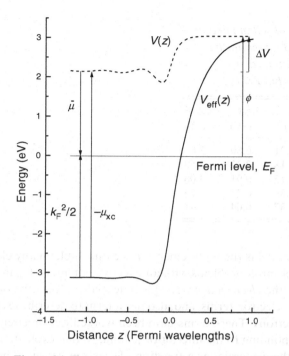

Figure 6.1. Energy diagram defining the terms ϕ, ΔV, $\bar{\mu}$ and the effective potential $V_{\text{eff}}(z)$ in relation to the Fermi level and the bottom of the conduction band of a metal. This diagram is drawn to scale from the data in table 1 of Lang & Kohn (1970) for $r_s = 4$. See text for discussion.

$$\phi + \bar{\mu} = V(+\infty) - V(-\infty) \equiv \Delta V. \tag{6.2}$$

From this simple manipulation we can understand the following points: (1) $\bar{\mu}$ is a bulk property, determined by the kinetic energy and exchange-correlation energy of the electron gas; (2) the fact that the work function ϕ depends on the surface face {hkl} means that ΔV has to be a surface property also. This has various consequences, which are spelled out in the next sections; but first, we need a bit of background theory. The details can be quite complicated, especially considering that there are (at least) two length scales in the problem, one connected with the electron gas, and another connected with the lattice of ions.

It is a good idea to understand the elements of density functional theory (DFT), even if only in outline, in the form that Lang and Kohn used in the early 1970s to derive values for the work function and surface energies of monovalent metals (Lang & Kohn 1970, 1971; Lang 1973). In order not to lose the thrust of the argument, this material is relegated to Appendix J. These calculations characterize free electron metals in general in terms of the radius (r_s) which contains one electron; in particular, their calculations spanned the range $2 < r_s < 6$ (in units of the Bohr radius a_0) which includes the alkali metals Li to Cs. Figure 6.1 is drawn to scale for $r_s = 4$, which is close to the value needed to describe sodium.

Table 6.1. *The work function of jellium and its components. Columns 2 and 3 represent the kinetic, and exchange-correlation energy respectively (after Lang & Kohn, 1971)*

r_s	$k_F^2/2$ (eV)	μ_{xc} (eV)	$\bar{\mu}$ (eV)	ΔV (eV)	ϕ (eV)
2	12.52	-9.61	2.91	6.80	3.89
3	5.57	-6.75	-1.18	2.32	3.50
4	3.13	-5.28	-2.15	0.91	3.06
5	2.00	-4.38	-2.38	0.35	2.73
6	1.39	-3.76	-2.37	0.04	2.41

The main aspect of this model is the replacement of the (insoluble) many electron N-body problem by N one-electron problems with an effective potential, V_{eff} in figure 6.1, which is a *functional* of the *electron density*. This potential contains the original electron–nuclei and electron–electron terms, and also has a term to describe exchange and correlation between electrons. These terms have been worked out precisely for a uniform electron gas, corresponding to the interior of jellium, so that explicit, numerical values can be given to these energies as a function of electron density. The trick now is to apply these same numerical recipes to non-uniform densities, whence the term local density approximation (LDA). There are many further methods which try to correct LDA for non-local effects and density gradients, such as the generalized gradient approximation (GGA), but it is not clear that they always produce a better result. In any case, we are now getting into the realm of arguments between specialists.

Some results of Lang & Kohn's work on jellium are indicated on figures 6.1 to 6.3. The electron density (figure 6.2(a)), electrostatic potential and effective potential (figure 6.1) have oscillations normal to the surface in the self-consistent solution obtained; there are substantial cancellations between the various terms. The work function of these model alkali metals (figure 6.3) varies weakly from Li (r_s about 3.3) to Cs (r_s about 5.6), whereas the individual components of the work function vary quite a lot. This model was the first to get the order of magnitude, and the trends with r_s correct: a big achievement. Note that the position of the ions do not enter this model at all: everything is due to the electron gas, and the importance of the exchange-correlation term μ_{xc}, and the variation of the electrostatic contribution, are evident in table 6.1.

In the quarter century since Lang & Kohn's initial work, there have been major developments within the jellium model. As computers have improved, this method has also been applied to clusters, especially of alkali metals, of increasing size. Figure 6.2(b) shows the comparison of the electron density in a spherical sodium atom cluster of more than 2500 atoms, modeled as jellium, compared with the free planar jellium surface on the same scale (Brack 1993). The only difference of note between the two curves is that the oscillations in the cluster produce a standing wave pattern at the center of the cluster, whereas they die away from the planar surface. This central peak

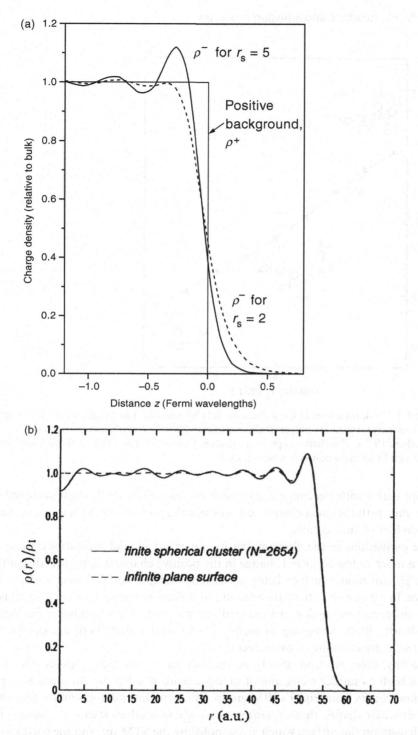

Figure 6.2. Electron density at a metal surface in the jellium model: (a) Lang & Kohn (1970) for $r_s = 2$ and 5; (b) comparison between a spherical cluster of 2654 simulated Na atoms ($r_s = 3.96$) and a planar surface for $r_s = 4$ (after Genzken & Brack 1991, and Brack 1993, reproduced with permission).

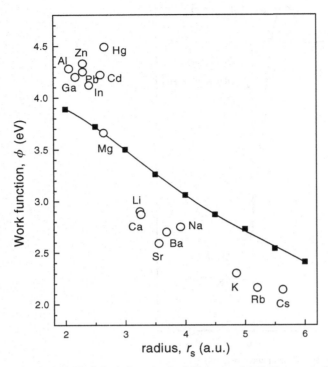

Figure 6.3. Work functions in the jellium model (full squares, Lang & Kohn 1971), compared with experimental data for polycrystalline alkali and alkaline earth metals (open circles: Michaelson 1977). The elements plotted are after Lang (1973) and the solid line fourth-order polynomial fit to these points has been added.

(or dip) varies with electron energy and is dominated by the highest occupied states which vary with the exact cluster size, whereas the oscillations close to the surface are independent of such details.

The oscillations in the electron density are called Friedel oscillations; these occur when a more or less localized change in the positive charge density (the discontinuity at the jellium model surface being an extreme case) is coupled with a sharp Fermi surface. In other words, they are a feature of defects in metals in general, not just surfaces, and are an expression of Lindhard screening, which is screening in the high electron density limit. Screening in metals is so effective that there are ripples in the response, corresponding to overscreening.

Recently, these electron density oscillations have been seen dramatically in STM images both of surface steps, and of individual adsorbed atoms on surfaces, reported in several papers from Eigler's IBM group. By assembling adatoms at low temperature into particular shapes, these 'quantum corrals' can produce stationary waves of electron density on the surface which are sampled by the STM tip, and the corresponding Friedel oscillations are energy dependent; two examples from a circular assembly of 60 Fe atoms on Cu(001) are shown in figure 6.4.

Whether or not these effects can be explained in detail as yet (Fe and Cu are both

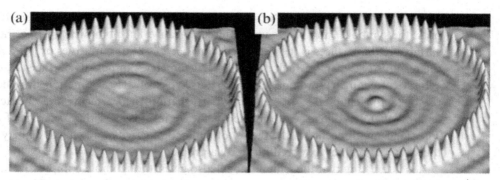

Figure 6.4. A 'quantum corral' of 60 Fe atoms assembled and viewed on Cu(001) by STM at 4 K. The tip imaging parameters are (a) $V_t = +10$ mV and (b) -10 mV, with current $I = 1$ nA (after Crommie *et al.* 1995, reproduced with permission).

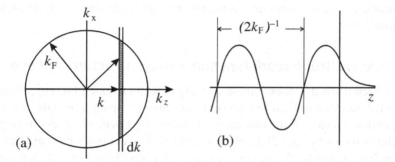

Figure 6.5. (a) Cross section of the free electron Fermi surface, radius k_F; (b) the combination of traveling wave states $\pm k$ near a surface. See text for discussion.

transition metals with important d-bands), these oscillations are present in free electron theory. To see how such effects arise, one needs to do as simple a calculation as possible, and try to understand how the physics interacts with the mathematics. The calculation done by Lang & Kohn goes roughly as follows, using figure 6.5 as a guide.

Consider pairs of states, ordered by their **k**-vector perpendicular to the surface, k and $-k$. Their wavefunction is $\psi \sim \psi_k(z) \exp i(k_x x + k_y y)$, and when $\pm k$ are combined to vanish in the vacuum (outside the surface), $\psi_k(z) \sim \sin(kz - \gamma_F)$, where γ_F is a phase factor, dependent on k_F, since the origin doesn't have to be exactly at $z = 0$. Draw a Fermi sphere, radius k_F, with the k-axis (perpendicular to the surface) as a unique axis, as in figure 6.5. Make a slice at k, dk thick; the density of states $g(k)$ is just the area of this slice which is $\pi(k_F^2 - k^2)$. Now we can write

$$\rho^- = n(z) = \pi^{-2} \int g(k) |\psi_k|^2 dk, \tag{6.3}$$

where the limits of integration are 0 and k_F, and with a bit of manipulation you should get the result

$$n(z) = \bar{n}[1 + 3\cos\{2(k_F z - \gamma_F)\}/(2k_F z)^2 + O(2k_F z)^{-3}], \tag{6.4}$$

where the O-notation means 'of order $(2k_F z)^{-3}$'. Here $\bar{n}$ is the electron density in the bulk; the symbols $\bar{n}$ and ρ^- are used interchangeably. The point which is specific to 2D surfaces and interfaces is the dependence on $(2k_F z)^{-2}$. For impurities or point defects, the result is $O(2k_F z)^{-3}$, which is due to 3D geometry. For corrals on the surface with cylindrical geometry, we encounter various types of Bessel function, for the same reasons as in chapter 5. In scattering/perturbation theory terms, the characteristic length, $(2k_F)^{-1}$, is due to scattering across the Fermi surface without change of energy. The same length occurs in the theory of superconductivity and charge density waves; these features can be explored further via problem 6.1.

It is interesting that the jellium model also gives, though not so impressively, values for the surface energy of the same metals as shown later in figure 6.10. The agreement is again excellent for the heavier alkali metals, but fails dramatically for small r_s. This arises from the need to include the discreteness of the positive charge distribution associated with the ions, a point which was recognized in Lang & Kohn's original paper. With a suitable choice of pseudopotential, agreement is much improved (Perdew et al. 1990, Kiejna 1999).

6.1.2 Beyond free electrons: work function, surface structure and energy

There have been many developments since Lang & Kohn to extend this approach, first to s-p bonded metals and then to the complications of transition metals involving d-electrons, and in the case of the rare earths, f-electrons as well. The d-electrons give an angular character to the bonding, often resulting in structures which are not close-packed, e.g. b.c.c. (Fe, Mo, W, etc.) or complex structures like α-Mn. This is in contrast to s-p bonded metals which typically are either f.c.c. or h.c.p. There are many challenges left for models of metallic surfaces.

To start we need a few names of the methods, for example 'nearly-free electron' method, pseudopotentials, orthogonalized plane waves (OPW), augmented plane waves (APW), Korringa–Kohn–Rostoker (KKR), tight-binding, etc. These long-standing methods are described by Ashcroft & Mermin (1976). For surfaces, an intro-ductory account of electronic structure is given by Zangwill (1988), which contrasts with a highly detailed version from Desjonquères & Spanjaard (1996). Typically tight-binding (where interatomic overlap integrals are thought of as small) is taken as the opposite extreme to the nearly free electron model (where Fourier coefficients of the lattice potential are thought of as small). However, this is more apparent than real, in that both pictures can work for arbitrarily large overlap integrals or lattice potentials; the only requirement is that the basis sets are complete for the problem being studied. This of course can lead to some semantic problems: methods which sound different may not in fact be so different; in particular, when additional effects are included they are almost certainly not simply additive.

The basic feature caused by including the ions via any of these methods is that the electron density near the surface is now modulated in x and y with the periodicity of the lattice; an early calculation which shows this for the lowest atomic number metal lithium is given in figure 6.6. So there are now two length scales in the problem which

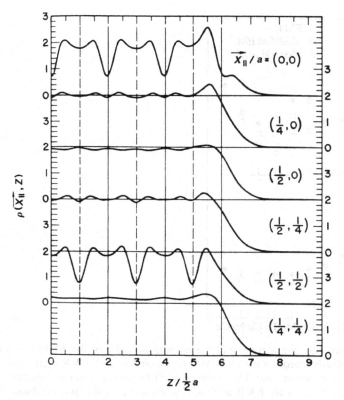

Figure 6.6. Valence electron density at several $x_{//}$ points for Li(001) in a pseudopotential calculation (from Alldredge & Kleinman 1974, reproduced with permission, after Appelbaum & Hamann, 1976).

compete; surface states have oscillation periods with no simple relation to the lattice period in the z-direction.

Contrary to the free electron starting point, it has more recently proved fruitful to consider models based on wavefunctions relatively localized in real, rather than reciprocal, space, and to construct interatomic potential functions arising from atomic-like entities interacting with the electron gas in which the 'atom' is embedded (Sutton 1994, Pettifor 1995, Sutton & Balluffi 1995). The resulting methods are known as embedded atom models (EAM) or effective medium theories (EMT); in these models the embedding energy ΔE is expressed in terms of the cohesive function $E_c(n)$, as

$$\Delta E = E_c(n) + \Delta E_c, \tag{6.5}$$

where the correction energy ΔE_c differs between the various schemes, but is relatively small.

The cohesive function $E_c(n)$ is a function of the homogeneous electron gas density n in which the atom is embedded (Jacobsen et al. 1987, Jacobsen 1988, Nørskov et al. 1993). The cohesive energy, and the component $E_c(\bar{n})$ at the optimum density $\bar{n}$ is shown for the 3d transition metals in figure 6.7. A major effect of these models is to

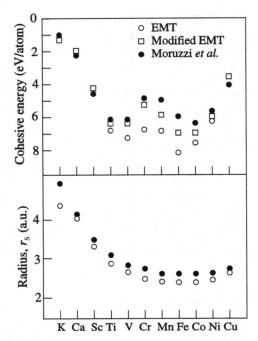

Figure 6.7. Calculated cohesive energies and the equilibrium radius r_s for the 3d transition series, comparing effective medium theory (EMT) (open circles) with KKR methods (Morruzzi *et al.* 1978, closed circles). The modified EMT (open squares) corresponds to EMT applied at the density given by the KKR method (Jacobsen *et al.* 1987, and Jacobsen 1988; redrawn with permission).

show clearly that metallic binding is strongly non-linear with coordination number. The first 'bonds' to form are strong, and get progressively weaker as extra metal atoms are added to the first coordination shell. Some of these effects were exhibited by the experimental examples described in section 5.4. There are many subtleties in the 3d series resulting from magnetism; here the overall cohesion peaks before and after the middle of the transition series, unlike the 4d and 5d series, where cohesion from the d-bands peaks in the middle. Note that this particular calculation does not include spin-correlation effects, but some of these are discussed in relation to magnetism in section 6.3.

Many metal surface relaxations and reconstructions are due to competing electronic and vibrational effects of a quite complex kind. For example, reconstructions of transition metals are often subtle, such as the W(001)2×1 and the 'almost 2×1' incommensurate Mo(001) structures at low temperatures mentioned in section 1.4.3. These structures are driven by (angular) bonding instabilities at low temperature and by anharmonic lattice dynamics at high temperature (Inglesfield 1985, Estrup 1994, Titmus *et al.* 1996). F.c.c. noble metal surfaces can be strongly affected by their d-electrons, interacting with the ions and the other electrons. Although Ag(111)(1×1) is unreconstructed, Au(111) has a uniaxially compressed herringbone (roughly 23×1) reconstruction, Au(001) has a quasi-hexagonal surface layer giving a (roughly 20×5) diffraction pattern (Van Hove *et al.* 1981, Barth *et al.* 1990), and Pt(111) has a 7×7 reconstruction which can be removed by depositing Pt adatoms (Needs *et al.* 1991, Bott *et al.* 1993). This last case shows that the surface structure and lattice parameter of a metal can be a function of the supersaturation $\Delta\mu$ of its own vapor, and that adatoms and surface reconstructions can change the surface stress. This possibility is also well known from adsorption studies of rare gases on graphite, as discussed in section 4.4.

An increasing number of theorists are sufficiently practical and public-spirited that they collaborate closely with experimentalists, and make their computer codes available to others for work on specific problems. It is a welcome recent development that theorists have addressed the problem of 'understanding'. By this I mean that they acknowledge that the 'true' solution is obtained by keeping all the terms in the Schrödinger equation that they can think of, but that this doesn't necessarily help one understand trends in behavior, or help one make predictions. Pettifor (1995), for example, starts with a quote from Einstein: 'As simple as possible, but not simpler'. This is excellent: with such an attitude there is real prospect that we can 'understand' a higher proportion of theoretical models than we would be able to otherwise.

Increasingly what counts is the speed of the computer code; if this speed scales with a lower power of the number, N, of electrons in the system, then more complex/larger problems can be tackled; $O(N)$ methods are *in!* For example, because the interactions between atoms and the electron gas are parameterized initially, EMT calculations are fast enough that they can be used to simulate dynamic processes such as adsorption, nucleation or melting on metal surfaces; here an approximate electronic structure calculation is being done for each set of positions of the nuclei, i.e. at each time step (Jacobsen *et al.* 1987; Stolze 1994, 1997). This requires computer speeds that would have been inconceivable just a few years ago. It is now feasible to download EMT programs from a website in Denmark (see Appendix D) in order to run them for a class project in Arizona! There are real possibilities for experiment–theory collaborations here which were impractical just a few years ago.

6.1.3 Values of the work function

There are several methods of measuring the work function, as described by Woodruff & Delchar (1986, 1994), by Swanson & Davis (1985) and by Hölzl & Schulte (1979) amongst others. The work function varies with the surface face exposed, as shown for several elemental solids in table 6.2. Note that for b.c.c. metals, the surfaces decrease in roughness in the order (111), (100), (110) presented, whereas for f.c.c. the same order corresponds to an increase in roughness. These variations are responsible for several interesting effects, as described here and in the next section.

A polycrystalline material, with different faces exposed, gives rise to fields *outside* the surface, referred to as patch fields. Such fields are very important for low energy electrons or ions in vacuum, and can thereby influence measurement accuracy in surface experiments. Molybdenum is often used for such critical parts of UHV apparatus,

Table 6.2. *Experimental work functions for metals assembled by Michaelson (1977), compared with calculations by Perdew, Tran & Smith (1990; PTS), Skriver & Rosengaard (1992; SR) and Methfessel, Hennig & Scheffler (1992; MHS), plus others as indicated in the last column*

Metal/ structure	Face {hkl}	Experiment* (eV)	Model (PTS)	Model (SR)	Model (MHS) + others*
Li	111		2.90		
b.c.c.	100	2.9 (poly)	2.92	3.15	3.03[b]
	110		3.09	3.33	3.27[b]
Na	111		2.54		
b.c.c.	100	2.75 (poly)	2.58	2.76	2.66[b]
	110		2.75	2.94	2.88[b]
K	111		2.17		
b.c.c.	100	2.30 (poly)	2.21	2.34	2.27[b]
	110		2.37	2.38	2.44[b]
Cs	111		1.97		
b.c.c.	100	2.14 (poly)	2.01	2.03	2.04[b]
	110		2.17	2.09	2.19[b]
Al	110	4.06	3.81		
f.c.c.	100	4.41±0.03[a]	3.62		4.50[a]
	111	4.24	3.72	4.54	4.09[b]
Cu	110	4.48		4.48	
f.c.c.	100	4.59		5.26	
	111	4.98		5.30	
Ag	110	4.52		4.40	4.23
f.c.c.	100	4.64		5.02	4.43
	111	4.74		5.01	4.67
Au	110	5.37		5.40	
f.c.c.	100	5.47		6.16	
	111	5.31		6.01	
Nb	111	4.36			
b.c.c.	100	4.02			3.68
	110	4.87		4.80	4.66
Mo	111	4.55			
b.c.c.	100	4.53			4.49
	110	4.95		5.34	4.98
W	111	4.47			
b.c.c.	100	4.63			
	110	5.25		5.62	

Note: *Error bars and other calculations by: (a) Inglesfield & Benesh (1988); (b) Perdew (1995).

because the work function doesn't vary more than 0.4 V between the low index faces (table 6.2), whereas Nb and W, which are otherwise similar, have variations of around 0.8 V.

The origin of this face-specific nature of the work function can be seen qualitatively by considering jellium again. First, we can see from figures 6.1 and 6.2 that the negative charge spills over into the vacuum, causing a dipole layer, whose dipole moment is directed into the metal. Now we use Gauss' law and show that

$$\Delta V(\text{volts}) = \sigma d/\varepsilon_0 = pN/\varepsilon_0, \tag{6.6}$$

where the sheets of charge, surface charge density σ, are separated by a distance d. To get an idea of how big the potential change is, think of each atom on the surface ($N\text{m}^{-2}$) having a charge of 1 electron separated by 0.1 nm (1 ångström). With $N = 2 \times 10^{19}$ m^{-2}, $p = 1.6 \times 10^{-29}$ Cm, and $\varepsilon_0 = 8.854 \times 10^{-12}$ Fm^{-1}, we get $\Delta V = 36.14$ V. This value is perhaps 2–5 times as large as most voltage (energy) differences between the vacuum level and the bottom of the valence band (which is also the conduction band in monovalent metals).

So a charge separation of <0.5Å is needed to produce the desired effect. Is it a coincidence that this is the same order of magnitude as the Bohr radius, $a_0 = 0.0529$ nm? Not really: the reasons for both effects, the spill over of electrons due to the need to reduce kinetic energy, are the same! This is, of course, a zero order argument: to get the numbers right we have to go back to exchange and correlation energies, and the details. However, models may contain rather arbitrary parameters. For example, the 'corrugation factor' introduced into the 'structureless pseudopotential model' (Perdew et al. 1990) sounds rather dubious, although it moves the model in the right direction (Perdew 1995). Brodie (1995) has proposed a model, building on the idea of corrugation, which is 'too simple' in Einstein's sense; this model should be ignored since it is incapable of further elaboration.

While on this subject, we can note the unit to describe dipole moments, the Debye (D). This is 10^{-18} esu·cm $= 3.33 \times 10^{-30}$ Cm. Thus 1 electron charge $\times$ 1Å $= 4.81$ D. Adsorbed atoms change the work function considerably, but only alkalis give rise to dipole moments this large; for example Cs adsorbed on W(110) at low coverages has been calculated to have a dipole moment of at least 9D (see e.g. table 2.2. in Hölzl and Schulte 1979); in this case the single electron charge distribution would be shifted by about $d = 0.2$ nm. This simple picture is illustrated in figure 6.8(a), and corresponds to (partial) ionization of the alkali, a model first introduced by Langmuir in 1932 and developed by Gurney in 1935. But we need to be careful about inclusion of the image charge, and the nature of bonding, which varies with coverage and is the subject of ongoing discussion (Diehl & McGrath 1997).

The same arguments about electron spillover tell us that stepped, or rough surfaces will have lower work function than smooth surfaces, due to dipoles associated with steps, *pointing in the opposite direction* to the dipole previously considered for the flat surface. A schematic (top view) of this situation, referred to sometimes as the Smoluchowski effect after a 1941 paper, is shown in figure 6.8(b). Experiments on vicinal surfaces, close to low index terraces, do indeed show that the work function

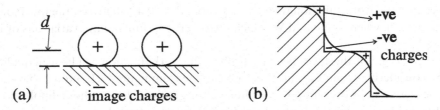

Figure 6.8. Origins of face- and adsorbate-specific work function: (a) dipoles due to charge transfer from adsorbates; (b) top-view of a stepped surface showing smoothing of the charge distribution around the steps (after Gomer 1961, and Woodruff & Delchar 1986, redrawn with permission).

decreases linearly with step density, as shown in figure 6.9; this implies that there is a well defined dipole moment per ledge atom, around 0.3 D for steps parallel to [001] on W(110) and varying with step direction on Au and Pt(111) surfaces (Krahl-Urban *et al.* 1977; Besocke *et al.* 1977; Wagner 1979).

Fascinatingly, work function changes as a function of temperature can be used to define 2D solid–gas phase changes via these same effects. An adatom has a dipole moment which depends both on its chemical nature and on its environment. In a solid ML island the dipole moment per atom is considerably smaller than in the 2D gas. This effect has been used to map out the gas–solid phase boundary for Au/W(110) at high temperatures (Kolaczkiewicz & Bauer 1984). Phase changes in adsorbed layers are discussed in more detail in chapter 4.

6.1.4 *Values of the surface energy*

While the work function is a sensitive test of electronic structure models, particularly exchange and correlation, surface energies are sensitive tests of our understanding of cohesion. Surface energies clearly increase, in general, with sublimation energies, and this has led to many studies trying to embody such relationships into universal potential curves, or into other semi-empirical models (Rose *et al.* 1983, deBoer *et al.* 1988). But the effective medium and density functional models have proven to be more durable, and we may well now have arrived at the point where these models are more accurate than the experiments, which were almost all done a long time ago on polycrystalline samples, sometimes under uncertain vacuum conditions. An example of a microscope based sublimation experiment for Ag, which has stood the test of time, and which agrees with recent calculations within error, is that by Sambles *et al.* (1970). Some experimental and theoretical values for a range of metals are given in table 6.3.

As in the case of the work function, the starting point for models of alkali and other s-p bonded metals has been the jellium model. As was shown in the original Lang & Kohn papers, jellium has an instability at small r_s values, which is due to the neglect of the ion cores. This topic has been pursued by Perdew *et al.* (1990), Perdew (1995) and Kiejna (1999), who have been interested in exploring the simplest feasible pseudopotential models for such metals, and in particular obtaining trends in calculations as a

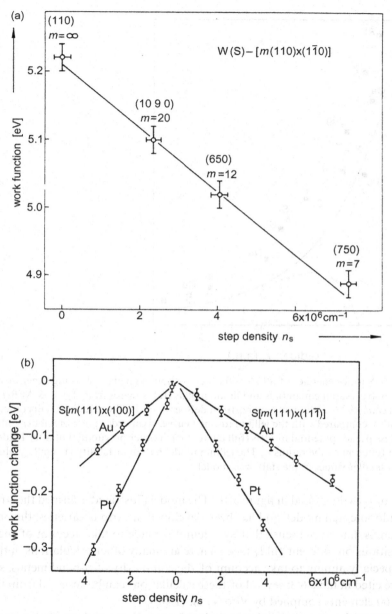

Figure 6.9. Work function of stepped (vicinal) surfaces: (a) vicinals of W(110) in the [001] zone as a function of step density; (b) vicinals of Au and Pt(111) with two different step directions (Krahl-Urban *et al.* 1977; Besocke *et al.* 1977; Wagner 1979, reproduced with permission).

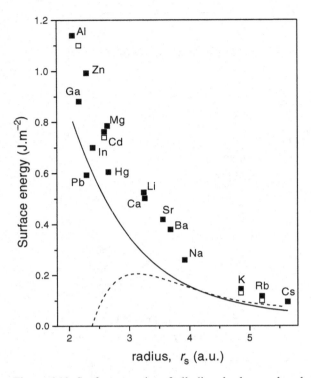

Figure 6.10. Surface energies of alkali and other s-p bonded metals, showing experimental values on polycrystalline materials and liquid metals (full squares from Tyson & Miller 1977 and Vitos *et al.* 1998, with open squares from deBoer *et al.* 1988 where the results differ significantly), compared with the jellium (dashed curve, sixth order polynomial fit) and flat structureless pseudopotential models (full curve, fourth order polynomial fit after Perdew *et al.* 1990) as a function of the radius r_s. The jellium model has an instability at small r_s, which is pushed to smaller values in the stabilized model.

function of r_s as illustrated in figure 6.10. The model illustrated is termed the structureless pseudopotential model, and has been developed in various varieties; the one illustrated here is 'flat' in the sense that no attempt is made to take account of the actual ionic positions on different {hkl} faces. There are many other calculations in the literature, especially aiming to take account of the complexity of d-band metals, some of which are cited in table 6.3; several of these calculations can be accessed from the data base for 60 elements compiled by Vitos *et al.* (1998).

In section 1.2.3, we noted that pair bond models overestimated the anisotropy of surface energy in comparison with the classic experiments of Heyraud & Métois (1983) on Pb, which were shown in figures 1.7 and 1.8. The modern calculations shown in table 6.3 are the only ones getting the anisotropy of surface energy low enough to be even close to experiments on small metal particles. However, it is ironic that the full charge density (FCD) calculation, carried out with the GGA approximation by Vitos *et al.* (1998) still gives too high an anisotropy, especially for Pb, which is almost the only case (plus In and Sn, see Pavlovska *et al.* 1989, 1994) for which there

Table 6.3. *Experimental surface free energies for metals assembled by Eustathopoulos et al. (1973), Tyson & Miller (1977), Mezey & Giber (1982) and deBoer et al. (1988), compared with calculations by Perdew, Tran & Smith (1990; PTS), Skriver & Rosengaard (1992; SR), Methfessel, Hennig & Scheffler (1992; MHS) and Vitos, Ruban, Skriver & Kollár (1998; VRSK)*

Metal/ Structure	Experiment* (J/m^2), at T and at 0 K	Face {hkl}	Model (PTS)	Model (SR)	Model (MHS) + others*	Model (VRSK)
Li		111	0.433			
b.c.c.	0.525	100	0.371	0.436	0.412[b]	
		110	0.326	0.458	0.362[b]	
Na		111	0.252			
b.c.c.	0.260	100	0.216	0.236	0.237[b]	
		110	0.190	0.307	0.208[b]	
K		111	0.134			
b.c.c.	0.130	100	0.115	0.129	0.126[b]	
		110	0.134	0.116	0.111[b]	
Cs		111	0.092			
b.c.c.	0.095	100	0.079	0.092	0.080[b]	
		110	0.069	0.072	0.070[b]	
Al	1.14	110	1.103			1.271
f.c.c.	175 °C	100	0.977			1.347
	1.14–1.16	111	0.921	1.27	1.096[b]	1.199
Cu	1.60	110		2.31		2.237
f.c.c.	900–1070 °C	100		2.09		2.166
	1.78–1.83	111		1.96		1.952
Ag	1.20 ± 0.06[a]	110		1.29	1.26	1.238
f.c.c.	800–830 °C	100		1.20	1.21	1.200
	1.25	111		1.12	1.21	1.172
Au	1.40 ± 0.05[c]	110		1.79		1.700
f.c.c.	950–1000 °C	100		1.71		1.627
	1.50	111		1.61		1.283
Nb	2.30	111				3.045
b.c.c.	1900 °C	100			2.36	2.858
	2.67–2.70	110		1.64	2.86	2.685
Mo	2.00	111				3.740
b.c.c.	1800 °C	100			3.14	3.837
	2.95–3.00	110		3.18	3.52	3.454
W	2.80	111				4.452
b.c.c.	1700 °C	100				4.635
	3.25–3.68	110		3.84		4.005

Note: *Error bars and other calculations from (a) Sambles *et al.* (1970); (b) Perdew (1995); (c) Heyraud & Métois (1980).

are reliable experimental data. One important point in comparing experiment and theory is often only mentioned in passing: the electronic structure model typically refers not only to zero temperature, but indeed to a solid without zero-point vibrations. The experiment, on the other hand, is done at high temperature to avoid kinetic limitations, and exhibits a substantial temperature dependence due to entropic effects, as shown in figure 1.8, and discussed in sections 1.1 and 1.2. Thus the point of comparison is often quite difficult to establish.

Theorists can now calculate not only the surface energies of the different low index {hkl} faces, but also the step energies in particular directions on these surfaces, enabling vicinal surfaces, the shape of 2D nuclei, and unstable facets to be explored, complementing the results for atomic diffusion on surfaces described in section 5.4 (Jacobsen *et al.* 1996, Ruggerone *et al.* 1997, Vitos *et al.* 1999). Current research is exploring the reliability of such models, and their usefulness in interpreting crystal growth experiments, which are strongly influenced by kinetics. An example is the study of fluctuations in the shape of ML-deep pits on the Cu(111) surface by dynamic STM observations (Schlösser *et al.* 1999), leading to a value of the step energy of 0.22 eV/atom along the close-packed directions; this value is close to that calculated by EMT models. These results and others have lead to the realization that entropic effects associated with steps on metals are important, even at room temperature and below (Frenken & Stoltze 1999).

6.2   Electron emission processes

Electron emission processes are central to many effects at surfaces and interfaces, and to many techniques for examining the near-surface region. Most obviously we have emission from the solid into the vacuum, the electron overcoming the work function barrier in the process. This happens both in *thermal* emission, as described in section 6.2.1 below, and in photoemission and Auger electron spectroscopy, described in section 3.3. In a high electric field, the barrier height can be substantially reduced, resulting in cold or thermally assisted *field* emission, as discussed here in sections 6.2.2 and 6.2.3. Finally an incoming beam can result in *secondary electron* emission, as described in section 6.2.4, and hot electrons can penetrate internal barriers by *ballistic* emission, as described in connection with the microscopy of semiconductors in section 8.1.3. All of these effects are connected with electron sources for various types of microscopy. Consequently, one can think of this section as providing a complement for those sections which deal with (electron) microscope techniques.

6.2.1   Thermionic emission

The Richardson–Dushman equation, dating from 1923, describes the current density emitted by a heated filament, as

$$J(T) = AT^2 \exp(-\phi/kT), \tag{6.7}$$

so that a plot of $\log(J/T^2)$ versus $1/T$ yields a straight line whose negative slope gives the work function ϕ. This value of ϕ is referred to as the 'Richardson' work function, since there is an intrinsic temperature dependence of the work function, whose value $d\phi/dT$ is of order 10^{-4} to 10^{-3} eV/K, with both positive and negative signs (Hölzl & Schulte 1979, table 4.1). When data is taken over a limited range of T, this temperature dependence will not show up on such a plot, but will modify the pre-exponential constant. This constant, A, can be measured in principle, but is complicated in practice by the need to know the emitting area independently, since what is usually measured is the emission current I rather than the current density, J.

The form of this equation can be derived readily from the free electron model, by considering the Fermi function, and integrating over all those electrons, moving towards the surface, whose 'perpendicular energy' is enough to overcome the work function. In this calculation, ignoring reflection at the surface by low energy electrons, the value of A is $4\pi mk^2e/h^3 = 120$ A/cm^2/K^2. Where absolute values of current densities have been measured, values of this order of magnitude have been found. This derivation is quite suitable as an exercise (problem 6.2) but is also available explicitly in the literature (Modinos, 1984).

Thermionic emitters in the form of pointed wires or rods are used as electron sources in many electron optical devices such as oscilloscopes, TV and terminal displays, and both scanning and transmission varieties of electron microscopes. A good thermionic emitter has to have a combination of a low work function and a high operating temperature. However, as can be seen from tabulations such as table 6.2, higher melting point metals typically have higher work function. Thus the search is on for metals with a moderate work function which are sufficiently strong, or creep-resistant, near to their sublimation temperature, which in many cases is a long way below the melting temperature. Note that an additional possibility is to take a high melting point material and to coat or impregnate it with a thin low work function layer. This is done for high current applications (TV and computer terminals) in sealed vacuum systems as described by Tuck (1983). For specialists, updates on current practice can be found in conference proceedings published in *Applied Surface Science* **111** (1997) and **146** (1999).

The standard material for comparison is a polycrystalline tungsten 'hairpin' filament with ϕ around 4.5 V, made of drawn wire a few tenths of a millimeter in diameter, bent, and situated in a triode structure, using a gate electrode called a Wehnelt. The competition is between the *brightness* of the source and its *lifetime*, which decreases markedly as the operating temperature is increased. For example, standard W-filaments used as electron microscope sources may have a lifetime of around 15 h when operated at 2800 K, but this extends to maybe 50 h when the operating temperature is dropped to 2700 K (Orloff, 1984).

The brightness, B, is typically the parameter which matters most in electron optical instruments, the current density per unit solid angle (J/Ω); B is conserved if the energy of the beam is constant and geometrical optics applies. Tungsten filaments have an effective source diameter around 50 μm, an emission current around 50 μA, resulting in $B \sim 5 \times 10^4$ A/cm^2/sterad at 100 kV electron energy; the brightness scales linearly with energy.

A material which has replaced tungsten filaments very successfully for high brightness applications is LaB_6, lanthanum hexaboride, which has ϕ around 2.5 V, grown as small single crystal rods in [001] orientation with a square pointed end made of natural facets. When operated at around 1700 K, the lifetime is around 500 hours, and the brightness around 3×10^6 A/cm²/sterad at 100 kV, which is a major improvement, despite the increased cost and vacuum requirement. This increase in B mostly comes from a decrease in the emission diameter to around 5 μm; the actual current emitted is typically lower than the tungsten hairpin.

In instruments such as analytical SEM, TEM and STEM, we need to force as much current into a small spot as possible, in order to extract a high spatial resolution signal which has a sufficient signal to noise ratio (SNR). This means that there has been an intensive search for materials with better performance as thermionic emitters than LaB_6. It is clear that the desired material must be very stable at high temperature, and moreover must have a stable surface. Borides, carbides and nitrides are natural candidates, which have strong (largely ionic) bonds and can be, or can be made, adequately conducting.

Futamoto *et al.* (1980, 1983) investigated mixed rare earth borides ($La_xM_{1-x}B_6$), where several metals M were tried out. They found that these additions made the emission go down rather than up, but that after some use, they improved somewhat, but never exceeded the performance of pure LaB_6. Using a microprobe AES apparatus, they investigated the surface composition of the tips, and found that the other metallic elements evaporate faster, leaving a surface layer, a few nm thick enriched with La; emission properties thus remained remarkably similar across the series. Swanson *et al.* (1981) changed the surface plane away from (001), measuring the lifetimes for a given emission current: no luck, (001) was the best!

Electron microscopy conferences typically have a few papers on carbides and nitrides; some of these have promising properties, but they have not proved to be stable enough to be used routinely. Thus LaB_6(001) stays! The competition has come from field emission as described below.

6.2.2 Cold field emission

A high electric field near the emitter lowers the work function barrier; the barrier height can be sufficiently reduced to increase emission substantially, as drawn for a field $F = 4$ V/nm in figure 6.11. When the field is this strong, the width of the barrier is of order 1nm, and electrons can escape even at low (room) temperature by tunneling. This is (cold) *field emission*.

The field F plays a similar role to the temperature in thermionic emission, and the governing equation is that by Fowler & Nordheim, derived in 1928 from free electron theory. The current density J, in the simplest case without the image force correction, is given by

$$J = AF^2 \exp(-B\phi^{3/2}/F), \tag{6.8}$$

where the constants are

$$A = 6.2 \times 10^6 (\mu/\phi)^{1/2}/(\mu + \phi) \text{ A/cm}^2 \text{ and } B = 6.83 \times 10^7 \text{ V}^{-1/2}/\text{cm}, \tag{6.9}$$

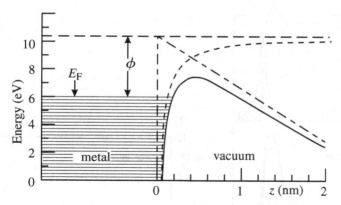

Figure 6.11. Electron energy diagram for field emission as a function of distance z drawn for a field $F=4$ V/nm.

μ being the Fermi level with respect to the bottom of the valence band, i.e. $\mu=\bar{\mu}$ as expressed in section 6.1.1, with F measured in V/cm (Gomer 1961, 1994, Modinos 1984). Free electron theory is also able to calculate the electron energy distribution, as shown in figure 6.12(a), as a product of the Fermi–Dirac distribution for perpendicular energies, and the barrier transmission function, as discussed in problem 6.3. At low F, the distribution is sharp, but the intensity is weak, and vice versa.

Experimentally, cold field emission requires a sharp tip, radius r, and UHV conditions. A voltage V_0 is applied to a first anode with a small hole in it, but most of the field in generated very close to the tip, giving $F=V_0/kr$, with k dependent on the tip shape, but typically $k\sim5$. With $V_0=3$ kV and $r=100$ nm, a field $F=3000/(5\times10^{-7})=0.6\times10^{10}$ V/m, or 6 V/nm is obtained. Field emission tips are usually operated with V_0 from 1 to 5 kV, and radii around 100 nm. The linear dependence of F on the voltage V_0 means that a Fowler–Nordheim plot of $\log(I/V_0^2)$ versus $1/V_0$ gives a straight line, and is a good check on the field emission mechanism.

A single crystal W wire emitter is used, in a low work function orientation. Both (310) and (111) orientations have been widely used in high performance SEM, TEM and STEM instruments; the ultimate single atom tip on W(111) has been demonstrated, and checked by comparison with FIM (Fink 1988). Improving the performance of CFE in an analytical STEM instrument used for electron energy loss spectroscopy (EELS) is described by Batson *et al.* (1992), with the measured field emission spectrum shown in figure 6.12(b), which includes the Fermi tail at room temperature. Here the technical limits are at full stretch. The authors want to study the composition and electronic structure of materials, such as strained Ge/Si quantum wells, using the Si 2p energy loss edge at 100 eV, with nm spatial resolution (Batson & Morar 1993). They need a high current in order to get enough SNR in the spectrum, but if more current is drawn from the tip, both the energy and spatial resolution degrade. The trick is to achieve a modest improvement in energy resolution by deconvolution, using the Fermi tail of the emission (broadened by the spectrometer resolution as in figure 6.12(b)) as a sharp feature which enables the deconvolution to succeed.

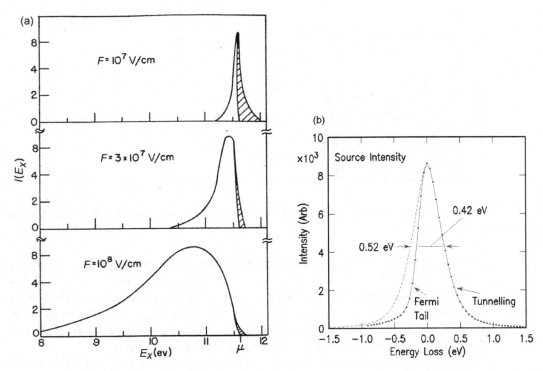

Figure 6.12. Field emission energy distributions from tungsten tips (a) as a function of applied field F, where the shaded area indicates those electrons having the energy component normal to the surface $E_x > \mu$; the total emission increases strongly with F, the curves are scaled for easier comparison (after Gomer 1961); (b) energy distribution of a field emission source with half-width 0.42 eV measured with an EELS spectrometer operating at 120 keV with 0.2 eV energy resolution, showing the Fermi tail due to emission at room temperature (after Batson *et al.* 1992, both diagrams reproduced with permission). Note that the energy scale is inverted right to left in these figures, corresponding to normal practice in these fields.

At the other end of the commercial spectrum, there have been widespread developments in field emission for use in flat panel displays for TV and computer screens. This has now been demonstrated as prototype in industrial laboratories, so the real issues become manufacturability, reliability and of course cost relative to competitive schemes (Slusarczuk 1997). The systems have to work at relatively low voltage, which makes light output from the phosphors also an issue. Two very similar schemes are in competition: the first is based on the Spindt cathode, a lithographically etched assembly based on micrometer-sized structures containing arrays of field emission tips, as shown in figure 6.13(a); this technology is reviewed by Brodie & Spindt (1992). Other specific thin film materials are in contention as the source, most notably diamond-like carbon (DLC) films with specific nanometer scale structures, using the setup shown in figure 6.13(b). This is a very competitive area; recent progress is reviewed and possible mechanisms are discussed by Robertson & Milne

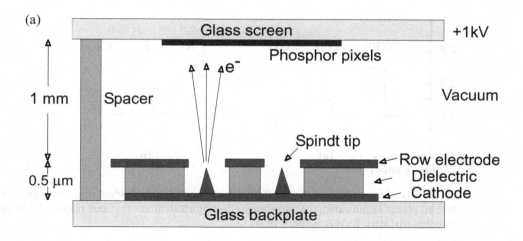

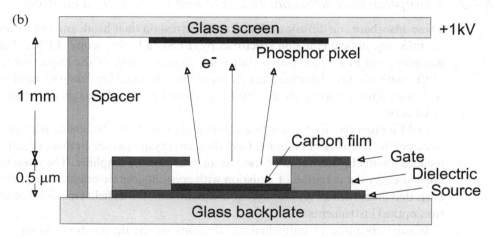

Figure 6.13. Field emission display geometries using (a) Spindt cathodes and (b) DLC films (after Robertson 1997, reproduced with permission).

(1997, 1998) and Robertson (1997). One of the main issues is how to synthesize nanometer-scale structures reliably over large areas; we return to this topic in section 8.4.2.

Field emission requires a very good vacuum, and often, even in UHV, emission is not due to the clean surface. A typical field emitter tip needs to be 'flashed' to clean it, usually by passing a current through a loop on which it is mounted. After flashing the emission current is high, but rather unstable; the current decays with time, and becomes more stable as it does so. This is due to contamination of the tip, either from the vacuum, or more often from diffusion of adsorbed surface species to the tip. Thus the nature of real field emission tips during use, and indeed of real STM tips, is somewhat shrouded in mystery.

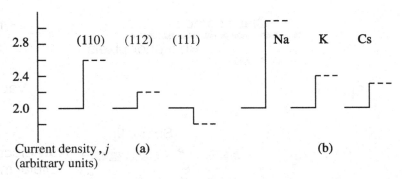

Current density, j (a)   (b)
(arbitrary units)

Figure 6.14. Current jumps associated with the arrival of (a) individual tungsten atoms on W(hkl) planes as indicated; (b) the average effect of alkali atoms on these planes of a W field emission tip (after Todd & Rhodin 1974).

6.2.3 Adsorption and diffusion: FES, FEM and thermal field emitters

These adsorbate and diffusion effects can be turned on their head, and put to good use scientifically, in field emission spectroscopy (FES) and microscopy (FEM). This field was pioneered by Gomer, whose 1961 book contains many of the important features of the methods: his 1990 review article should be consulted for details of methods and results on diffusion using the current fluctuation method. Three types of example are given here.

FEM images the tip itself, with a plate anode which may be coated with phosphor to detect the intensity of emission from different crystal planes; in more recent experiments a channel plate would be used as an intermediate amplifier. The main features are caused by the variation of emission with crystallographic orientation. It is on this basis that faces such as W(310) were subsequently chosen as field emission tips for electron optical instruments.

In situ deposition of individual metal atoms on the tip has been shown to cause jumps in the emitted current, as illustrated in figure 6.14. Todd & Rhodin (1974) showed that they could distinguish 1, 2 and 3 W-atoms arriving on individual W (hkl) faces, and their subsequent desorption when the field remained on. Then they investigated the response to adsorption of different alkali adatoms (Na, K, Cs), which all increase emission markedly via lowering ϕ.

A sophisticated technique was developed to measure diffusion coefficients due to diffusion of these adatoms. A probe hole, or slot, is cut out of the screen, and the current through this hole is measured as a function of time. If no adatoms move in or out of the 'hole area', then the current stays constant; on the contrary, if they do, it changes. This can be expressed as a current–current correlation function $\langle \delta I(0) \cdot \delta I(t) \rangle$, which in normalized form is shown to decay with the delay time t ($\delta I = I - \bar{I}$, the deviation from the average).

Rigorous results can be derived from the fluctuation-dissipation theorem, to show that the decay time for such correlations scales as the radius r of the probe hole squared, divided by the diffusion coefficient. This makes sense qualitatively: the fluctuations with this characteristic decay time τ are caused by adatom diffusion in or out of the probe hole, and $\tau \sim r^2/4D$. So measuring the decay time yields the diffusion coefficient. The subtlety can be increased further: using a slot in different orientations allows one to explore diffusion anisotropy, since the measurement is dominated by diffusion parallel to the *short* axis of the slot of half-width x; now $\tau \sim x^2/2D$. An example of O/W(110) is shown in figure 6.15; in this case the work function of the oxygen covered surface is greater than the clean metal, so the adsorbate *reduces* emission. Note that O-diffusion was found to be anisotropic in the ratio about 2:1, faster parallel to $[\bar{1}10]$ (Tringides & Gomer, 1985).

Adsorption is useful for an electron source if the adsorbate *increases* emission. The stringent vacuum requirements of field emission can be reduced somewhat if one both increases the operating temperature, and also uses an adsorbed layer which reduces the work function. This is thermal field emission (TFE). Typical thin layers, which have long been used in TV and other sealed tube applications, are refractory (Ba, Sr) oxides, pasted onto, or indirectly heated by, the W filament; however, these coatings degrade badly if let up to atmosphere.

For high performance SEM applications, W(001) tips coated with Zr/O have been used as TFE sources; this cathode has $\phi \sim 2.6$–2.8 V (Orloff 1984, Swanson 1984). When operated at $T \sim 1800$ K, the molecules adsorbed from the vacuum, or diffused from the support, do not remain on the surface long enough to cause the current to decay with time; the tips can also have a larger radius than for CFE at a given anode voltage V_0, due to the lower ϕ. In higher current applications, it is the angular current density (I/Ω) which is more important than the brightness: the angular current density can be in excess of 1 mA/sterad. TFE is a successful compromise for many applications: a reasonable current which is stable in moderate ($<10^{-8}$ mbar) vacuum, an infinite lifetime barring accidents, requiring only routine preparation after bakeout: all of which is just as well, considering the initial cost!

Research into alternative CFE/TFE emitters also continues, and papers occasionally appear in the journals. An example detailing the TFE properties of LaB_6 (Mogren & Reifenburger, 1991) emphasized that the emission process is rather more complex than the simplest Fowler–Nordheim treatment presented here, and needs to take into account the actual density of states in the material just above the Fermi level. This effect is indicated by the difference between the full and dashed lines in figure 6.16; additionally, this paper showed that the current decay after flashing may be due to the removal of emission from surface states. Note that the narrow energy distribution is a feature of CFE which gets lost to some extent in TFE. The highest performance analytical STEM and STEM-spectroscopy instruments use CFE primarily for this last reason. Once the vacuum in the source region has been improved to $<10^{-10}$ mbar, the advantages of CFE can be realized in such applications.

6.2.4 *Secondary electron emission*

When a sample is bombarded with charged particles, the strongest region of the electron energy spectrum is due to secondary electrons. We have discussed this extensively

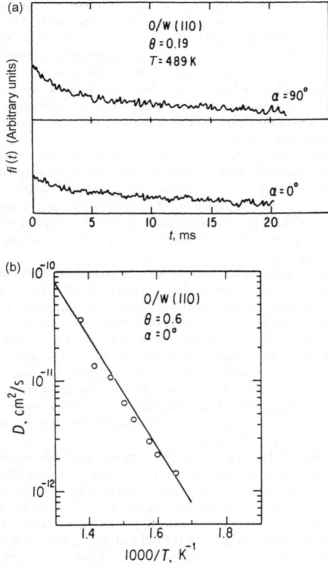

Figure 6.15.(a) Current–current correlation functions for two orientations, diffusion perpendicular and parallel to [$\bar{1}$10]; (b) diffusion coefficients for O/W(110) parallel to [$\bar{1}$10] (from Tringides & Gomer 1985, reproduced with permission).

in chapter 3, for example in relation to figure 3.7 in section 3.3, and to the operation of the SEM. The secondary electron yield depends on many factors, and is generally higher for high atomic number targets, and at higher angles of incidence. There is a lot of information in this secondary electron 'background', but, unlike Auger and other electron spectroscopies, it is not directly chemical or surface specific in general.

However, there are cases where the secondary electrons can be seen to convey more

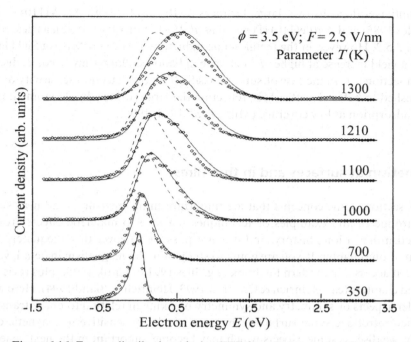

Figure 6.16. Energy distributions during TFE from LaB_6 as a function of temperature. Full lines are a calculation for a detailed field emission model including the effect of a peak in the density of states at energies just above E_F, compared with the Fowler–Nordheim expression for $\phi = 3.5$ eV and $F = 2.5$ V/nm (Mogren & Reifenburger, 1991, reproduced with permission).

specific surface information. Under clean surface conditions, a change of surface reconstruction, or an adsorbed layer, will change the work function, the surface state occupation, and may also, in a semiconductor, change the extent of band bending in the surface region. A technique developed from this effect is biased-secondary electron imaging (b-SEI), since biasing the sample negatively (anywhere from -10 to ~ -500 V) causes the low energy electrons to escape the patch fields at the surface. This signal is much more sensitive for imaging than the corresponding Auger microscopy, as discussed in section 3.5. It has been shown that this technique is sufficient to detect sub-ML deposits with good SNR, as illustrated in figure 3.21 for Ag/W(110). The case of Cs/W(110), studied earlier by Akhter & Venables (1981), showed that phase transitions could be observed and the activation energy for Cs surface diffusion measured, complementing original values by Taylor & Langmuir in the 1930s. The patch field effect is very strong in this case, extending for distances of more than 0.1 mm away from the Cs/W boundary at low bias fields.

Once again, the corresponding spectroscopy is useful in determining the origin of the contrast (Janssen *et al.* 1980, Futamoto *et al.* 1985, Harland *et al.* 1987). For 2 ML Ag deposited onto W(110) an increase in the secondary yield at the lowest electron energies E was observed, which is readily explained by a decrease in the work function. This form of surface microscopy has been exploited to measure diffusion of sub-ML and multilayer deposits over large distances (>10 µm) for both Ag/W(110) (Jones & Venables 1985) and Ag/Fe(110) (Persaud et al. 1994, Noro et al. 1996), as described in section 5.5.2. However, in the metal–semiconductor case of sub-ML Ag/Si(111) there is also a yield increase at higher E, a change in band-bending is involved, as discussed later in section 8.1. In the case of sub-ML Cs/Si(100), the detection sensitivity of b-SEI was pushed to below 1% of a ML, reflecting the large surface dipole moment caused by Cs adsorption at low coverage (Milne et al. 1994, 1995).

6.3 Magnetism at surfaces and in thin films

In this section, some concepts that are important in magnetism at and near surfaces are introduced, and examples of techniques sensitive to magnetic effects are given. Magnetism has a long history, and it is not possible to cover this adequately in one section of one chapter. If you are going to study thin film magnetism in detail, you will also need access to a modern textbook (e.g. Jiles 1991 or Craik 1995), plus review articles and chapters (e.g. Heinrich & Cochran 1993, Heinrich & Bland 1994). Here we first consider aspects of symmetry and symmetry breaking in relation to phase transitions, and then introduce some surface techniques which are sensitive to magnetic effects. Finally, we discuss some aspects which may become important in the next generation of thin film devices based on magnetism.

6.3.1 Symmetry, symmetry breaking and phase transitions

Arguments about symmetry, and symmetry breaking, figure strongly in the magnetic literature. Magnetism is often the prototype system in discussions of phase transitions. The Weiss theory of ferromagnetism published in 1908 (Ashcroft & Mermin, chapter 33, or Kittel, chapter 15) for example, is one of the earliest examples of a mean field theory applied to a (second order) phase transition. Onsager's exact solution of the 2D Ising model in 1944 was couched in terms of a magnetic transition. In the book series edited by Domb, *Phase Transitions and Critical Phenomena*, many of the articles and theorems stem from magnetic interests. There is a discussion of Onsager's papers in volume 1 of this series by Temperly (1972).

One such theorem is that due to Mermin & Wagner, which shows that magnetic long range order (in the absence of anisotropy) is impossible for a Heisenberg spin system in 2D or 1D, whereas it is clearly possible in 3D systems. The argument goes as follows. In an ordered lattice of magnetic spins which can have any orientation, such as exists in a ferro- or antiferro-magnet, the excitations are spin waves at low temperature. In these waves, the spins on neighboring lattice sites twist with respect to each other, giving rise to a magnetic energy $\omega(k)$ proportional to k^2; these quantized excitations are called magnons. Then we count the number of magnons, using Bose–Einstein statistics and obtain

$$n_{\mathrm{m}} = \int g(k) \langle n(k) \rangle dk, \qquad (6.10)$$

integrating from zero to infinity, with the density of states $g(k) = \mathrm{d}n/\mathrm{d}k = k/2\pi$ appropriate for $\omega = Ak^2$. The number of magnons, n_{m}, then is given by substituting for $\langle n(k)\rangle$, and in the high temperature limit, this gives

$$n_{\mathrm{m}} = \int (k\, k_{\mathrm{B}} T/hAk^2)\mathrm{d}k \sim \int \mathrm{d}k/k \sim [\ln (k)], \qquad (6.11)$$

which diverges at the lower limit.[1] This means that theoretically we cannot have long range order in 2D, because long wavelength (low k) excitations are possible in these systems with negligible energy. The same Mermin–Wagner theorem applies to positional order in 2D, due to the divergence, also logarithmic, of long range positional correlations; thus the corresponding theorem has also been invoked in theoretical studies of adsorption, as discussed in chapter 4.

This theorem has been shown to be of interest in some situations, but usually the length scales are too long to be of practical interest, and what happens first has to do with symmetry breaking. For instance, you can't make a free standing monolayer, or a truly 2D magnetic system. Once we have a monolayer or a magnetic system on a substrate, we have broken the symmetry. Logarithmic divergences are very easy to break; examples are the finite energies in the core of dislocations due to atomic structure, or the inductance of a finite, versus an infinitesimal diameter, wire. The breakdown of the Mermin–Wagner theorem for such practical reasons is another case.

6.3.2 Anisotropic interactions in 3D and '2D' magnets

In 3D magnetic systems we have many examples of symmetry breaking. The basic magnetic interaction is the exchange interaction related to the spins on a lattice as

$$E = -\Sigma J \mathbf{S}_i \cdot \mathbf{S}_j, \qquad (6.12)$$

where the summation is typically limited to neighboring sites only; this is referred to as the Heisenberg Hamiltonian between spins $\mathbf{S}_i$ and $\mathbf{S}_j$ with exchange coupling constant J. In the presence of an external magnetic field $\mathbf{H}$, a unique axis is imposed (orientational symmetry breaking), because $E_{\mathrm{H}} = -\Sigma \mathbf{m}_i \cdot \mathbf{H}$. The combination of these two terms for $S_i = \pm 1/2$ is the Ising model which Onsager solved exactly to find the magnetization M as a function of T. The approach to the Curie temperature, above which the system is paramagnetic, goes like

$$M \sim (T - T_{\mathrm{c}})^{1/8}, \qquad (6.13)$$

rather than the mean field exponent of 1/2. These critical exponents are characteristic of the models as T_{c} is approached, and the dimensionality (two, three or higher dimensions), but are not dependent on the details of the interactions. This is the basis of interest in universality classes, within which the critical exponents are the same: impress your friends with '… as in the 2D XY model, we can see that …'! An introduction to these critical exponents is given by Stanley (1971), and the details for 2D systems are

---

[1] Note that we use k_{B} for Boltzmann's constant in this section and the next, to distinguish it from the wave-number k.

described by Schick (1981), Roelofs (1996) and others, as discussed in section 4.5.1.

Magnetism has many other symmetry breaking interactions, and we can't realistically discuss them all here. But one very important case is the magneto-crystalline anisotropy (MCA) energy E_K, which is due to the anisotropic charge distribution in the crystal field, and orients the magnetic moments along specific crystalline axes. The form of this energy depends on the crystal symmetry, the most often encountered being uniaxial anisotropy, which, for example, makes the c-axis in h.c.p. cobalt the easy axis of magnetization; the leading term has the form $K\sin^2\theta$. In a cubic crystal, such as b.c.c. Fe, we have cubic anisotropy, which is expressed in terms of the direction cosines $\alpha_1, \alpha_2, \alpha_3$ to the three cube axes as

$$E_K = K_1(\alpha_1^2\alpha_2^2 + \alpha_2^2\alpha_3^2 + \alpha_3^2\alpha_1^2) + K_2\alpha_1^2\alpha_2^2\alpha_3^2 + \ldots \tag{6.14}$$

We can see that there won't be a second order term because $\alpha_1^2 + \alpha_2^2 + \alpha_3^2 = 1$. The easy axis for Fe is along the $\langle 100 \rangle$ directions, and this corresponds to $K_1 > 0$ and $K_1 > -K_2/9$; for Ni the easy axis is $\langle 111 \rangle$ (Craik 1995, section 1.11). A practical example is the use of Fe–4% Si for transformer cores. Why? Not because Si does anything wonderful for the magnetization of Fe, but because it gives polycrystalline Fe a {100} texture, making it easy to magnetize in the plane of transformer laminations, leading to small energy losses when used with alternating currents.

There are several other anisotropic terms, which can be important in particular circumstances. A very important term is the demagnetizing energy, which is a macroscopic effect caused by the shape of the sample, and derives from the magnetic self energy, E_S. This self energy can be expressed as either the interaction of the demagnetization field inside the sample with the magnetization, or equivalently, the integral of the energy density of the stray field over all space. If, for example, the magnetization is perpendicular to a thin film, there is a large energy due to the dipolar field outside the film; but if the magnetization is in the plane of the film, this effect is minimized. In real films, this causes the formation of domains. These domains can be seen in transmission, even in quite small samples (not necessarily single crystals), by Lorentz microscopy (coherent Fresnel, Foucault, and differential phase contrast imaging), as described in several papers from Chapman's group in Glasgow (e.g. Chapman et al. 1994, McVitie et al. 1995, Johnston et al. 1996, Chapman & Kirk 1997). They can also be seen using electron holography as developed initially by the Möllenstedt school in Germany and Tonomura's group in Japan, and further developed and reviewed by Mankos et al. (1996). In uniaxial crystals such as h.c.p. cobalt, there will still be a small field outside the film, connecting two oppositely oriented domains. In cubic crystals, even this can be avoided, by the formation of small closure domains at the ends of the film. The price paid for these domains is the energy of the interfaces between oppositely magnetized regions: these are known as Bloch or Néel walls, depending on the details of how the spins rotate from one domain to the other (Kittel 1976).

Another term relevant to thin films is magnetoelastic anisotropy, or magnetostriction. In this effect, the crystal parameters change *because* of the magnetism; this also implies that structure and symmetry changes will influence the magnetism, as exemplified by the much studied system Fe/Cu(001). Fe is b.c.c. in bulk at temperatures below the b.c.c.–f.c.c. transition at 917 °C; the Curie temperature in b.c.c. is at 770 °C, and f.c.c. Fe is overall non-magnetic, although the calculated details depend very sensitively on the lattice constant. However, when Fe is deposited on a cold (77 K) Cu(001) substrate, and warmed to room temperature, the magnetization is perpendicular to the film for a coverage < 5 ML, and is parallel for > 5 ML. But below 10 ML, Fe is not b.c.c., but is pseudomorphic with the Cu(001) and is nominally f.c.c.

The detailed structure is actually f.c.t. (face-centered tetragonal), where the expansion parallel to the film plane causes compression along the normal direction. This type of distortion is very common, occurring in the opposite sense for Ge/Si(001) as discussed in section 7.3.3. In the magnetic case, the tetragonality induces uniaxial anisotropy favoring perpendicular magnetization, which overcomes the shape effects favoring parallel alignment, if the film is thin enough. The particular system Fe/Cu(001) is in fact rather complicated, because in deposition even at room temperature, we can get exchange diffusion of Fe into the Cu, since, on surface energy grounds, the Cu wants to cover the Fe layer. There are now several similar examples (Fe/Ag, Fe/Au, Co/Cu, etc.), which have been seen by Auger spectroscopy, STM and other methods. The extreme sensitivity of the magnetism to the exact lattice parameter and micro-structural condition of the film means that there are (too many) contradictory results in the literature. These points have started to become clear in recent research papers; they are discussed further here in sections 5.5.3 and 8.3.

6.3.3 Magnetic surface techniques

Investigation of magnetic surfaces and thin films proceeds at two levels: structural and microstructural examination can be done using the same techniques as for non-magnetic materials, as described in chapter 3, where some relevant examples were given. Some techniques which are specific to magnetism are described here in outline. These include optical rotation (Faraday and Kerr effects plus magnetic circular dichroism (MCD)) and spin-polarized electron techniques. For analysis of domain structures, several microscope based methods have been developed which display magnetic contrast. These include SEMPA, SMOKE microscopy, TEM (Lorentz or holography, described in the last section), spin-polarized LEEM and magnetic force microscopy (MFM), which are explained below.

Optical techniques work in magnetic materials via the rotation of the plane of polarized light. The dielectric constant ε is a tensor, with off-diagonal terms of the form $\pm i\varepsilon_{xy}$, in addition to the usual diagonal terms, so that right- and left-handed circularly polarized light behave differently. The effects are called the Faraday effect in transmission and the Kerr effect in reflection. The most commonly used technique is called MOKE (magneto-optic Kerr effect) and the acronym SMOKE is used when this technique is applied to surfaces. Depending on the light polarization with respect to the magnetization of the sample, one can measure different Kerr signals which have components perpendicular and parallel to the sample. By varying the magnetic field cyclically, one can obtain hysteresis loops to characterize the magnetic state of the sample.

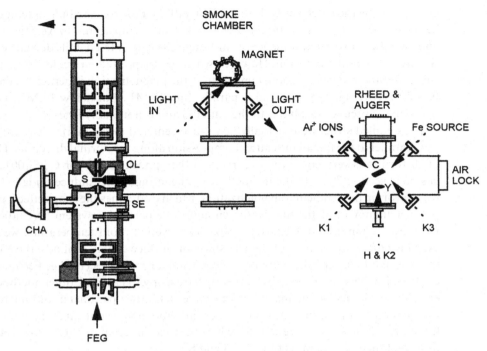

Figure 6.17. The MIDAS column and preparation chamber arranged for magnetic studies. The column shows the position of the sample (S), objective lens (OL), electron parallelizers (P) and secondary electron detector (SE), all of which are shown in more detail in figure 3.24, plus the field emission gun (FEG). The sample can be transported from the airlock to the sample preparation station, which has multiple ports used as shown plus those for sample heating (H) and extra Knudsen cell evaporators (K1–K3); the YAG screen (Y) is for viewing the RHEED pattern. After preparation, the copper sample (C) can be transported to the SMOKE station before being examined by high resolution SEM and analysis (after Heim *et al.* 1993, reproduced with permission).

A diagram of the geometry of the preparation chamber of MIDAS at Arizona State University, and typical Kerr loops, are shown in figures 6.17 and 6.18. This configuration has enabled *in situ* comparisons of structural and magnetic properties of a range of thin film magnetic systems.

Magnetic circular dichroism (MCD) is a powerful recent technique which is a cross between photoemission and MOKE. By using spin-orbit split core levels, separated by 10 eV or more, the magnetism of thin films, including internal interfaces, can be studied. The core levels are specific to particular elements, and the rotation of the plane of polarization is specific to the magnetism at the sites of these elements. A special merit of MCD is that it enables spin-specific and element-specific measurements to be made concurrently (Bader & Erskine 1994). This is particularly powerful, e.g. (a) in ferrimagnetic systems where differing spin sublattices have different orientations; (b) in trilayers and multilayers such as Fe/Cr/Fe(001), where the magnetic alignment is different in the various layers (Idzerda *et al.* 1993). MCD is typically performed using

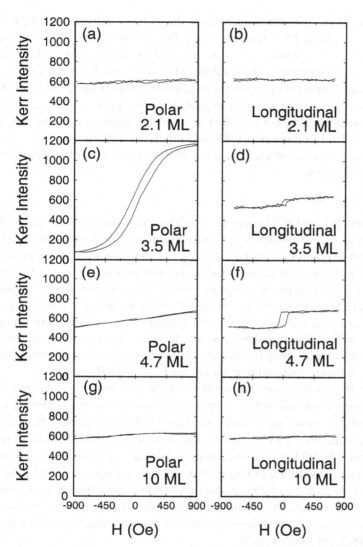

Figure 6.18. Polar and longitudinal hysteresis loops from Fe/Cu(001) grown and measured at room temperature, taken with the SMOKE setup in MIDAS shown in figure 6.17 with an angle of incidence of 45°: (a) and (b) 2.1 ML, no remanent magnetization; (c) and (d) 3.5 ML, remanent, mostly out of plane; (e) and (f) 4.7 ML remanent, in plane; (g) and (h) 10 ML, non-magnetic f.c.c. Fe film (after Hembree *et al.* 1994, reproduced with permission).

a synchrotron radiation source, and is a powerful application of display analyzers, as shown in figure 3.9(d) (Daimon *et al.* 1995).

Electron spectroscopy can be used to study magnetic domains. Electrons emitted from a magnetic material are spin polarized, because the spin up and spin down bands are populated differently. This is a strong effect for low energy secondary electrons, where the polarization can reach ±40%, with ~±20% for Fe and ~±10% for Co at higher energies, where the spectrum reflects the polarization of the valence band (Kirschner 1985, Landolt 1985, Hopster 1994). Combined with SEM, this has lead to the development of scanning electron microscopy with polarization analysis (SEMPA). The extra element is the addition of a spin polarization detector. These detectors detect left–right (or up–down) asymmetries caused by spin-orbit (Mott) scattering, typically in a heavy target such as gold. Clear views of the domain structure, completely different from the normal SEM image of the same area, can be obtained as shown in figure 6.19 (Hembree et al. 1987; Scheinfein et al. 1990).

There are several versions of such detectors, and the polarization P is determined by an algorithm of the form

$$P = C(L - R)/(L + R), \tag{6.15}$$

using the two signals L and R. The sensitivity of this technique depends on the effective Sherman function $S_{eff} \sim 0.1$–0.3 typically, increasing with incident electron energy in the range 10–100 keV, and the constant $C \sim S_{eff}^{-1}$. The figure of merit for the detector is proportional to the ratio of the detected to the incident current $(I_D/I_0) \cdot S_{eff}^2$ which is typically small ($< 10^{-4}$). Typically, magnetic effects can be enhanced by reversing the field and taking difference signals, but this is not always necessary if one is prepared to live with an offset signal arising from possible alignment errors in the detector system. Spin-polarized AES is also possible using an electron spectrometer in addition to a spin-polarized detector. As may be imagined, SPAES signal levels are very small, and long collection times are required to achieve an adequate SNR.

The above techniques use unpolarized electron sources, but spin polarized sources can be made using circularly polarized photoemission from spin-polarized valence bands. The most commonly used source is p-type GaAs(001), selectively exciting the heavy hole ($p_{3/2}$) band with 1.4 eV photons. This puts spin-polarized electrons into the conduction band. The trick is then to activate the surface to negative electron affinity, by coating the surface with a Cs/O layer. This strongly reduces the work function of GaAs, such that the bottom of the conduction band is above the vacuum level; electrons therefore spill out into the vacuum, and are sufficiently intense to form a source, even for a microscope. Comprehensive reviews of this technique have been given by Pierce et al. (1980) and Pierce (1995).

Spin-polarized LEEM is a technique which is being developed for magnetic materials. Phase sensitive detection to eliminate unwanted background signals is possible, by modulating the laser polarization, and detecting the electrons in synchronism. This work is in its infancy at present, but progress has been reviewed (Bauer 1994); more recent results are given by Bauer et al. (1996). An example of a SPLEEM image from the last reference is shown in figure 6.20.

Magnetic force microscopy is a development of AFM which measures the field gradient above ferromagnets; the force on a magnetic moment μ in the z-direction $F_z = -\mu dB_z/dz$, and lateral force measurements are also possible. Typically, a cantilever with a etched Si tip of radius $r \sim 10$ nm is sputter coated with a ferromagnetic material. If the coercivity is high, this allows the magnetization distribution of the tip to remain fixed as the fields on the sample are changed. The field emanating from the tip falls off

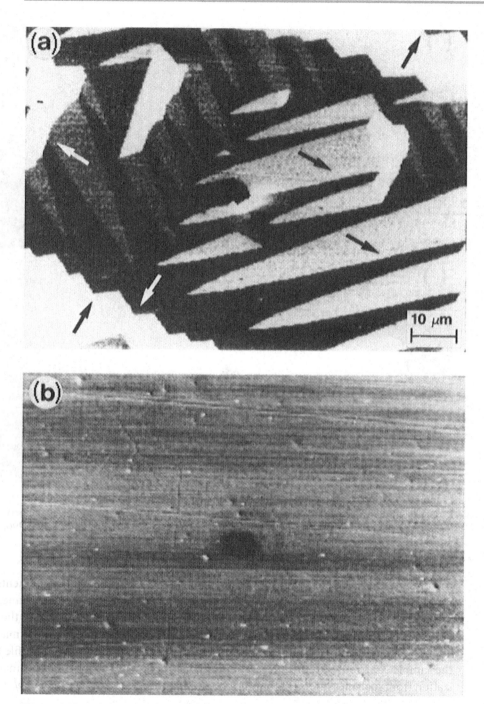

Figure 6.19. (a) SEMPA spin polarization image, contrasted with (b) SEM intensity image of an Fe–3% Si single crystal. The gray levels in (a) give the four different magnetization directions in the domains as marked by arrows (Hembree *et al.* 1987, reproduced with permission).

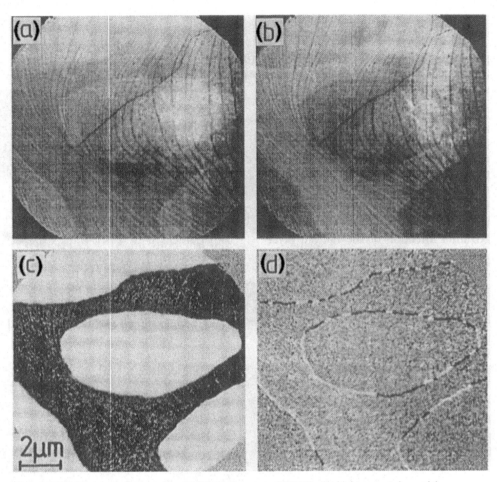

Figure 6.20. Magnetic image of a 6 ML Co layer on W(110): (a, b) images taken with
P parallel and antiparallel to **M** in the two domain orientations; (c) difference image between
(a) and (b); (d) difference image between two images with **P** ⊥ **M** (from Bauer *et al.* 1996,
reproduced with permission).

rapidly and the sharper the tip, the faster the field gradient decay. Measurements are
usually made in an a.c. detection scheme where the tip is vibrated at some resonance frequency, and the departure from that resonance due to the tip's interaction with the field
gradient is detected with lock-in amplifiers. In this fashion, an image can be made at
about 50 nm resolution. But inevitably, the entire integrated field gradient profile from
the sample contributes to the image, so the reconstruction of the local sample magnetization from such measurements may not be entirely straightforward (Rugar *et al.* 1990).

6.3.4   Theories and applications of surface magnetism

Magnetic interactions in 3d metals are dominated by the d-electrons and perturbed by
s-d hybridization. The 3d-electrons, responsible for the magnetism of Fe, Ni and Co,

form a relatively narrow band which overlaps with the wide 4s band. The question of why these members of the 3d series are ferromagnetic, while others are antiferromagnetic, and why the 4d and 5d series are not magnetic, is a typically subtle problem in cohesive energy, in which several terms of differing sign are closely balanced (Moruzzi *et al.* 1978, Sutton 1994, Pettifor 1995). The magnetism of the parent atoms is a result of Hund's rule, which asserts that the first five d-electrons are populated with parallel spins, and the remaining five then fill up the band with antiparallel alignment. This is due to the reduced electron–electron Coulomb interaction between pairs with parallel spins, because the exchange-correlation hole which accompanies each electron (see Appendix J) keeps these electrons further apart on average. The rare earth elements are an important class of magnetic materials based on 4f-electrons, but are not discussed here.

When these atoms are assembled into solids, several effects occur which we should not try to oversimplify. The d-band is very important for cohesion, and the simplest model is that due to Friedel (1969), which predicts a parabolic dependence of the bond energy as the number of d-electrons N_d is increased across the series. This model leads to the contribution of d-d bonding to the pair-bond energy, E_b

$$- E_b = 2 \int^{E_F}(E - \varepsilon_d)\,(5/W)\mathrm{d}E = - (W/20)N_d(10 - N_d),\qquad(6.16)$$

where ε_d is the unperturbed atomic d-level energy and W is the d-band width in the solid. This parabolic behavior with N_d is quite closely obeyed by the 4d and 5d series, leading to surface energies displaying similar trends (Skriver & Rosengaard 1992). In terms of the second moment of the energy distribution μ_2, the overlap integrals between d-orbitals of strength β, the band width are related by

$$W = (12z)^{1/2}|\beta|,\qquad(6.17)$$

with z nearest neighbors; this can be derived for a rectangular d-band, where the second moment $\mu_2 = W^2/12$ (Sutton 1994). However, when magnetic effects are considered, the *shape* of the d-band is also very important, and ferromagnetism only results when both the d-d nearest neighbor overlap is strong *and* the density of states near the Fermi energy is large. These conditions are fulfilled towards the end of the 3d series, aided by the two-peaked character of the density of states, sketched in figure 6.21(a); this energy distribution has a large fourth moment μ_4, which is also implicated in the discussion of why Fe has the b.c.c. structure, points which can be explored further via project 6.4.

When detailed band structure calculations are done including magnetic interactions, we have to account separately for the majority spin-up ($\rho\uparrow$) and minority spin-down ($\rho\downarrow$) densities. By analogy to LDA, there is a corresponding local spin density (LSD) approximation. This is illustrated in figures 6.21(b, c) and 6.22 by the calculations for b.c.c. Fe by Papaconstantopoulos (1986); the up and down spins bands are shifted by almost 2 eV. Above the ferro-paramagnetic transition at $T = 770\,°\mathrm{C}$ these spins lose long range order, but short range order is still present.

These spin density methods have been pursued intensively by Freeman & co-workers (Weinert *et al.* 1982, Freeman *et al.* 1985), particularly in the version known as the FPLAPW (full potential, linearized APW). Several features of thin film magnetism have been studied by this method as described by Wu *et al.* (1995). Comparisons of

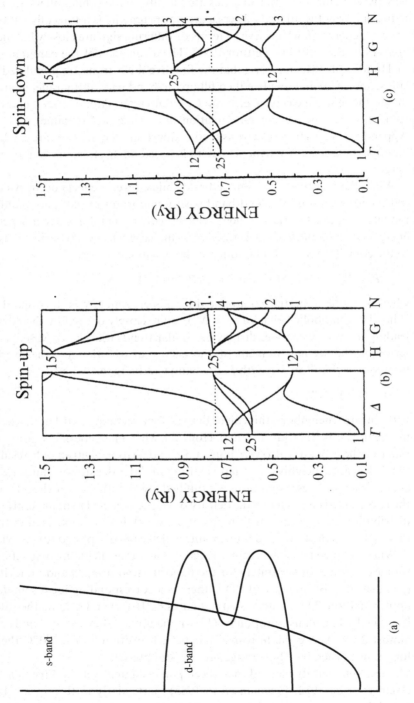

Figure 6.21. (a) Schematic distribution of s-d band overlap with the d-band having a double-peaked density of states; (b) the calculated spin-up and (c) spin-down band structures of Fe along $\Delta = [100]$, after Papaconstantopoulos (1986, reproduced with permission). The symmetry points at Γ labelled 1 are s-like; the d-like states are 12 ($e_g = 3z^2 - r^2$, $x^2 - y^2$) and 25 ($t_{2g} = yz$, zx, xy).

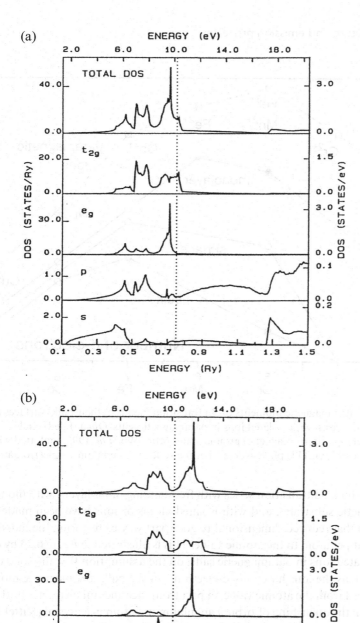

Figure 6.22. (a) Majority and (b) minority spin density of states, as calculated for b.c.c. Fe by Papaconstantopoulos (1986, reproduced with permission); the vertical dashed line corresponds to the Fermi energy. Note that both the 3d-bands have large fourth moments, with the t_{2g} band having a large DOS at E_F, and that the s- and p-bands are much broader than the d-bands.

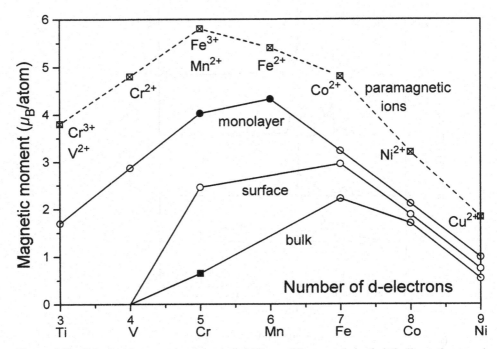

Figure 6.23. Calculated magnetic moments of 3d transition metals in their bulk, surfaces and monolayers, in comparison with isolated ions in paramagnetic salts. Open (filled) circles denote ferromagnetic (antiferromagnetic) ground states (after Wu *et al.* 1995, replotted with permission, and Jiles 1991). This plot assumes that the solid states contain 1 s-electron/atom.

magnetism in the bulk 3d transition series with freestanding monolayers, with monolayers on non-magnetic substrates, and with isolated atoms or ions have been made. The general feature is that reduced dimensionality goes part way to restoring the individual magnetic moment per atom to the atomic value. This is illustrated in figure 6.23 by comparison with isolated ions in paramagnetic salts, on the assumption that the solids have 1 s-electron, whereas the ions have only d-electrons. In the bulk, the magnetic moment per atom is reduced from the atomic value, in part from the itinerant character of d-electrons, in part from the quenching of orbital angular momentum in a crystal (Kittel 1976, Jiles 1991). These reductions are less marked at the surface and in monolayers.

Perhaps the most dramatic effect is that these changes may be sufficient to change the sign of the coupling between layers from ferromagnetic (F) to antiferromagnetic (AF) or vice versa. Some of these effects have been seen over the last few years in magnetic multilayers, in which thin magnetic layers are separated by non-magnetic spacers. The coupling between the layers can be either F or AF, and can be changed, both by the thickness of the spacer layers, and by the application of a magnetic field.

The coupling between magnetic layers separated by noble or transition metals, as in Fe–Cr, Fe–Ag or –Au, or Co–Cu superlattices, have all the magnetic interactions we have discussed, plus a coupling due to the conduction electrons in the non-magnetic spacers. The phenomenon is best thought of as a quantum size effect with magnetic

Monolayers

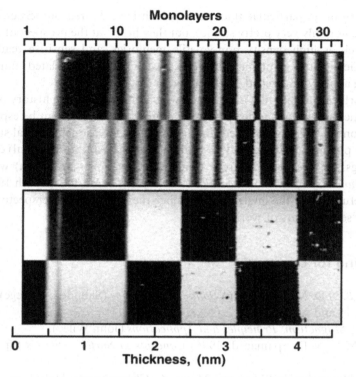

Thickness, (nm)

Figure 6.24. SEMPA magnetization images of the magnetic coupling of Fe layers in an Fe/Cr/Fe sandwich grown on a wedge shaped Cr layer whose thickness increases from left to right, on a single crystal Fe(001) substrate. There are two domains in the substrate with magnetization to the left and right, giving the sharp horizontal demarcation in both panels of size $\sim 300\ \mu$m square viewed obliquely. In the lower panel, the rough Cr layer was grown at room temperature giving long period reversals between antiferro- and ferro-magnetic coupling, whereas in the upper panel, layer by layer growth at $T = 300\,°$C reveals additional short, ~ 2 ML, period reversals (after Unguris *et al.* 1991, reproduced with permission).

complications (Stiles 1993, 1996). In effect, there are Friedel oscillations at each metal interface, and in the case of magnetic materials these are spin-dependent; there are standing waves in the spacer layer, and reflection and transmission amplitudes at the interfaces. The period is given by $(2k_F)^{-1}$ of the *spacer* layer in the direction perpendicular to the layers, but in noble and transition metals there can be more than one value of k_F due to the topology of the Fermi surface.

The competition between these length scales and the ML period produces complex magnetic patterns in superlattice 'wedges' which have been seen by SEMPA, as shown in figure 6.24 (Unguris *et al.* 1991, 1994, Pierce *et al.* 1994). These studies show that observing the finer periods is dependent on the quality of the interfaces, i.e. on crystal growth processes. The use of wedged samples is a clever way of studying several different thicknesses in the same experiment, by using a microscope to pinpoint the place where the multilayer is being sampled. It can even be done in 2D to probe two thickness variables at once (Inomata *et al.* 1996); this is clearly very advantageous as a means of homing in on particular thicknesses which have desired properties. These observations are not only very pretty science, but they hold out the prospect of device applications, such as high density non-volatile memories, and sensitive read/write devices. In particular, the giant magneto-resistance (GMR) and related multilayer effects are being actively researched, as described in section 8.3.

A field like surface and thin film magnetism, which builds on a long history of electric and magnetic properties, surface physics and growth processes, can be especially difficult for anyone trying to get started. In this situation, a reasonable initial strategy is to skip all the preliminary work, and go straight to the latest (international) conference proceedings. One conference (from which some examples are taken) was the *Second International Symposium on Metallic Multilayers (MML'95)* (Booth 1996); it is especially useful to read the invited papers, since these have more perspective, and typically survey several years of work.

Further reading for chapter 6

Ashcroft, N.W. & N.D. Mermin (1976) *Solid State Physics* (Saunders College) chapters 8–11, 14, 15 and 33.

Craik, D. (1995) *Magnetism: Principles and Applications* (John Wiley).

Desjonquères, M.C. & D. Spanjaard (1996) *Concepts in Surface Physics* (Springer) chapter 5.

Jiles, D. (1991) *Magnetism and Magnetic Materials* (Chapman and Hall).

Kittel, C. (1976) *Introduction to Solid State Physics* (6th Edn, John Wiley) chapters 14 and 15.

Pettifor, D.G. (1995) *Bonding and Structure of Molecules and Solids* (Oxford University Press) chapters 2, 5, 7.

Stanley, H.E. (1971) *Introduction to Phase Transitions and Critical Phenomena* (Oxford University Press).

Sutton, A.P. (1994) *Electronic Structure of Materials* (Oxford University Press) chapters 7, 8 and 9.

Sutton, A.P. & R.W. Balluffi (1995) *Interfaces in Crystalline Materials* (Oxford University Press) chapter 3.

Woodruff, D.P. & T.A. Delchar (1986, 1994) *Modern Techniques of Surface Science* (Cambridge University Press) chapters 6 and 7.

Zangwill, A. (1988) *Physics at Surfaces* (Cambridge University Press) chapter 4.

Problems and projects for chapter 6

Problem 6.1. Why is $(2k_F)^{-1}$ the characteristic length for electrons at surfaces?

A whole series of problems can be devised to get a feel for the size and relevance of this characteristic length; the following questions require access to a standard solid state textbook (e.g. Ashcroft & Mermin 1976, or Kittel 1976).

(a) For b.c.c Li, find the value of the lattice parameter a, and hence of the nearest neighbor distance. Assuming one electron per atom, evaluate the radius of the Wigner-Seitz sphere r_s, the magnitude of $(2k_F)^{-1}$, and hence the periodicity Δz of Friedel oscillations. Indicate these lengths (a and Δz) in relation to figure 6.6, and identify the corresponding periods in the electron density oscillations.

(b) Consider the discussion based on figure 6.5, and use this to derive equation (6.4), which is correct asymptotically for z well inside jellium. Note that we cannot use (6.4) near $z = 0$. Use the positions of maxima and minima in Lang & Kohn's calculation for $r_s = 5$ to estimate the value of the phase factor γ_F, and plot this prediction of $n(z)$ for comparison with figure 6.2(a).

(c) Study in outline the ingredients of the Bardeen–Cooper–Schrieffer (BCS) theory of superconductivity, noting how electrons of opposite momenta with energies within $\hbar\omega_D$ of E_F form the BCS ground state, via coupling by phonons of wave-vector $2k_F$.

(d) Similarly note how scattering of electrons across the Fermi surface by phonons of wavevector $2k_F$ is thought to contribute to structural transitions (charge density or static distortion waves) at metal surfaces, for example in the case of Mo(001).

(e) Estimate the wavelength of the ripples seen in the quantum corral of figure 6.4, and relate this length to the Fermi surface of copper, incorporating the cylindrical geometry. Consult the paper by Petersson *et al.* (1998) to see how such measures can be made quantitative.

Problem 6.2. Derivation of the Richardson–Dushman equation from free electron theory

Consider the Fermi–Dirac distribution in the form $f(E) = (1 + \exp(E - \mu)/k_B T)^{-1}$, where the chemical potential of the electron gas, μ is the same as the Fermi energy E_F.

(a) Use this form, together with the density of states shown in figure 6.5 and the concept of the *perpendicular* energy, $E_z = (\hbar k_z)^2/2m$, to derive the free electron form of the Richardson–Dushman equation (6.7) when barrier transmission occurs for all $E_z > \mu$.

(b) By performing a 1D calculation, matching electron waves at the surface at $z = 0$, investigate how the reflection coefficient for electrons incident on the surface changes as a function of E_z when $E_z \geq \mu$. Use this result to show how energy-dependent barrier transmission affects the formula derived in part (a).

Project 6.3. Barrier transmission, the Fowler–Nordheim equation and models of STM operation

The Fowler–Nordheim equation can be derived from free electron theory in a similar manner to problem 6.2, except that we now *need* the probability of transmission through the finite barrier as a function of E_z. This is typically given by the Wentzel–Kramers–Brillouin (WKB) approximation, where the transmission coefficient

$T \sim \exp[-(2/\hbar)\int(V(z)-E_z)^{1/2}dz]$. In this formula, the potential $V(z)$ is as in figure 6.11, and the limits of the z-integration are set by the perpendicular energy E_z.

(a) Show that the result (6.8) arises in the limit of a triangular barrier at zero temperature, and that the energy distribution will be broadened at elevated temperature as in figure 6.12.

(b) Show that the same set of arguments applied to the operation of the STM, predict qualitatively the observed exponential dependence of tip-sample voltage on tip separation at constant tunneling current, in the limit of small voltages.

(c) Alternatively, you may prefer to start from the quantum-mechanical expression for the particle current in terms of gradients of the (1D) wavefunction, and show using first order perturbation theory that the tunneling current I is given by

$$I = (e/\hbar) \sum_{\mu\nu} [f(E_\mu) - f(E_\nu)] |M_{\mu\nu}|^2 \delta(E_\nu + V - E_\mu),$$

where the matrix element $M_{\mu\nu}$ is given by

$$M_{\mu\nu} = (\hbar/2m) \int dS \cdot (\psi_\mu^* \nabla \psi_\nu - \psi_\nu \nabla \psi_\mu^*),$$

where dS is the element of surface area lying in the barrier region. In these formulae, $f(E)$ is the Fermi function and $\delta(E)$ the Dirac δ-function.

Project 6.4 Interactions between d-electrons, magnetism and structures in transition metals

Starting from the relevant sections of Sutton (1994) and/or Pettifor (1995), and other literature cited in section 6.3.4, investigate models of b.c.c. and f.c.c. Fe, and neighboring elements in which you are interested.

(a) Describe how magnetism stabilizes the b.c.c. structure at low temperature, and find out about f.c.c. Fe at high temperatures, and the magnetism of neighboring elements in the 3d series.

(b) What features of the 4d and 5d series make the elements in corresponding columns of the periodic table non-magnetic?

(c) Investigate ideas of bond-order potentials, and describe how 'embedding bonds' can create angular dependent interactions, and thus help to stabilize crystal structures which are not close-packed.

(d) Show that the angular dependent interactions can be formulated in terms of higher moments (μ_i with $i > 2$) of the valence electron density of states, and that the b.c.c. stability is largely attributable to μ_4, while the h.c.p.–f.c.c. difference is affected by μ_6. From this viewpoint, investigate the contribution such effects can make to the surface energy of transition metals.

7 Semiconductor surfaces and interfaces

This chapter gives a description of semiconductor surfaces, and the models used to explain them. Section 7.1 outlines ideas of bonding in elemental semiconductors, and these are used to discuss case studies of specific semiconductor surface reconstructions in section 7.2, building on the survey given in section 1.4. If you are not familiar with semiconductors and their structures, you will also need access to sources that describe the diamond, wurtzite and graphite structures, and which also describe the bulk band structures; these points can be explored via problem 7.1. It is also very helpful to have some prior knowledge of the terms used in covalent bonding, such as s and p bands, sp^2 and sp^3 hybridization. Section 7.3 describes stresses and strains at surfaces and in thin films, including the thermodynamic discussion delayed from section 1.1; the importance of such ideas in the growth of semiconductor device materials is discussed, especially those based on the elements germanium and silicon, with references also given to the 3–5 compound literature.

7. 1.   Structural and electronic effects at semiconductor surfaces

The first thing to realize is that the reconstructions of semiconductor surfaces are not, in general, simple. In section 1.4 reconstructions were introduced via the (relatively simple) Si(100) 2×1 surface. This introduced ideas of symmetry lowering at the surface, domains, and the association of domains with surface steps. At the atomic cell level we saw the formation of dimers, organized into dimer rows. If all this can happen on the simplest semiconductor surface, what can we expect on more complex surfaces? More importantly, how can we begin to make sense of it all? This is a topic which is still very much at the research stage. But enough has been done to try to describe how workers are going about the search for understanding, which is what is attempted here.

7.1.1   Bonding in diamond, graphite, Si, Ge, GaAs, etc.

The basis of understanding surfaces comes from considering them as intermediate between small molecules and the bulk. In the case of the group 4 elements, there is a progression from C (diamond, with four nearest neighbors), through Si and Ge with the same crystal structure, then on to Sn and Pb. The last two elements are metallic at

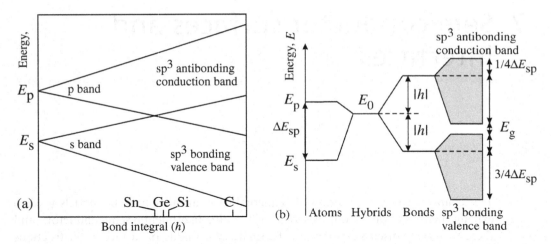

Figure 7.1. (a) Hybridization gap in due to sp³ bonding in diamond, Si, Ge and gray Sn; (b) stages in the establishment of the valence and conduction bands via s-p mixing, involving ΔE_{sp} and the overlap integral h (after Harrison 1980, and Pettifor 1995, replotted with permission).

room temperature, Pb having the 'normal' f.c.c. structure with 12 nearest neighbors. We might well ask what is giving rise to this progression, and where Si, Ge, GaAs, etc. fit on the relevant scale. A frequent answer is to say something about sp³ hybrids, assume that is all there is to say, and move on. However, there is much more to it than that; the extent to which one can go back to first principles is limited only by everyone's time (Harrison 1980, Sutton 1994, Pettifor 1995, Sutton & Balluffi 1995, Yu & Cardona 1996).

In lecturing on this topic, I have typically started with a two-page handout, the essence of which is given here as Appendix K. This connects bonding and anti-bonding orbitals in σ-bonded homonuclear diatomic molecules with the overlap, or bonding integral, h. (Note that h is *not* Planck's constant, and the symbol often used for overlap integral is β which is *not* $(kT)^{-1}$.) For heteronuclear diatomic molecules where ΔE is the energy difference of levels between the molecules A and B, the splitting of the levels w_{AB} combines as

$$w_{AB} = \sqrt{(4h^2 + \Delta E^2)}. \tag{7.1}$$

This leads to ideas, and scales, of electronegativity/ionicity, based on the relevant value of $(\Delta E/h)$: for group IV molecules this is zero, increasing towards III–V's, II–VI's etc., roman numerals being the convention for the different columns of the periodic table; these scales try to establish the relevant mixture of covalent and ionic bonding in the particular cases: 3–5's are partly ionic, and 2–6's are clearly more so.

In the diamond structure solids, the tetrahedral bonding does indeed come from sp³ hybridization, but it is not obvious that this will produce a semiconductor, and the question of the size of the band gap, and whether this is direct or indirect, is much more subtle, as indicated in figure 7.1. The s-p level separation in the free atoms is about 7–8 eV, but the bonding integrals are large enough to enforce the s-p mixing and to open up an energy gap (valence–conduction, equivalent to bonding–antibonding) within the sp^3 band, largest in C (diamond) at 5.5 eV, and 1.1, 0.7 and 0.1 eV for Si, Ge and (gray) Sn respectively. Sn has two structures; the semi-conducting low temperature form, alpha or gray tin (with the diamond structure), and the metallic room temperature form, beta or white tin (body centered tetragonal, space group $I4_1/amd$).

The question of phase transitions in Si as a function of pressure is also a fascinating test-bed for studies of bonding (Yin & Cohen 1982, Sutton 1994). Even at normal pressure, there is some discussion of bonding in these group IV elements, especially in the liquid state (Jank & Hafner 1990, Stich *et al.* 1991). For example, liquid Si is denser than solid Si at the melting point, and interstitial defects are present in solid Si at high temperature. In this state, the bonding is not uniquely sp^3, but is moving towards s^2p^2. Pb has basically this configuration, but, as a heavy element, has strong spin-orbit splitting. This relativistic effect is also important in Ge, being the cause of the difference between light and heavy holes in the valence band. You can see that all these topics are fascinating: the only danger is that if we pursue them much further here, we will never get back to surface processes!

7.1.2   *Simple concepts versus detailed computations*

Simple concepts start from the idea of sp^3 hybrids as the basic explanation of the diamond structure. These hybrids are linear combinations of one s and three p electrons. Their energy is the lowest amongst the other possibilities, but as seen in the arguments given by Pettifor, Sutton and others, it can be a close run thing. The hybrids give the directed bond structure along the different $\langle 111 \rangle$ directions in the diamond structure, so that

$$\psi[111] = 1/2 \{s + p_x + p_y + p_z\}, \; \psi[1\bar{1}\bar{1}] = 1/2 \{s + p_x - p_y - p_z\}$$

$$\psi[\bar{1}\bar{1}1] = 1/2 \{s - p_x - p_y + p_z\}, \text{ and } \psi[\bar{1}1\bar{1}] = 1/2 \{s - p_x + p_y - p_z\}, \tag{7.2}$$

which has a highly transparent matrix structure, exploited in the tight binding and other detailed calculations. The key point is that these *bonds* are directed at the tetrahedral angle, 109° 28′. This is the angle preferred by the group IV elements, not only in solids and at surfaces, but also in (aliphatic) organic chemistry (i.e. from CH_4 onwards).

We can contrast this with the planar arrangement in graphite, where three electrons take up the sp^2 hybridization, leaving the fourth in a p_z orbital, perpendicular to the basal (0001) plane. The in-plane angle of the graphite hexagons is now 120°, with a strong covalent bond, similar to that in benzene (C_6H_6) and other aromatic compounds, and weak bonding perpendicular to these planes. The binding energies of carbon as diamond and graphite are almost identical (7.35 eV/atom), but the surface energies are very different – basal plane graphite very low, and diamond very high. The combination of six- and five-membered rings that make up the soccer-ball shaped Buckminster-fullerene, the object of the 1996 Nobel prize for chemistry to Curl, Kroto and Smalley, is also strongly bound at ~ 6.95 eV/atom. All these are fascinating aspects of bonding to explore further.

The next level of complexity occurs in the III–V compounds, of which the archetype is GaAs. This is similar to the diamond structure (which consists of two interpenetrating f.c.c. lattices), but is strictly a f.c.c. crystal with Ga on one diamond site and As on the other; with the transfer of one electron from As to Ga, both elements adopt the sp^3 hybrid form of the valence band, and so GaAs resembles Ge. However, there are differences due to the lack of a center of symmetry (space group $\bar{4}3m$), which we explore in relation to surface structure in the next section. In addition, many such III–V and II–VI compounds have the wurtzite structure, which is related to the diamond structure as h.c.p. is to f.c.c. These two structures often have comparable cohesive energy, leading to stacking faults and polytypism, as in α- and β-GaN, which are wide-band gap semiconductors of interest in connection with blue light-emitting diodes and high power/ high temperature applications, as described in section 7.3.4.

7.1.3   Tight-binding pseudopotential and ab initio models

Professional calculations of surface structure and energies of semiconductors typically consider four valence electrons/atom in the potential field of the corresponding ion, in which the orthogonalization with the ion core is taken into account via a pseudopotential. This yields potentials which are specific to s-, p-, d-symmetry, but which are much weaker than the original electron–nucleus potential, owing to cancellation of potential and kinetic energy terms. All the bonding is concentrated outside the core region, so the calculation is carried through explicitly for the pseudo-wavefunction of the valence electrons only, which have no, or few nodes;[1] overlaps with at most a few neighbors are included. There are many different computational procedures, and there is strong competition to develop the most efficient codes, which enable larger numbers of atoms to be included. In particular, the Car–Parinello method (Car & Parinello 1985), which allows finite temperature and vibrational effects to be included as well, has been widely used. This method is reviewed by Remler & Madden (1990), while Payne *et al.* (1992) give a review of this and other *ab initio* methods.

The tight binding method is described in all standard textbooks (Ashcroft & Mermin 1976); a particularly thorough account is given by Yu & Cardona (1996). Tight binding takes into account electrons hopping from one site to the next and back again in second order perturbation theory, which produces a band structure energy which is a sum of cosine-like terms, as can be explored via problem 7.2. Zangwill (1988) applies this method in outline to surfaces; he shows that the local density of states (LDOS) is characterized by the second moment of the electron energy distribution. As a result, the second moment of $\rho(E)$ is proportional to the number of nearest neighbors Z, and to the square of the hopping, or overlap, integral (h or β). At the surface, the number of neighbors is reduced, and so the bandwidth is narrowed as $Z^{1/2}$. But this simple argument on

---

[1] The pseudo-wavefunction, being 'smoother', requires fewer (plane wave) coefficients to compute energies to a given accuracy. However, this can lead to difficulties in comparing different situations, e.g. between solids and atoms, which may require different numbers of terms. Some of these points are mentioned in Appendices J and K.

Table 7.1. *Lattice constants, binding energies and vibrational frequencies of Si and Ge*

Material		Spacing (nm)	Energy (eV)	Frequency (THz)
1. Bulk Si	calc.	0.545[c], 0.550[e], 0.537[g]	4.67[c], 5.03[e], 4.64[g]	15.16[c], 15.6[e]
	expt.	0.5430	4.63 ± 0.04	15.53[c]
2. Si$_2$	calc.	0.225[b], 0.227[e], 0.228[f]	4.18[b], 3.62[f]	15.0[b], 15.9[e], 14.4[f]
	expt.	0.224[e]	3.21 ± 0.13	15.3[b,e]
3. Bulk Ge	calc.	0.556[c], 0.558[g]	4.02[c], 3.86[g]	8.90[c]
	expt.	0.5658	3.83 ± 0.02	9.12[c]
4. Ge$_2$	calc.	0.234[a], 0.242[d], 0.2326–0.2385[h]	4.14[a], 2.50–2.67[h]	8.57[a], 8.48[d], 8.42–8.82[h]
	expt.		2.70 ± 0.07[d]	

References: (a) Northrup & Cohen (1983); (b) Northrup *et al.* (1983); (c) Yin & Cohen (1982), Cohen (1984); (d) Kingcade *et al.* (1986); (e) Sankey & Niklewski (1989); (f) Fournier *et al.* (1992); (g) Krüger & Pollmann (1994, 1995); (h) Deutsch *et al.* (1997). Where not referenced, experimental values are from table 1.1; others can be traced via the papers cited.

its own is not sufficient to reproduce the band structure of Si and Ge in any detail, either in bulk or at the surface. The various approximations for bulk band structures are discussed by Harrison (1980), Kelly (1995), Yu & Cardona (1996) and Davies (1998).

One of the first workers to pioneer tight binding methods was D.J. Chadi (for a short review see Chadi 1989), but there are many others who have made realistic calculations on semiconductor surfaces, and atoms adsorbed on such surfaces (e.g. in alpha-order C.T. Chan, M.L. Cohen, C.B. Duke, R.W. Godby, K.M. Ho, J. Joannopoulos, E. Kaxiras, P. Krüger, R.J. Needs, J. Northrup, M.C. Payne, J. Pollmann, O. Sankey, M. Scheffler, G.P. Srivastava, D. Vanderbilt, A. Zunger, to mention only a few). The most frequent use of tight binding methods is as an interpolation scheme, fitting *ab initio* LDA/DFT methods of the type discussed in chapter 6 or more chemical multiconfiguration calculations, but computationally much faster.

Examples of the level of agreement with lattice constants, dimer binding energies and vibrational frequencies from the *ab initio* work are given in table 7.1. Note that the spacing is the lattice spacing of the solid, or the internuclear distance in the dimer. The energy represents the sublimation energy at 0 K, including the zero point energy, for the bulk solid; for the dimer it is the dissociation energy of the molecule in its ground state. The frequency is the optical phonon or stretching frequency. The argument is that if one gets both bulk Si (Ge) and the dimer Si$_2$ (Ge$_2$) correct, then surfaces and small clusters, which are in between, must be more or less right. If tight binding schemes can bridge this gap, then large calculations can be done with more confidence. One may note from table 7.1 that the early *ab initio* LDA calculations tended to be overbound, sometimes by as much as 1 eV, but this improved over time. Many more details of tight binding methods in the context of surfaces are explained by Desjonquères & Spanjaard (1996); however, work is still proceeding on schemes which really can span the range of configurations which are encountered in molecules, in solids and at surfaces (Wang & Ho 1996, Lenosky *et al.* 1997, Turchi *et al.* 1998).

Some of the named authors have spent time in establishing principles by which such surfaces can be *understood*. This is possible because a large data base of solved structures now exists; one can therefore discuss trends, and the reasons for such trends. In particular, Duke has enunciated five principles in several articles, which can help us understand the following examples (Duke 1992, 1993, 1994, 1996). Zhang & Zunger (1996) and Kahn (1994, 1996) have looked at *structural motifs* which occur at III–V surfaces, regarding surfaces as special arrangements of these motifs. A useful point to note is that a Ga atom, being trivalent, would prefer sp^2 bonding, which has the 120° angle, but that the pentavalent As atom prefers s^2p^3 bonding, with an inter-bond angle of 94°. Atoms at the surface have some freedom to move in directions which change their bond angles, and do indeed move in directions consistent with the above arguments.

7.2   Case studies of reconstructed semiconductor surfaces

While studying this section, one needs to take enough time with a model or models to get as much of a three-dimensional 'feel' of the structures discussed. Two-dimensional cuts of various low index unreconstructed surfaces can be found in Zangwill (1988) and Lüth (1993/5) along with the corresponding 2D Brillouin zones. Not all of you will need to know all the details referred to: I have found during teaching this material that any one of these sections is suitable for elaboration via a mini-project on 'understanding surface reconstructions'.

7.2.1   GaAs(110), a charge-neutral surface

In the f.c.c. III–V semiconductors, (110) is the cleavage face which is charge-neutral, the surface plane containing equal numbers of Ga and As atoms. Figure 7.2 shows the top view of the unit cell (a), and two side views, the dashed lines indicating dangling bonds. The unrelaxed surface (b) has the form of a zig-zag chain As–Ga–As, though, as seen in the top view, the atoms are not in the same plane. This structure is (1×1), so it does not introduce any further diffraction spots; however, LEED and other experiments have shown convincingly that the surface relaxes as in diagram (c): the As atom moves outwards and the Ga moves inwards, corresponding to a rotation of the Ga–As bond away from the surface plane. LEED I–V intensity analysis has been used to show that best fits are obtained with a rotation of 29 ± 3°, with small shifts in the outer plane spacings, remarkably consistently across several III–V and even II–VI compounds. This large body of work has been reviewed by Chadi (1989), Duke (1992, 1993, 1994, 1996) and Kahn (1994, 1996). I do not give any details of 2-6 compound surfaces, nor of adsorbed atoms on any of these surfaces, but discussions of such structures and associated theoretical models are given by Mönch (1993) and Srivastava (1997).

The rotation is important for several aspects. First, the unrelaxed surface would be metallic. This arises because the cleavage results in one dangling bond per atom; thus the surface band is half-filled. The rotation results in a semiconducting surface, in

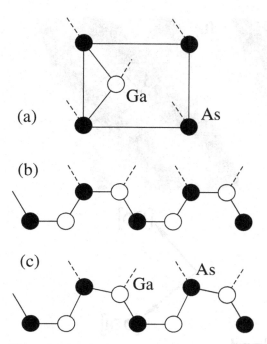

Figure 7.2. Surface structure and bond rotation in GaAs (110), with broken bonds shown as dotted lines: (a) top view of the unit cell, with [1$\bar{1}$0] vertical and [001] horizontal, the Ga and As forming a zig-zag chain; (b) side view of (a) without rearrangement; (c) with bond rotation of about 28°, so that Ga moves towards the planar sp^2 and As towards the pentagonal s^{2}p^3 configurations (after Chadi 1989, redrawn with permission).

which electrons are transferred to the outer As atom and away from the Ga. Second, and intimately related, the filled As state is lower in energy, near the valence band edge, and its environment and angles are closer to the s^{2}p^3 configuration. The unfilled Ga state moves up in energy, above the conduction band minimum, with its environment and angles closer to the sp^2 configuration. This is real cluster chemistry in action at the surface.

Finally, we can see that this means that the filled (valence band-like) and the empty (conduction band-like) surface states will have the same periodicity, but will be shifted in phase, to be located over the As and Ga atoms respectively. The amazing feat of visualizing this arrangement was first achieved by STM and spectroscopy in 1987, as shown in figure 7.3. Tunneling from the sample into the tip showed the filled As atom states, whereas reversing the sample bias showed up the unfilled Ga states. Suitably colored in red and blue, this made an impressive cover for *Physics Today* in January 1987; tunneling spectroscopy was then used to verify these assignments in detail (Feenstra *et al.* 1987). This work was also correlated with extensive previous work on UPS and surface band structure, some of which is described by Lüth, Mönch and Zangwill. More images are given by Wiesendanger (1994), and an update on STM/STS for studying semiconductor surfaces and surface states is given by Feenstra (1994).

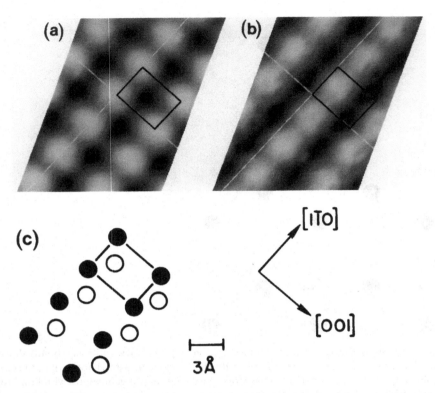

Figure 7.3. Constant current STM images of the GaAs(110) surface acquired at sample bias voltages (a) $+1.9$ V and (b) -1.9 V. Image (a) shows the unoccupied state images, dominated by the Ga sp^2 configuration, while (b) shows the filled states associated with the As s^{2}p^3 configuration; (c) the corresponding unit cell and crystallography (after Feenstra *et al.* 1987, reproduced with permission, and Wiesendanger 1994).

7.2.2   GaAs(111), a polar surface

There are many examples of polar semiconductor surfaces, but the archetype is GaAs (111). Viewed along the [111] direction we have layers: Ga As space Ga As space, so that along the [$\bar{1}\bar{1}\bar{1}$] direction is not the same, it is As Ga space .... This results from the lack of a center of symmetry in the GaAs lattice, ($\bar{4}$3m), not m3m as the normal f.c.c., or the diamond lattice.

If now the Ga layers are somewhat positive, and the As somewhat negative, then there are indeed alternating sheets of charge, as discussed in problem 7.3. Consider a test charge moving through this material. It will undergo a net (macroscopic) change of potential energy as it goes through the crystal. In fact this change is *HUGE!* We calculated in section 6.1.4 that a dipole layer consisting of 1 electron/atom separated by 1 Å caused a potential change of about 36 V; but this case has a dipole sheet of similar magnitude on each lattice plane, and gives rise to a really large dipole – of order 1 electron/atom times the thickness of the crystal. Anyway, this cannot be what happens in reality; nature does not like long range fields, which

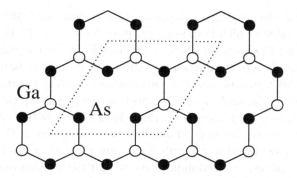

Figure 7.4. Top view of the 2×2 vacancy reconstruction of Ga-rich GaAs(111). The six-fold ring of atoms surrounding the corners of the unit cell consist of alternating three-fold coordinated Ga and As atoms, closely resembling the zig-zag chains of the (110) surface (after Chadi 1989, redrawn with permission).

store large amounts of energy. There must be an equal and opposite dipole due to the surfaces somehow.

The two opposite faces are referred to as Ga-rich (111A) or As-rich (111B), and they may well not have the stoichiometric composition. If they don't, they will carry a surface charge density (opposite on the two faces), which will produce a compensating long range dipole and hence no long range field. The most common solution is thought to be the 2×2 vacancy reconstruction, shown here in figure 7.4 for the Ga-rich surface. It is an interesting exercise to do the bond counting and show that it works out correctly (problem 7.3). You should also note the changes in bond angles, which take the Ga towards the sp^2, and the As towards the s^{2}p^3 configurations, which these elements would like. The As moves into the vacancy and towards five-fold coordination, and the Ga uses the extra space so created to move into the surface and to a more planar, three-fold configuration.

What is perhaps difficult to comprehend is the fact that the changes in *electronic* energy involved are so large, that they are sufficient to create *atomic* structural defects such as surface vacancies. In this case, we have removed one Ga atom in four; so the cost of this has to be about three Ga–As bonds, of order $3 \times 1.7 = 5.1$ eV per surface unit cell, the excess Ga typically existing in the form of small (liquid) droplets on the surface. But instead of the metallic surface, we have four filled As-derived states, gaining of order $4E_g \sim 5.6$ eV, where the energy gap of GaAs is $E_g = 1.42$ eV; we also have to pay for the bond (and other forms of elastic) distortion, but against that we get rid of the long range electric field completely. There are delicate balances involved, but the result is clear. The arguments in favor of vacancy formation at II–VI surfaces in such situations are even stronger because of larger band gaps and lower bond energies (Chadi 1989).

7.2.3 Si and Ge(111): why are they so different?

In section 1.4, we introduced the various reconstructions of Si(111), and the fact that the famous 7×7 structure was solved by a combination of STM, THEED and LEED.

The crucial breakthrough was the proposal of the dimer-adatom-stacking fault (DAS) model by Takayanagi *et al.* (1985) which built upon the prior STM and LEED work, and a detailed analysis of THEED intensities. Since the diffraction pattern contains 49 beams, a truly quantitative analysis of the diffraction pattern was thought to be impossible. But once this model had been articulated, detailed surface X-ray diffraction and LEED *I–V* analyses were successful, and the refinements lead to a very complete set of atomic positions in the structure (Robinson *et al.* 1988, Tong *et al.* 1988). The 7×7 structure is shown in figure 7.6; versions in color can be found via Appendix D.

This is the hallmark of a really extraordinarily successful piece of science: long fought for, but worth every penny. Understanding why we get these structures, and what are the competing structures, is equally fascinating. First, Si and Ge(111) are the lowest energy surfaces of these elements at low temperatures, but when we cleave the crystals at room temperature, we get a (2×1) reconstruction. This has been found to have a π-bonded chain structure; it is illustrated and discussed in detail by Lüth (1993/5). On annealing this structure to around 250 °C, it transforms *irreversibly* to the 7×7. The DAS structure is therefore more stable energetically; but it requires atom exchange, which is not possible at low temperatures. At 830 °C, the 7×7 pattern disappears, to be replaced *reversibly* by a simple 1×1 pattern. But Ge(111) has a quite different sequence: $c(2 \times 8)$ at room temperature, with a reversible transition to 1×1 at 300 °C.

What on earth is going on, you might well ask. More detective stories, good ones too; should the plot be spelled out, or should you be left to find out? Difficult question; the detailed history is a good topic for a mini-project during a course. Early work using a semi-empirical tight binding model showed that 7×7 was more stable than 1×1 by around 0.4 eV/(1×1) cell (Qian & Chadi 1987). But the 7×7 structure is only one of a family of DAS structures of the form $(2n + 1) \times (2n + 1)$; the smallest of these is 3×3. The elements of the 1×1 and 3×3 structure are shown in figure 7.5. When *ab initio* theorists first calculated the energy of DAS structures, they naturally started with this one (Payne 1987, Payne *et al.* 1989). The basic adatom unit is in a 2×2 arrangement, so that was another possible approach (Meade & Vanderbilt 1989).

There was then an enormous effort to calculate the energy of the 7×7 *ab initio*, a huge task, resulting in two groups publishing back to back in *Physical Review Letters* volume 68: Stich *et al.* (1992) on page 1351 from Cambridge, England, and Brommer *et al.* (1992) on page 1355, from Cambridge, MA. Both these groups showed that the 7×7, illustrated in figure 7.6, indeed has a lower energy than both the 3×3 and 5×5, and also by a margin of only 0.06 eV/(1×1)cell than the 2×1, very close to the 0.04 eV previously estimated by Qian & Chadi (1987) – especially considering the likely errors in the calculations. The values they quote for these energies are shown in table 7.2. To show that 7×7 is really the most stable structure, one should surely also calculate the 9×9 and 11×11 and show that the energy goes up: yes, but one must remember that these calculations were at the limit of massively parallel supercomputer technology. At the time of writing, such a calculation is definitely feasible; but is it now anyone's first priority to do it again, and be *really* careful? Probably not!

The stacking fault in the DAS structures enables dimers to form along the cell edges,

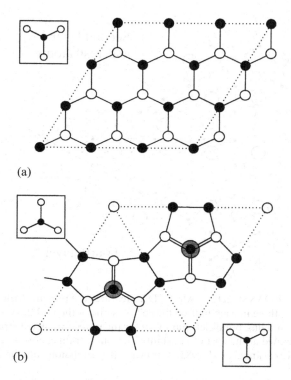

(a)

(b)

Figure 7.5. Simple Si(111) surface structures (a) bulk terminated, showing a 3×3 cell for comparison with (b) 3×3 DAS structure. In (a) the open and full circles show unrelaxed atoms of the first layer (three-fold) and second layer (four-fold) coordinated. In (b) the second layer atoms form the dimers along the (dotted) cell edges, which is coupled to the existence of the stacking fault in the lower left hand half of the cell (small boxes). The larger shaded circles are the adatoms, which are three-fold coordinated, each replacing three dangling bonds by one (after Chadi 1989, redrawn with permission).

and the ring at the corners at the intersection of cell edges. Without the stacking fault, we simply have the adatoms, which are arranged in a 2×2 array. The Ge(111) structure is thought to be based simply on these adatoms; within the cell there are two local geometries, subunits of 2×2 and $c2 \times 4$; together they make the larger $c2 \times 8$ reconstruction as determined by X-ray diffraction (Feidenhans'l et al. 1988); reviews of this technique plus many structural details are given by Feidenhans'l (1989) and Robinson & Tweet (1992). In case you think it is always easy for great scientists, it isn't; for example, Takayanagi & Tanashiro (1986) generalized their Si(111) 7×7 model to produce a model of Ge(111)$c2 \times 8$ based on dimer chains – too bad, wrong choice!

The high temperature 1×1 structure is often written '1×1', meaning 'we know it isn't really'; both Si and Ge are thought to form a disordered structure of mobile adatoms which may locally be in 2×2 or similar configurations. Diffuse scattering from these adatoms has been seen for both Si and Ge(111), e.g. using RHEED (Kohmoto & Ichimiya 1989) and medium energy ion scattering (MEIS) (Denier van der Gon et al. 1991). Similar structures are expected on Ge/Si mixtures, where Ge segregates to the

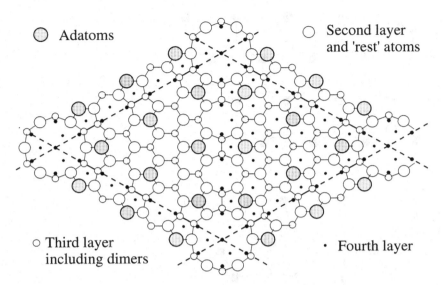

Figure 7.6. The equilibrium 7×7 DAS structure. All the DAS features are present, with 12 adatoms plus seven 'rest atoms', three in each half of the cell plus one in the middle of the corner hole. There are thus 19 dangling bonds left out of 49 for the unreconstructed structure. This reduction in energy is obtained at the cost of substantial internal strain energy, stretching over at least four layers (after Takayanagi *et al.* 1985, redrawn with permission, and later authors).

surface because the lower binding energy. These details are also fascinating and are discussed in section 7.3.3.

Are there any further checks on these models, and can we make sense of them? STM has been invaluable; adatoms were seen in the original pictures by Binnig *et al.* (1982), and subsequent work by many people showed up back bonds and other features of the electronic structure; i.e. one gets different pictures as a function of bias voltage, because different states are active. The most ambitious, yet relatively simple, attempt to understand the various structures is that by Vanderbilt (1987), where he tries to estimate the energy costs of the stacking fault (f) and of the corner holes (c), expressed as a ratio to the dimer (domain wall) energy. He then draws a phase diagram, shown in figure 7.7. This exhibits a series of DAS structures if f is small, which have increasing $(2n+1)$ periodicity as c increases. At larger values of f, the stacking fault is unfavorable, and there is a transition to an ordered adatom structure, notionally the $c2 \times 8$.

This simple diagram explains how Si and Ge could be close together on such a diagram, and yet have such different structures. It also explains (in the same sense) how the surface stress, quenching, or Ge addition to the surface can give rise to $5 \times 5, 9 \times 9$, and mixed surfaces. Beautiful STM pictures illustrating all these possibilities have been published by Yang & Williams (1994); one example is shown here in figure 7.8. The important point to note is that these different reconstructions do not have the same areal density of atoms; so the change from one structure to another requires a lot of

Table 7.2. *Calculated energies of reconstructed Si and Ge(111) surfaces (eV/1 ×1 cell)*

Material	1×1 relaxed	2×1 cleaved	2×2 adatom	3×3 DAS	5×5 DAS	7×7 DAS
Si(111)	1.39[b]	1.239[d]	1.12[b]	1.196[c]	1.168[c]	1.153[c], 1.179[d]
Ge(111)	1.15[a]		1.04[a]	0.88[a]		
	~1.34 <1.40[b]		~1.08 <1.20[b]			

References: (a) Payne (1987), Payne *et al.* (1989); (b) Meade & Vanderbilt (1989). The Ge calculations were not well converged, and should be lower than the Si values; the ~values assume that the changes on convergence are the same as for Si; (c) Stich *et al.* (1992); (d) Brommer *et al.* (1992). The only experimental values available date from 1960 and are 1.24 J·m⁻² for cleaved Si(111) (0.998 eV/1×1 cell) and 1.10 J·m⁻² for Ge(111) (0.952 eV/1×1 cell) (Kern *et al.* 1979); these values are not unreasonable, but they have unknown error bars.

adatom movement, and can nucleate 2D islands or pits; the microstructure thus becomes very complex, depending on the kinetics in detail.

7.2.4 Si, Ge and GaAs(001), steps and growth

The geometry of the basic 2×1 reconstruction of Si(001) was fully described in section 1.4. We need to recall the formation of the dimers, their organization into dimer rows (perpendicular to the dimers), and the correlation with surface steps (Chadi 1979, 1989, Griffith & Kochanski 1990). There has been much debate as to whether the 2×1 reconstruction is symmetric, or asymmetric; by now you will realize that this is the same question as whether the surface is metallic or semiconducting. A consensus has emerged that the Si dimer is asymmetric, but that the energies are so close that the dimer flips between two equivalent states – either the left-hand or the right-hand atom is up at any one time. At high temperature, this is like having a low frequency anharmonic vibrational mode; at low temperature, ordered arrays of up and down dimers can give various superstructures, such as p2×2 or c2×4.

There are a host of such calculations in the literature: one (Ramstad *et al.* 1995) gives the c4×2 as the lowest energy structure, and calculates by how much it is stable. The dimerization gives a large energy gain over the unreconstructed 1×1 structure, about 2 eV per dimer. The asymmetric dimer is favored by a further 0.2 eV; ordering these dimers into either the p2×2 or c4×2 gains a further 0.02 eV, and the eventual stability of the c4×2 is a mere 0.002 eV/dimer. It is not entirely clear whether we should believe this slender margin, but it is clear why the complex superstructures will not be stable at high temperatures. Finite temperature molecular dynamics simulations have mapped out the timescale on which the dimers flip between the two positions; an example which takes account of the actual p2×2 or c4×2 structure is shown in figure 7.9 (Shkrebtii *et al.* 1995). Here it can be seen that dimers flip, and also twist, causing

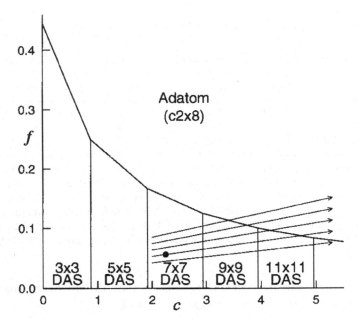

Figure 7.7. Model phase diagram of surface structures for the Si(111) surface; c and f are measures of the effective corner hole and stacking fault energies relative to the dimer (domain wall) energies. The arrows show possible trajectories that can occur by increasing the adatom density, and hence the chemical potential of Si which acts to compress the surface phase (after Vanderbilt 1987, and Yang & Williams 1994, reproduced with permisison).

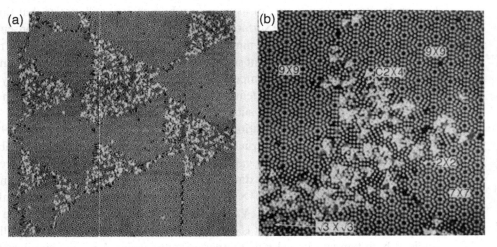

Figure 7.8. Coexistence of reconstructions on Si(111) (a) area approximately 100×100 nm^2 showing 7×7, 9×9 and regions with higher densities of adatoms; (b) atomically resolved image of a high adatom density area 43×43 nm^2 showing 7×7, 9×9 and other reconstructions where adatoms are seen individually (after Yang & Williams 1994).

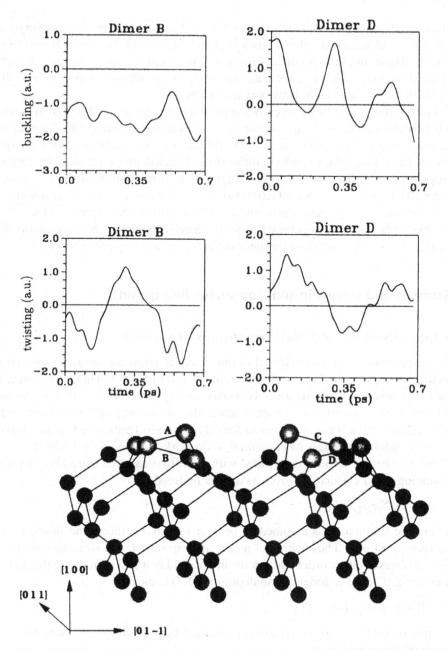

Figure 7.9. Dimer buckling (*z*-motion) and twisting (*y*-motion) on Si(001) revealed in a finite temperature molecular dynamics simulation at 900 K, showing two of the four dimers in a (4×2) cell (after Shkrebtii *et al.* 1995, reproduced with permission).

changes of structure, roughly every 0.5–1 ps. The amplitudes are suprisingly large, namely >0.05 nm up and down (buckling) and sideways (twist plus shift) ~ 0.03 nm at 900 K. These amplitudes correspond to tilt angles in the range 16–19°, though this value does vary somewhat between the different calculations (Krüger & Pollmann 1994, 1995, Ramstad *et al.* 1995, Srivastava 1997).

For Ge (001), and more recently Ge/Si (001) also, there has been great interest in whether these surfaces are also asymmetric, in what way, and in establishing the trends in bond angles (Tang & Freeman 1994). All these reconstructions reduce the symmetry of the surface, which results in diffusion and growth properties that are very anisotropic, and alternate across single height steps. The growth of devices based on Ge/Si(001), or GaAs on Si or Ge(001) gives rise to many fascinating problems. There is a vast literature on these topics, and even more unpublished empirical knowledge in the firms who make devices based on such materials. Some of these issues are aired in section 7.3; but the last word on these subjects is a long way in the future.

7.3   Stresses and strains in semiconductor film growth

7.3.1   Thermodynamic and elasticity studies of surfaces

In the previous section we referred to the effects of stress on surface reconstructions. Before proceeding further it is time to return to section 1.1.3, where it was noted that surface stress had the same units as surface energy, but that the stress σ_{ij}, sometimes written f_{ij}, is a second rank tensor, whereas the surface energy γ is a scalar quantity. This complication is the main reason for delaying consideration of surface stress until now. Somehow, the stress has potential to do work, but doesn't actually do so unless there is a surface strain ε_{ij} associated with it. In that case the stored internal energy, assuming linear elasticity, is given as in bulk material by

$$\Delta U = \tfrac{1}{2}\Sigma \sigma_{ij}\varepsilon_{ij}, \tag{7.3}$$

where the summation is assumed to be over repeated suffices, and in the case of a surface $i, j = 1$ or 2. The creation of a new surface, area A, in a strained solid proceeds in two stages in either order: we can strain the solid first and then create the surface, or vice versa. The work done, dW analogous to (1.3) is then

$$\mathrm{d}W = \mathrm{d}(\gamma A) = \gamma \mathrm{d}A + A\mathrm{d}\gamma. \tag{7.4}$$

Noting that $\mathrm{d}W = A\Sigma\sigma_{ij}\mathrm{d}\varepsilon_{ij}$ we can then deduce the relation for the individual components of the stress tensor

$$\sigma_{ij} = \gamma\delta_{ij} + \partial\gamma/\partial\varepsilon_{ij}, \tag{7.5}$$

since $\mathrm{d}A = A\delta_{ij}\mathrm{d}\varepsilon_{ij}$, where δ_{ij} is the Kronecker delta (i.e. 1 for $i=j$ and zero otherwise). If the surface has three-fold rotation symmetry or higher, then the off-diagonal terms of (7.5) are zero and the diagonal terms are equal, so the surface stress σ is isotropic in the surface plane.

The thermodynamic background leading to (7.5) is given by Cammarata (1994), who acknowledges that there has historically been much confusion on this topic. In addition, his paper describes calculations on the values of surface and interface stresses in particular materials. For example, noble metals are under *positive* surface stress if they have unreconstructed surfaces, because the surface atoms want to immerse themselves in a higher electron density to compensate for the loss of neighbors caused by the surface. The stress of the unreconstructed surfaces is calculated to be much larger for Au than for Ag; this is consistent, qualitatively at least, with the fact that Au(111) has the contracted 23×1 herringbone structure, and Au(001) the more close-packed 5×20 structures described in section 6.1.2, whereas Ag(111) and (001) are both unreconstructed (Needs *et al.* 1991). For semiconductors such as Si and Ge(111), the different reconstructions are calculated to have different surfaces stresses as well as energies, and indeed it is thought that the stability of the 7×7 reconstruction results from partial compensation of positive and negative stresses between different layers (Meade & Vanderbilt 1989). However, the point to remember is that the stability of these surfaces depends on the surface *energy*, not the stress; the existence of the stress is only a reason why the surface might want to adopt a different structure.

This situation changes if we apply a stress to the surface by external means; now work can be done by and on the surface, and the configuration of the surface may change in response to the applied stress. The 2×1 reconstruction on Si(001), discussed in detail in section 1.4.4, is only mirror (2mm) symmetric, and so the surface stress tensor is not isotropic. Since single-height steps are associated with a switch in domain orientation, there is a change in surface stress across each step, and this can be portrayed as a force monopole F_0 at each step, alternating in sign between S_A and S_B steps and numerically equal to the difference in stress tensor components $(\sigma_{//} - \sigma_{\perp})$. Calculations for Si(001) indicate that the value of $\sigma_{//}$ is positive parallel to the dimer bond direction, and $\sigma_{\perp}$ is negative in the direction perpendicular to it, thus parallel to the dimer rows. If the steps can move, F_0 couples to the external strain, work is done, and the equilibrium domain configuration of the surface changes.

The classic experiment was the observation of changes in domain population on the Si(001) surface at elevated temperature (~ 625 K) in response to bending a Si wafer, studied by LEED and STM by Webb and co-workers (Webb 1994) illustrated in figure 7.10. With a surface strain of only 0.1%, the domain population as observed by LEED half-order intensities was shifted from equal areas to more than a 90–10 distribution (Men *et al.* 1988); follow-up studies by STM showed not only this distribution of areas, but also the statistics of kinks along ledges (Swartzentruber *et al.* 1989, Webb *et al.* 1991). A model developed by Alerhand *et al.* (1990) was among the first to describe the elastic and entropic interactions between the steps, and to fit such experiments so that energies for the direct step–step interactions, and for kink energies on S_A and S_B steps could be extracted. We should note in passing that the original straining experiments were unsuccessful, since the steps cannot move at room temperature because of insufficient surface mobility. It is necessary that kinks can move, and that adatoms and/or ad-dimers can diffuse along and detach from steps for local equilibrium to be established.

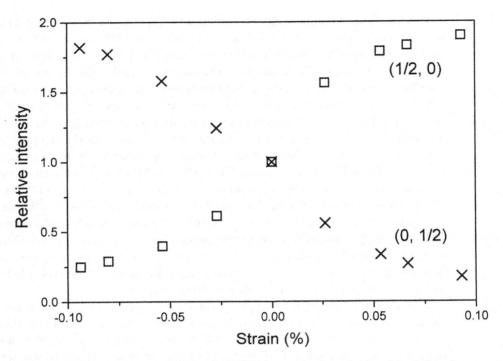

Figure 7.10. The asymmetry of half-order LEED intensities from 2×1 and 1×2 domains on Si(001) surfaces as a function of surface strain. The domain compressed along the dimer bond is favored (after Men *et al.* 1988, replotted with permission).

A very interesting development consisted of depositing sub-ML amounts of Ge on the Si(001) surface and repeating the same series of straining experiments (Wu & Lagally 1995). By this means it was shown that the surface stress anisotropy could be made to change sign as a function of Ge concentration, passing through zero at around $\theta = 0.8$. The anisotropy was measured for pure Si as $F_0 = 1.0 \pm 0.3$ eV/Ω, where Ω is the area per 1×1 surface unit cell. For 2ML Ge/Si (001) F_0 had decreased to -0.9 ± 0.3 eV/Ω. Comparison with calculations, including those by García & Northrup (1993), indicated that the main change is the strong decrease in $\sigma_{//}$ with increasing Ge coverage, with $\sigma_{\perp}$ staying essentially constant. There is clearly scope for studies on other surfaces with low symmetry such as (113), which is a possible substrate for quantum wires and grooves (Knall & Pethica 1992, Baski *et al.* 1997).

All these experiments should be seen in the context of the limited amount of data which exist on the equilibrium form of Si and Ge crystals. These have either been obtained on crystals of a few micrometers in size at relatively high temperatures $\sim 1050\,^\circ$C, which parallel the data on metals explored in sections 1.2.3 and 6.1.4 (Bermond *et al.* 1995, Suzuki *et al.* 1995), or via the formation of ~ 10 nm size helium-filled voids after ion implantation by annealing to temperatures between 600–800 $^\circ$C (Follstaedt 1993, Eaglesham *et al.* 1993). For silicon, the latter authors agree that γ_{111} is the minimum in the free energy, and estimate $\gamma_{001}/\gamma_{111} = 1.09 \pm 0.07$, and $\gamma_{110}/\gamma_{111} =$

1.07 ± 0.03 (Follstaedt 1993); Eaglesham *et al.* (1993) give $\gamma_{001}/\gamma_{111}$ in the range 1.11 ± 0.03, $\gamma_{113}/\gamma_{111} \sim 1.12$ and $\gamma_{110}/\gamma_{111} \sim 1.16$ at $\sim 800\,°C$.

However, at the higher temperature of $1050\,°C$ (1323 K, which is above the 7×7 to 1×1 transition), Bermond *et al.* (1995) find a smaller anisotropy ($\sim 4\%$), as one might expect, but they also find that the ordering has changed to $\gamma_{111} \gtrsim \gamma_{110} > \gamma_{113} > \gamma_{001}$. A curious feature is that the (001) face does not have a true cusp in the equilibrium form, which consists largely of $\{111\}$ and $\{113\}$ facets and rounded regions. This is probably due to the long range stress field, and the associated strain energy, due to the (2×1) surface domains, which results in the spontaneous formation of steps discussed by Alerhand *et al.* (1990) and observed by using LEEM by Tromp & Reuter (1992, 1993). A reminder that impurities can be influential in such measurements was shown by the (reversible) segregation of 0.3% carbon to create extra facets, which are not in the clean equilibrium form.

Many studies on facetting transitions on Si and Ge surfaces vicinal to (111), including the effect of the 7×7 to 1×1 transition for Si, have been made using LEED, LEEM and STM by Bartelt, Williams and co-workers, as reviewed by Williams *et al.* (1993), Williams (1994) and Jeong & Williams (1999). All this work suggests that at the higher temperatures the free energy differences between the various faces $\{hkl\}$ are quite small, due to a subtle balance of substantial energetic and entropic factors. Large amplitude motion of steps on Si(111) has been observed also by REM (Pimpinelli *et al.* 1993, Suzuki *et al.* 1995), the analysis of which also implies a sizable adatom population on the terraces to mediate the step movement.

The surface entropies involved in these transitions are unknown, but microscopy is beginning to visualize directly some configurations (and motion) involved on the timescale of 1 s and upwards. Molecular dynamics studies, as illustrated here by figure 7.9, can be used to estimate the entropy associated with configuration and motion on the pico-second time scale. Another point to note is that the latent heats of melting of Si and Ge are almost a factor of 2 higher than that of close-packed metals which melt at similar temperatures. The interrelation of these apparently isolated facts can be explored further via project 7.4; a link is the angular nature of sp³ bonding in semiconductors, and its (partial) disappearance in the liquid state.

7.3.2 Growth on Si(001)

The classic substrate for semiconductor growth is Si(001), since this has the simplest structure, and is used for growth of most practical devices. Typically device growers use a surface which is tilted off-axis by about 2–4°, to form a vicinal surface which contains a regular step array. The reason for this is to promote layer by layer growth, sometimes referred to as step-flow, and to suppress random nucleation on terraces; nucleation is typically not wanted, because it increases the possibility of incorporating defects (e.g. threading dislocations) which have bad electrical properties. Thus the fact that higher miscut angles favor double-height steps (Chadi 1987, Tong & Bennett 1991, de Miguel *et al.* 1991, Men 1994) is of major importance for growing compound semiconductors such as GaAs (on Si), since single steps produce anti-phase boundaries, across which Ga and As are misplaced.

Figure 7.11. STM images of Si/Si(100) showing diffusional anisotropy of adatoms, and the effects of S_A and S_B steps, after 0.1 ML deposition at $R = 0.15$ ML/min, $T =$ (a) 563 K, (b) 593 K. The surface steps down from upper left to lower right. In (a) anisotropic islands can be seen on all terraces; the underlying dimer rows are orthogonal to these islands. In (b) diffusion is more rapid, so denuded zones are observed only on (2×1) terraces (after Mo & Lagally 1991, reproduced with permission).

Moreover, materials grown on such a substrate typically have a sizable misfit, which may be accommodated initially by strain, but eventually by missing dimer rows at the atomic level, or by dislocations. Most growers search for low misfit systems, so that dislocation introduction is delayed beyond the so-called critical thickness (Matthews 1975, Matthews & Blakeslee 1974, 1975, 1976, People & Bean, 1985, 1986). The bonding changes during semiconductor growth are extremely complex, since any surface reconstruction has to be undone in order for growth to proceed; at low temperatures there is the possibility of creation of many, largely unwanted, metastable structures. On the other hand if growth of complex multilayer structures, such as multiple quantum wells (MQWs) with different compositions, is conducted at too high temperatures then they will be degraded by surface segregation and interdiffusion. Device engineers are always treading a fine line in trying to grow crystals at the lowest practicable temperature – reducing the 'thermal budget'; many of the more technical methods described in section 2.5 have been introduced solely for this reason.

There have been many studies of Si/Si(001) growth primarily using STM, in addition to spot profile analysis using LEED (Heun et al. 1991, Falta & Henzler 1992) and RHEED (Tong & Bennett 1991). Large area STM pictures such as figure 7.11 are very helpful (Mo & Lagally 1991, Liu & Lagally 1997); one can identify both S_A and S_B single height steps which have very different roughness due to different edge energies and the anisotropy of diffusion.

In one detailed study, the nucleation density N of 2D islands on the terraces was observed as a function of R and T, and an analysis similar to that of section 5.2 performed, but taking into account the diffusion anisotropy, and the anisotropy in binding at the edges of the monolayer islands (Mo *et al.* 1991, 1992). The low temperature region of this $N(R, T)$ data, with a slope of 0.165 eV, is consistent with a critical nucleus size $i = 1$, and a diffusion energy, in the 'easy' direction parallel to the dimer rows, $E_d = 0.67 \pm 0.08$ eV. At higher temperatures, a transition to a higher critical nucleus size was observed, probably involving the breakup, and coarsening of, larger clusters into (stable) dimers, via dimer motion. In the initial papers it was not entirely clear what mechanism (e.g. adatom or ad-dimer motion) was actually being discussed; further work showed that adatoms typically move too fast to be observed directly by STM, but that for $T < 500$ K, adatom motion is responsible for nucleation. Subsequently, many papers have been published on alternate diffusion mechanisms and their diffusion energies (Milman *et al.* 1996, Borovsky *et al.* 1997, Liu & Lagally 1997); some of these data are presented in table 7.3.

In addition to these observations of nucleation on the (001) terraces, the expected nucleus-free, or denuded, zones next to steps are seen in figure 7.11. In particular, the terraces to show denuded zones at lowest temperature have the (2×1) reconstruction, where the fast diffusion direction is towards the steps. A very elegant technique has shown the path of individual dimers can be tracked by an STM tip whose position is locked onto particular ad-dimers laterally (Swartzentruber 1996). This technique reveals preferred paths for the migrating dimers, and can also observe dimer rotation directly. The measurements, made as a function as both temperature and the field produced by the tip, coupled with detailed quantum calculations, have shown that dimer rotation has an activation energy around 0.7 eV, and that dimer diffusion takes place at somewhat higher temperatures with an activation energy just less than 1 eV (Swartzentruber *et al.* 1996). The work of Borovsky *et al.* (1997, 1999) has extended these measurements to include dimer diffusion both along and across the troughs; one should note that all these measurements are made over limited ranges of temperature, where STM observations are sensitive to particular activation energies on experimentally accessible timescales.

Another piece of the jigsaw is provided by LEEM measurements of the ad-dimer concentration as a function of temperature. Patterned substrates were first held between $1000 < T < 1300$ K and then quenched so that the ad-dimers form islands (Tanaka *et al.* 1997, Tromp & Mankos 1998). The total area, and hence the number of dimers contained in the ML islands was measured. The data are shown in figure 7.12, from which the activation energy 0.35 ± 0.05 eV was deduced for the formation of ad-dimers from the reconstructed Si(001) surface. It is clear from this low value that almost all of the energy gained during deposition is due to condensation into ad-dimers, and that relatively little is left to encourage the ad-dimers to incorporate into the growing (reconstructed) crystal. This makes it understandable that the critical nucleus size at normal growth temperatures has been found to be rather large, so that growth is typically quite close to 2D equilibrium, and the thermal population of dimers cannot be neglected (Theis & Tromp 1996). However, because of the large adsorption energies, equilibrium with the 3D vapor is far from being maintained. We can see that this general scheme is completely consistent with the range of energy values for Si and Ge binding and diffusion on Si(001) collected in table 7.3. Although there are gaps in

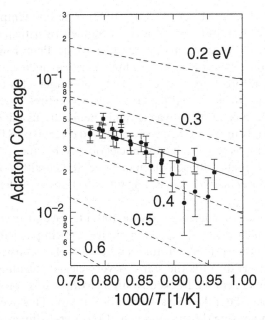

Figure 7.12. Concentration of Si ad-dimers on Si(001) at temperatures between 1000 and 1300 K, measured in quenching experiments (after Tromp & Mankos 1998, reproduced with permission). See text for discussion.

this table, one can expect these to be filled in quite rapidly over the next few years. Further thoughts along these lines can be explored via project 7.4.

There is a comparably detailed and complex history of the study of steps on Si(001) and their manipulation by external and internal stresses (Webb 1994, Cho et al. 1996); this work has been important in understanding the energies and stress fields of steps, and the interaction of steps with the 2×1 and 1×2 reconstructions. Most recent interest has centered on the role of steps in relation to incorporation of adatoms and dimers, which has been studied experimentally and theoretically both at the atomic (Roland & Gilmer 1992, Zhang et al. 1995, Swartzentruber 1997) and mesoscopic (Tsao et al. 1989, Swartzentruber et al. 1990, Zandvliet et al. 1995) scale.

For example, anisotropy in denuded zones on a single terrace, and the shapes of growing islands have been analyzed to show that S_B steps (the rough ones in figure 7.11) are at least 10 times better sinks for adatoms than the smooth S_A steps (Liu & Lagally 1997). In a few cases the energies of steps have been measured on low index faces, either directly via observations of step roughening (Swartzentruber et al. 1990) or derived from the equilibrium form (Eaglesham et al. 1993, Williams et al. 1993, Bermond et al. 1995). The former work estimated that S_B steps[2] have an energy per unit length $\beta = 0.09 \pm 0.01$ eV/a, whereas S_A steps have the much lower energy of $\beta = 0.028 \pm 0.002$ eV/a. This means there can be unstable step orientations where the step stiffness $(\beta + d^2\beta/d\theta^2)$

---

[2] The repeat distance along the step is a, in the same sense as used for the ledge energy e_1 in section 1.2.2.

Table 7.3. *Calculated and experimental surface and diffusion energies of Si and Ge adatoms on Si(001) substrates (eV). Adatom diffusion is either parallel (//) or perpendicular ($\perp$) to the dimer rows, and dimer diffusion has also been measured in the troughs (t) between the dimer rows*

Adatom/ substrate	γ_{100} eV/1×1	Adatom diffusion	Dimer rotation	Dimer diffusion	Dimer formation
Si/Si(001)	1.39±0.1[a,f]	0.67±0.08[b] (//) ~1.0[b] ($\perp$)	0.70±0.08[c]	0.94±0.09[d] (//) 1.09±0.05[g] (//) 1.27±0.08[g] (t) 1.22–1.26[h] (t) 1.36±0.06[g] ($\perp$)	0.35±0.05[e]
Ge/Si(001)		0.62[f] (//) 0.95±0.1[f] ($\perp$)			

References: (a) Northrup (1993); (b) Mo *et al.* (1991, 1992); (c) Swartzentruber *et al.* (1996); (d) Swartzentruber (1996); (e) Tromp & Mankos (1998); (f) Milman *et al.* (1994, 1996); (g) Borovsky *et al.* (1997, 1999); (h) Lee *et al.* (1999). References (a), (f) and (h) are calculations; other entries are experimental observations.

is negative, for the same reasons as there can be unstable facet orientations via equation (1.11).

From my own research perspective, that of trying to understand atomic processes at surfaces and determining the energies involved, nucleation and growth on these surfaces is intrinsically rather complicated. In principle, the dimer reconstruction has to be broken and reformed as each layer is grown, so there can be nucleation barriers at many stages of growth. At normal growth temperatures (400–650 °C), dimerization is not the rate-limiting step. However, in a complex system, experiments which cover a large range of the temperature or deposition rate variables must take into account the possibility that different atomic mechanisms may well become important under different conditions. Semiconductor surfaces have this feature in common with molecular biology: the problem of *rugged energy landscapes*, including multiple energy minima and reaction pathways. If this is true for the simplest semiconductor system, it may color how we think about the more complex processes which are discussed later.

7.3.3 Strained layer epitaxy: Ge/Si(001) and Si/Ge(001)

Ge and Si have the same structure, differing only in the lattice parameter by 4.2%, with the Ge slightly less strongly bound as seen in tables 1.1 and 7.1. Thus, the surface energy of Ge is expected to be lower than that of Si, and deposition of Ge will first occur in the form of layers. However, growth beyond the first few ML will build up substantial strain due to the mismatch, and after a certain thickness, the Ge prefers to grow as islands in which (if the islands are large enough) the strain has been relieved by misfit dislocations. The route to this state is quite complicated, but can be understood qualitatively by reference to figure 5.3(a). The equilibrium Ge layer thickness has been measured, after annealing, to be 3ML (Copel *et al.* 1989). But it is possible to grow much thicker coherent layers kinetically, or by using a surfactant; the first islands to form are also coherent with the underlying layers, not dislocated (Eaglesham & Cerullo 1990, Krishnamurthy *et al.* 1991, Williams *et al.* 1991).

The growth of this system, and the inverse Si/Ge(001), and the growth of SiGe alloys for practical devices are sufficiently important topics to warrant reprint collections, review articles and book chapters of their own (Stoneham & Jain 1995, Whall & Parker 1998, Hull & Stach 1999). For practical strained layer devices, there is a strong interest in suppressing island formation, which is practicable when alloys with low enough Ge content are used, or when alternating Si and Ge layers are thin enough. In the first few MLs $2 \times n$ reconstructions, with $n \sim 8$–12, are observed when monitoring the growth of Ge/Si by RHEED (Köhler *et al.* 1992). STM has shown that these structures consist of rows of dimer vacancies (Chen *et al.* 1994) which both relieve and respond to surface stresses.

But the growth process of most interest is the evolution of the islands, in competition with further growth of layers, and the instabilities which result at relatively high Ge content, or in the limit using pure Ge and Si layers. Here a large literature has been created, studying island densities and size distributions. Bimodal size distributions, some of them quite narrow can be created, which themselves may be of interest as quantum dot structures. As seen in TEM pictures, taken *ex situ* after UHV preparation, the smaller Ge islands are strongly strained as shown in figure 7.13(a). The strong black–white contrast is due to the bending of the substrate (Si) lattice caused by the Ge island, and indicates a radial strain, which also has a component normal to the substrate. This strain is relieved somewhat in the dislocated islands, which rapidly grow much larger. An individual dislocated island, observed in UHV SEM and UHV STEM is shown in figure 7.13(b) and (c). The higher secondary electron contrast from the ridges shows up the facetting in (b); moiré patterns in (c) indicate the presence of misfit dislocations between the island and the substrate.

The facets have been subsequently characterized principally by AFM and STM, and various shape transitions identified, both with and without surfactants (Horn-von Hogen *et al.* 1993, Floro *et al.* 1997, 1998), and by *in situ* TEM (Ross *et al.* 1998). At deposition temperatures above 500 °C, where surface diffusion is rapid, the size to which these coherent islands grow is markedly dependent on the presence of other sinks within the diffusion distance. Dislocated islands can be nucleated, preferentially from the larger coherent islands, or at impurity particles; once nucleated these islands form the strongest sinks, they grow rapidly and the supersaturation in the ($\theta > 3$ ML) Ge layer reduces. At a temperature of 500 °C, diffusion distances are of order 5 μm, whereas below 400 °C this figure drops below 0.5 μm. Assuming that the dimer energetics are similar on Ge to that on Si(001), then all these rearrangements on the surface are occurring via a substantial sea of migrating ad-dimers.

Similar effects are seen when Ge films, grown at room temperature to thicknesses above 3 ML are annealed at comparable temperatures, although the detailed

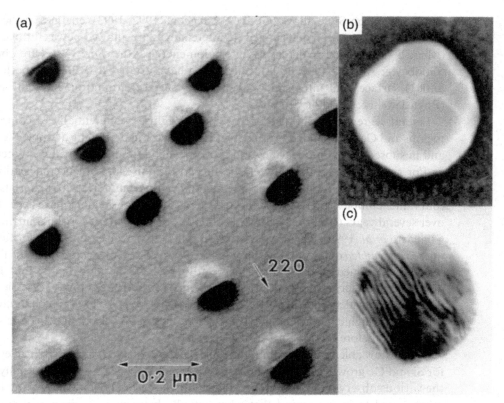

Figure 7.13. Island formation in vicinal Ge/Si(001). *Ex situ* bright field TEM image, showing (a) coherent islands. The strong black–white contrast parallel to the reflection $g = 220$ indicates a radial dilatational strain field; (b) UHV SEM and (c) UHV STEM images of a single dislocated island, showing (b) facets and (c) moiré fringes indicative of misfit dislocations (from Krishnamurthy *et al.* 1991, reproduced with permisison).

mechanisms and diffusion coefficients will be different. Initially, there are no large (> 10 nm radius) islands, but as annealing proceeds the bigger islands grow rapidly, while the size distribution of the smaller islands (< 10 nm radius) stays constant. This evidence suggests that the material for the rapid growth of the dislocated islands occurs primarily from the supersaturated layer rather than from the coherent islands, and in particular, that their strain fields are effective in keeping out migrating adatoms and/or dimers (Krishnamurthy *et al.* 1991, Drucker 1993, Tersoff *et al.* 1996).

Since Ge is less strongly bound than Si, growth of SiGe alloys leads to Ge segregation at the surface. This effect has been studied by several authors (Godbey & Ancona 1992, 1993, Li *et al.* 1995), and simple kinetic models have been developed to explain these results in terms of a segregation energy, E_s or E_{seg}. Segregation proceeds by an atomistic mechanism which is confined to essentially the surface layer, as diffusion within the bulk is quite negligible at typical growth temperatures; but within a two-layer model, segregation is almost complete for thin Si layers on Ge at $T \geq 500\,°C$. From the surface composition, as measured by AES on thin Si/Ge/Si layers (Li *et al.* 1995) or XPS work on SiGe alloys (Godbey & Ancona 1992, 1993), energies in the range 0.24–0.28 eV have been determined. If interchanges are allowed between three layers (where E_s is the sum of the layer segregation energies) the same values are obtained (Godbey & Ancona 1997, 1998). Calculation has retrieved $E_s \sim 0.25$ eV, both at the Si(001) surface, and interestingly also at the 'surface' around an internal vacancy (Boguslawski & Bernholc 1999).

Lateral segregation can also occur, since Si diffuses preferentially to compressed regions and Ge to expanded regions. This is one factor in producing a 'rippled' surface, and in subsequent non-linearities and growth instabilities in Ge–Si alloys. When dots and/or ripples interact there are elastic effects, and this can influence nucleation and subsequent growth. As such issues are potentially important for device materials, a full (but rather complex) literature has been created, and various regimes have been studied over several years (Cullis *et al.* 1992, Jesson *et al.* 1996, Deng & Krishnamurthy 1998, Chaparro *et al.* 1999). These types of effect, plus the perceived potential for fabricating *self-assembled* quantum dots, have sparked a great deal of related theoretical activity, and continued discussion of the relative role of thermodynamic and kinetic argument in understanding the structures formed (Tersoff & LeGoues 1994, Shchukin *et al.* 1995, Daruka & Barabási 1997, Medeiros-Ribeiro *et al.* 1998). Such discussions often go through complicated gyrations before they get resolved, but this one should get clarified before too long if we all keep at it!

We are now able to begin compiling tables of experimental and theoretical energies for Si and Ge growth systems, as in table 7.3, which one believes will eventually make the various observations comprehensible within a reasonably unified picture. But we always need to bear in mind that what actually happens in a given experiment or growth procedure may be a subtle combination of thermodynamic and kinetic effects, in which the competing effects of strain, adatom/dimer mobility and binding, surface and lateral segregation, facetting and coarsening play important parts. The subtleties of the transitions and the many competing structures, should make one rather wary of supposed clear-cut proofs of particular mechanisms; this is an area where metastabilty is very important, and where there are many routes to the supposedly final structures.

7.3.4   Growth of compound semiconductors

The properties of compound semiconductors such as GaAs, or more recently group III nitrides, grown by MBE and other techniques, are sufficiently important to have whole books devoted to the topic (Tsao 1993, Gil 1998). Often there is interest in growing such epitaxial layers on Si or GaAs substrates, in order to incorporate III–V, II–VI or IV–VI features with mature Si and GaAs-based device technology. Like Si and Ge, GaAs(001) surfaces also exhibit such higher order vacancy line structures, such as the 2×4 and 6×4 and various centered arrangements (Pashley *et al.* 1988, Chadi 1989, Biegelsen *et al.* 1990). Questions of layer growth versus nucleation on terraces have been addressed, as well as alloy segregation and pattern formation at steps (Arthur 1994, Gossard 1994, Joyce *et al.* 1994). The atomic (diffusion–reaction–incorporation) mechanisms are complicated, but calculations have developed to the point where the energies associated with some of the processes can be studied in some detail (Madhukar & Ghaisas 1988, Krishnamurthy *et al.* 1994, Kley *et al.* 1997).

Given the complexity of III–V chemistry, it is remarkable that the resulting microstructures of GaAs are rather similar to the elemental deposits described in the previous section. The main lesson to draw is that, in the presence of an As (i.e. group V) overpressure, the rate-limiting processes on the surface are typically associated with the incorporation of Ga (i.e. the group III cation). There are, of course, many subtle effects, particularly in relation to the incorporation of dopants. In general, II–VI and IV–VI compounds are prepared from compound sources, and then stoichiometry is less of an issue. However, in these cases, dopant incorporation presents particular challenges, as described by Han *et al.* (1999) and Springholz *et al.* (1999).

Several groups have monitored the growth of GaAs *in situ* using RHEED oscillations and a variety of light scattering techniques, such as ellipsometry or reflectance differential spectrometry. Of particular interest in the present context are those studies which correlate surface and step structures observed by a microscopic technique, typically STM or AFM, with the real time monitoring technique, such as RHEED or an optical technique. Nucleation of 2D islands and subsequent (rough) growth has been observed on wide terraces, for all the materials discussed in this section including GaAs, and the measured roughness correlated with the diffraction intensity (Johnson *et al.* 1993, 1994, Sudijono *et al.* 1993).

This roughness is thought to be caused by the Ehrlich–Schwoebel barrier, just as in the elemental case, though it is possible that selective adsorption at steps could also play an important role. In model computations, one can increase the strength of this barrier to the point that straight steps become wavy; for larger barriers, well developed mounds are seen. This is a fascinating example of pattern formation or self organization; in effect, the surface rearranges itself so that it (just) creates conditions for step flow, as illustrated in figure 7.14. There are also discussions in the literature of the role of adatom and ad-dimer concentrations in setting the chemical potential during growth (Northrup 1989, Tersoff *et al.* 1997) which parallel those discussed in the previous section for the silicon and germanium systems, but may contain further complexities yet to be explored. These compound systems are of particular interest for quantum dot structures.

One set of studies of GaAs(001) growth, combining RHEED oscillations and more recently STM observations, has been pursued over several years (Shitara *et al.* 1992, Smilauer & Vvedensky 1993, Joyce *et al.* 1994, Itoh *et al.* 1998). These studies are very interesting in respect of the relationship between model calculations, simulations and the growth of real materials. The presumption is that the RHEED intensity, measured at a particular well-chosen glancing incident angle, is most affected by the roughness of the surface, and thereby measures the step density. This is consistent with the ML oscillation period, and if true, would enable finer details of the waveform to be interpreted; however, this is not the key point. Most informative comparisons have been with studies on vicinal surfaces, in the relatively narrow temperature region where the transition from 2D island nucleation on the terraces to step flow takes place. Interest

Figure 7.14. (a) STM image of a GaAs buffer layer; (b) after termination of growth at the fourth RHEED maximum. Scan range for (a) and (b) 200×200 nm^2; (c) and (d) Monte Carlo calculation after 50 layers deposition, original scale 200×200 sites^2. Vicinal surface with slope $=0.1$ in (c), showing wavy steps, and on-axis surface in (d), showing large mounds (adapted from Johnson *et al.* 1993, with permisison).

was focused on the amplitude and phase of the initial transient which establishes the ML oscillation period, and in the relaxation to the smoother surface after the flux is switched off.

A set of experiment–model comparisons is given in figure 7.15, showing essentially perfect agreement with a conceptually simple solid-on-solid model containing no more than three important energy parameters. The key ingredients are: (1) the initial transient relaxation is caused by 2D nucleation and initial growth which establishes surface roughness at the ML level. The oscillations persist if the surface regains its smoothness each ML, but die out if growth is spread over several MLs, i.e. reaching a steady-state distribution. (2) The relaxation at the end of deposition has two components. These

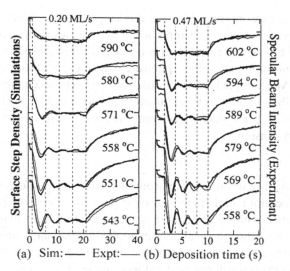

Figure 7.15. Experimental RHEED intensities during growth of GaAs(001) layers (miscut 2° towards [010], Shitara *et al.* 1992) in comparison with KMC simulations at a Ga flux rate of (a) 0.20 and (b) 0.47 ML/s (after Smilauer & Vvedensky 1993, reproduced with permission).

are a fast relaxation which is associated with processes taking place on the same level (island edge smoothing, loss of adatoms, some coarsening between islands) and a slow relaxation which requires mass transport between layers (long range diffusion including the ES barrier).

There have been several attempts to break down these processes into elemental steps, but they have not been without problems. From the description given here and from the discussion in chapter 5, it is clear that each process observed may well be composite. Moreover, we might well expect that any mechanisms deduced are only valid in a relatively narrow range of temperature and flux, so that the model cannot be used uncritically in other situations. In particular, the observed transition from 2D nucleation to step flow is remarkably sharp, which means that rather high activation energies would result from a direct comparison of the model with experiment. However, examination of the surface *ex situ* by STM shows that this transition does *not* imply that 2D nuclei are not produced at the higher temperatures, just that they are rapidly swept up by the steps. Models of the 2D nucleation process which take account of the need to reform the 2×4 structure as each layer is added give realism to the growth simulations, at the expense of several extra parameters (Itoh *et al.* 1998). But now we are beginning to see what specific features of the bonding are behind the growth of GaAs(001)! This is important and fascinating, but is rather a long way from the primary uses of RHEED oscillations, namely to count monolayers, and to provide a qualitative measure of surface and thin film quality.

The mathematics of step flow and mound formation is itself very interesting, both with and without impurities (Kandel & Weeks 1995, Siegert & Plischke 1996, Orme & Orr 1997). Models of multilayer growth inhabit a region where, although atomistic processes can be important in determining outcomes, it is simply not practical to follow them through many layer growth cycles. The approach is to take mesoscopic averages (along steps, or over several layers) and formulate (non-linear) differential equations to describe roughening and smoothing and other quantities which can be followed from initial starting conditions. An example is the work of Kardar, Parisi & Zhang (1986), whose KPZ equation describes differences from the average thickness h of the layer by a diffusion-like equation, with a nonlinear term proportional to the square of the surface slope, and an added constant, i.e.

$$\partial h / \partial t = \nu \nabla^2 h + 0.5\lambda \, (\nabla h)^2 + \eta. \tag{7.6}$$

The parameter ν describes surface tension which promotes smoothing, λ describes the fact that the growth front does not remain planar, and the parameter $\eta (\mathbf{x}, t)$ represents the noise involved in deposition which is random in space and time. Thus solving such equations becomes a series of challenging exercises in applied mathematics and statistical mechanics, which have been extensively described elsewhere (e.g. Krug & Spohn 1992, Barabási & Stanley 1995) and will not be developed further here. Many of these models are, as may be imagined, quite remote from the day to day concerns of the growers of actual devices. This does not mean, of course, that they do not have long, or even medium term significance.

Finally, in concluding this section, we should note that there is an enormous technologically driven thrust behind studies of these device growth processes, which we are only dimly seeing in the more scientifically oriented papers which have been highlighted in this section. An example is the effort to develop blue lasers and high power, high temperature electronics generally, based on the growth of wide band gap semiconductors such as nitrides, e.g. GaN with a band gap of 3.42 eV, and closely related ternary and quaternary compound multilayers. A particular success is the InGaN blue laser diode pioneered by Nakamura. These efforts are reviewed in several places (e.g. Ponce & Bour 1997, Ambacher 1998, Gil 1998, Nakamura 1998, Hauenstein 1999). It is clear from these accounts that there is much excitement in the field, and that control of growth and doping are central issues which give a lot of difficulty. Some of the background material needed to understand device issues involving thin film and surface processes is given in the next chapter.

Further reading for chapter 7

Ashcroft, N.W. & N.D. Mermin (1976) *Solid State Physics* (Saunders) chapters 10 and 28.

Davies, J.H. (1998) *The Physics of Low-dimensional Semiconductors: an Introduction* (Cambridge University Press), chapters 1–3.

Desjonquères, M.C. & D. Spanjaard (1996) *Concepts in Surface Physics* (Springer), chapter 5.

Gil, B. (Ed.) (1998) *Group III Nitride Semiconductor Compounds* (Oxford University Press).

Kelly, M.J. (1995) *Low-dimensional Semiconductors* (Oxford University Press) chapters 1–3.

Liu, W.K. & M.B. Santos (Eds.) (1999) *Thin Films: Heteroepitaxial Systems* (World Scientific).

Lüth, H. (1993/5) *Surfaces and Interfaces of Solid Surfaces* (2nd/3rd Edn, Springer) chapter 6.

Mönch, W. (1993) *Semiconductor Surfaces and Interfaces* (Springer) chapters 7–18.

Pettifor, D.G. (1995) *Bonding and Structure of Molecules and Solids* (Oxford University Press) chapters 3 and 7.

Sutton, A.P. (1994) *Electronic Structure of Materials* (Oxford University Press) chapters 6 and 11.

Sutton, A.P. & R.W. Balluffi (1995) *Interfaces in Crystalline Materials* (Oxford University Press) chapter 3.

Tsao, J.Y. (1993) *Materials Fundamentals of Molecular Beam Epitaxy* (Academic).

Yu, P.Y. & M. Cardona (1996) *Fundamentals of Semiconductors: Physics and Materials Properties* (Springer) chapter 2.

Zangwill, A. (1988) *Physics at Surfaces* (Cambridge University Press) chapter 4.

Problems and projects for chapter 7

Problem 7.1. Band structures of Si, Ge and GaAs

Look up the calculated band structures of bulk Si, Ge and GaAs, and explain the meaning of the following terms in relation to these three solids:

(a) the existence of a *direct* versus an *indirect* band gap, and the position of the conduction band minimum;

(b) spin-orbit splitting and the differences between *light* and *heavy* holes;

(c) the removal of *degeneracy* by stress in compressed thin films, e.g. of Ge on Si(001).

Project 7.2. Band structures in tight binding models

Tight-binding pseudopotential models of tetrahedral semiconductors such as Si or Ge treat the valence s- and p-electrons as moving in the potential field of the nuclei plus closed shell electrons. Consult selected references for section 7.1, and show one of more of the following.

(a) That eight electrons are required to describe the system, resulting in the need to diagonalize an 8×8 matrix to solve for the band structure as a function of the wave vector $\mathbf{k}$.

(b) That a possible approach to this problem is to use sp^3 hybrid states as the basis set, formed by linear combinations of 1 s- and 3 p-electrons as in equation (7.1). Find the relations between the overlap (or hopping) integrals expressed in the s, p_x, p_y, p_z system and the sp^3 system. What is the potential advantage of the sp^3 basis?

(c) That you can construct and solve for the energy bands of Si and or Ge using one

Table 7.4. *Melting temperatures and enthalpies of metals and semiconductors*

	Si	Ge	Fe	Ag
Atomic mass (u)	28.1	72.6	55.8	107.9
T_m (K)	1685 ± 3.0	1210.4	1809	1234
ΔH_m (kcal/mole)	12.0 ± 1.0	8.83	3.3 ± 0.1	2.7 ± 0.1

or other of these basis sets, and particular values of the matrix elements, using a matrix diagonalization package, display your results graphically, and compare your results to the literature.

(d) That the simplest model of band structure associated with the 2×1 reconstruction on Si or Ge(001) involves fixing the atoms below the surface plane in their bulk positions, and constructing a matrix as a function of wavevectors, k_x, k_y in the plane of the surface, and k_z perpendicular to the surface, again involving an 8×8 matrix, but now with some of the matrix elements set equal to zero. What is needed in addition to calculate the equilibrium position of the surface atoms and the resulting surface band structure and surface states?

Problem 7.3. Removal of long range fields in polar crystals via surface reconstruction

Draw the planar structure of GaAs to scale, viewed perpendicular to [111], and

(a) identify the planes containing Ga and those containing As, and the [111] versus the [$\bar{1}\bar{1}\bar{1}$] directions.
    If both Ga and As assume the sp^3 configuration,

(b) use Gauss's law to work out the average internal electric field along [111]. Note that a suitable choice of unit cell for the Gaussian surface may convince you that all the long range field is due to the surface layers.

(c) Show that removal of one quarter of the Ga atoms on the (111) A-face (as in figure 7.4) plus the same numbers of As atoms from the ($\bar{1}\bar{1}\bar{1}$) B face removes this internal field.

Project 7.4. Adatoms, surface vibrations, roughening and melting transitions on Si and Ge

The melting temperature and latent heat of melting of Si and Ge, Fe and Ag are listed by Honig & Kramer (1969) as in table 7.4.

(a) From these data evaluate the entropy of melting for the four elements, using Appendix C to express the results in units of k/atom.

(b) Do a literature search to find out whether any of the low index surfaces of these elements are known actually to undergo roughening or melting transitions, and if so under what conditions.

(c) The surface energy of Si(111) has been measured as ~10% less than Si(001) at 600 K, but experiments at $T = 1300$ K seem to suggest that their free energies are roughly equal (Bermond *et al.* 1995). Using the tables in the text, what can you deduce about the difference in surface excess entropies between the two faces?

   Estimate the surface excess entropy involved (in k/atom units), and compare with the melting entropy, in the following situations, using equations from section 4.2 as appropriate.

(d) We have a certain concentration of ad-dimers on Si(001). Evaluate the configurational entropy at $T = 1300$ K, assuming that the concentration shown in figure 7.10 corresponds to ad-dimers. What is the additional entropy if these ad-dimers are moving around as a 2D gas?

(e) We have asymmetric dimers on Si or Ge(001), each of which can be either up or down with equal probability. Show that this answer must also represent an upper limit to the entropy due to a single step moving freely across the surface. This is a (very rough) approximation to the entropy involved in the roughening transition, discussed in more detail by Desjonquères & Spanjaard (1996, chapter 2).

(f) The amplitude of z-motion of the surface atoms is increased from say 0.01 to 0.05 nm when the p2×2 or c4×2 ordered dimer phases on Si(001) disorder into to the '2×1' (Shkrebtii *et al.* 1995). Is this entropy the same as in (e) or are they additive?

8 Surface processes in thin film devices

This chapter discusses surface and near-surface processes that are important in the context of the production and use of various types of thin film device. In section 8.1 the role of band bending at semiconductor surfaces is considered along with the importance, and the perfection, of oxide layers and metal contacts on silicon surfaces. Section 8.2 describes models which have been developed to understand electronic and optical devices based on metal–semiconductor and semiconductor–semiconductor interfaces. Then section 8.3 describes conduction processes in both non-magnetic and magnetic materials, and discusses some of the trends which are emerging in new technologies based on thin films with nanometer length scales. The final section 8.4 discusses chemical routes to manufacturing, including novel forms of synthesis and materials development. The treatment in this chapter is rather broad; my aim is to relate the material back to topics discussed in previous chapters, so that, in conjunction with the further reading and references given, emerging technologies may be better understood as they appear.

8.1 Metals and oxides in contact with semiconductors

This section covers various models of metals in contact with semiconductors, and the oxide layers on semiconductors. Such topics are important for MOS (metal–oxide–semiconductor) and the widely used CMOS (complementary-MOS) devices. Models in this field have been extremely contentious, so sometimes one has felt that little progress has been made. However, recent developments and some new experimental techniques have shed light on what is happening.

8.1.1 Band bending and rectifying contacts at semiconductor surfaces

Band bending can occur just below free semiconductor surfaces, and when metals or oxides come into contact with semiconductors. Models of this effect are given in many places; the major effect is caused by the presence of surface states in the band gap, which pins the Fermi level and induces band bending. As indicated in figure 8.1 for an n-type semiconductor, the bands bend upwards towards the interface. This bending is associated with a dipole layer beneath the surface corresponding to the depletion

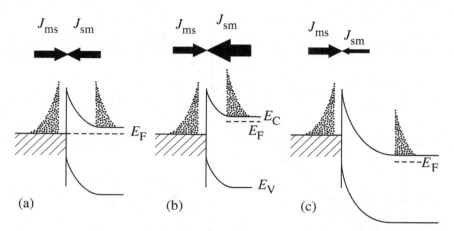

Figure 8.1. Band bending at the surface of an n-type semiconductor under different biassing conditions: (a) thermal equilibrium; (b) forward bias; (c) reverse bias.

region, where the n-type impurities are ionized, and the surface states are negatively charged. For a p-type semiconductor, band bending is reversed and surface states are charged positively, as indicated earlier in figure 1.23(a). If the semiconductor is in contact with a metal, the Fermi level is fixed by the metal, so that band bending or band flattening can be induced depending on changes from the previous surface state distribution.

The main electrical effects of this band bending arise from the asymmetry in the current flow under bias, as shown in figure 8.1. In the conduction band, the electron energy distribution corresponds to the tail of the Fermi function, so that at the Schottky barrier height ϕ_B, the number of electrons above the barrier scales as $\exp[-(q\phi_B/kT)]$, where q is the (effective, positive) electron charge. In the equilibrium case, when the Fermi levels are equal on both sides of the junction, the energy distributions above the barrier must be the same, and the current flow has to be zero, as indicated in figure 8.1(a). However, under bias voltage V, there is current flow. Using the formulae for thermionic emission over the barrier (see section 6.2), we have a current density J, given by

$$J = J_{sm} - J_{ms} = A^*T^2 \exp[-(q\phi_B/kT)]\{\exp(qV/kT)-1\}, \tag{8.1}$$

where the 'Richardson constant', A^* for the barrier, depends on the details of band structures and interface chemistry. The rectifying character of the contact is determined by the asymmetry with respect to $\pm V$, shown schematically in figures 8.1(b) and (c). For forward bias (positive V) the current J_{sm} can increase without limit, but under reverse bias, the current is limited to a low constant value J_{ms} by the Schottky barrier height ϕ_B. For higher reverse bias, catastrophic breakdown can occur.

This topic has a very long history, starting with the discovery of rectifying properties by Braun in 1874, and the use of 'cats' whiskers' as detectors in the early days of radio (Mönch 1990, 1994). The two classical means of checking (8.1) are given by Sze (1981). Figure 8.2 is a log–log plot of the forward bias current density J_F as a function

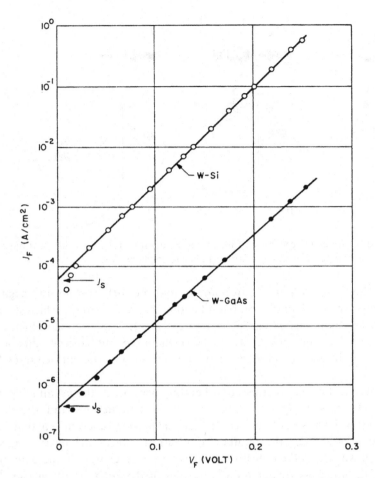

Figure 8.2. Logarithmic plot of forward current density J_F versus forward voltage V_F of W–Si and W–GaAs diodes (from Crowell *et al.* 1965, after Sze 1981, reproduced with permission).

of V_F, which gives a straight line whose slope is q/kT (Crowell *et al.* 1965). If q turns out not to be equal to e, the charge on the electron, then so be it. Device engineers replace q by q/n, where n is referred to as an 'ideality factor', with typical values of $n = 1.02$ to 1.04; i.e. the approximation that $q = e$ is pretty good, but 'non-ideality' can be used to cover several types of disagreement with this simple model. An Arrhenius plot of the reverse barrier current I_S, as $(\log I_S/T^2)$ versus T^{-1} shown in figure 8.3, shows that both A^* and ϕ_B for the important Al–Si diodes depend on processing conditions (Chino 1973). This sensitivity to cookery has plagued the field for a long time; both recipes and theories have on occasion become so complex as to be completely unbelievable. But, before discussing these points, the next topic is the width of the depletion region.

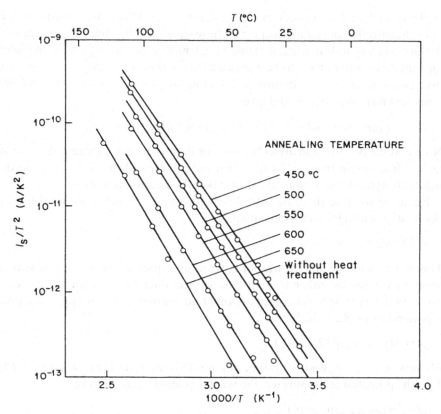

Figure 8.3. Arrhenius plot of the barrier current (I_S/T^2) of Al–Si Schottky barrier diodes under reverse bias for different processing conditions (from Chino 1973, after Sze 1981, reproduced with permission).

8.1.2 Simple models of the depletion region

Models of the depletion region are well developed in the literature, and to dwell on these here would take us too far from the topic of 'surface processes'. The important point to determine is the length scale over which (electron or hole) depletion is observed, and to establish the connection with electrical and optical properties. The key concept is screening, which was introduced briefly in section 1.5, and the closely related values of the (relative) dielectric constant ε. Screening is very effective in metals, with screening lengths $\sim (2k_F)^{-1}$, and moderately strong, but dependent on the doping level, in semiconductors. It is weak in insulators, where for example it depends on ionic defects in crystals such as NaCl, and photographic materials including AgBr. In metals, Lindhard screening is required to describe the resulting Friedel oscillations (see section 6.1), but in semiconductors and insulators, classical Thomas–Fermi screening provides a sufficient description.

Consider a potential $V(z)$ away from the surface, where the zero of V is in the bulk of the crystal, and an associated charge density $\rho(z)$. This charge density may consist of both ionized donors and acceptors, whose values are N_D^+ and N_A^- respectively, and the electron and hole density, $n(z)$ and $p(z)$, whose values are n_b and p_b in the bulk. In the bulk, the charges have to be compensated, so that $N_D^+ - N_A^- = n_b - p_b$. The electron and hole charge distribution is biased in the presence of $V(z)$, which is different from zero near the surface, and gives

$$\rho(z) = +q[n_b(\exp(-qV(z)/kT) - 1) - p_b(\exp(+qV(z)/kT) - 1)]. \tag{8.2}$$

Note that the positive sign arises because the donor and acceptor distributions stay the same, while the electron and/or hole distribution responds to $V(z)$; care is needed with signs throughout this argument, which takes the electron to have $q = -e$.

Equation (8.2) needs to be solved self-consistently, which is done within classical electrostatics using the Poisson equation

$$d^2V(z)/dz^2 = -\rho(z)/\varepsilon\varepsilon_0. \tag{8.3}$$

This can be solved numerically, but is typically expressed within one of two limiting approximations, for either n- or p-type semiconductors, i.e. when $N_D \gg N_A$ or vice versa. In the weak space charge approximation, we make a linear approximation to the exponentials in (8.2) which gives

$$\rho(z) = (n_b \text{ or } p_b)q^2V/(\varepsilon\varepsilon_0 kT), \tag{8.4}$$

where we use n_b or p_b for n- or p-type doping. This results in $V(z) = V_s\exp(-z/L)$, where V_s is the potential at the surface; the screening length L is given by

$$(Lq)^2 = (\varepsilon\varepsilon_0 kT)/(n_b \text{ or } p_b). \tag{8.5}$$

In the other limit we acknowledge that if V_s is large compared to kT, then $\rho(z)$ will approximate to a step function, such that all the charges are ionized up to a depth d below the surface, i.e. $\rho(z) = q(N_D \text{ or } N_A)$ for $0 < z < d$. Integrating (8.3) twice then gives a quadratic dependence:

$$V(z) = -q(N_D \text{ or } N_A)(d-z)^2/(2\varepsilon\varepsilon_0), \tag{8.6}$$

for $0 < z < d$, with $V(z) = 0$ for $z \geq d$, which is known as the Schottky approximation, see figures 8.4(b) and (c). A detailed discussion with examples is given by Lüth (1993/5, chapter 7).

The key point is to realize how the screening length L and depletion length d depend on the doping level in typical semiconductors. Inserting a set of values into (8.5), for Si with $\varepsilon = 11.7$ or Ge with $\varepsilon = 16$, a low doping level $n_b = 10^{20}$ m^{-3} (or equivalently 10^{14} cm^{-3}) gives $L = 410$ nm for Si and 480 nm for Ge. However, for a typical surface potential $V_s = 0.8$ V, the depletion length d is greater than 3 μm; since $d > L$ the Schottky model is most appropriate. These lengths are very long relative to atomic dimensions; although they will decrease as $(n_b \text{ or } p_b)$ increase, they are much greater than 10 nm, at least until samples are heavily doped, and have properties approaching those of metals. Thus it is not surprising that models of the electrical behaviour of semiconductors are typically not unduly concerned with atomic scale or surface properties. On the other

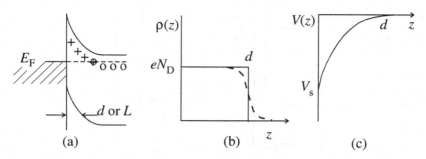

Figure 8.4. Depletion layer below an n-type semiconductor surface: (a) band bending showing ionized donors above the Fermi level E_F; (b) charge density $\rho(z)$; and (c) potential variation $V(z)$ in the Schottky approximation.

hand, as devices get smaller, such effects can be pervasive throughout the whole device, and the unwanted statistical distribution of impurities may well pose a limit to device dimensions in future.

8.1.3 Techniques for analyzing semiconductor interfaces

Classical experiments to determine the depth of the depletion layer and the barrier heights include C–V profiling, where the same types of arguments as used above show that C^{-2} is linearly proportional to the bias voltage V, with an intercept which gives ϕ_B, and the photo-electron yield or photocurrent, where the square root of the yield versus photon energy gives a straight line with intercept ϕ_B (Sze 1981, Schroder 1998). By biasing the sample with a d.c. offset, and using a.c. techniques for probing and detecting, more subtle techniques including deep level transient spectroscopy (DLTS) have been developed, and used to study the distribution of electrically active defects with depth. An example related to point defects produced during InGaAs thin film growth is given by Irving & Palmer (1992); but it is noticeable that deductions about the nature of the defects responsible are at best rather indirect.

Semiconductor devices are very demanding in terms of analytical techniques, since one would like to know the density and depth distribution of the dopants, and of other impurities (Schroder 1998). Of the wide range of surface analytical techniques available, so far only secondary ion mass spectrometry (SIMS) has been widely applied; although it is destructive, it does have the necessary sensitivity, and it specifically identifies the elements in question. The calculation in section 8.1.2 above can be used to estimate the sensitivity needed: to detect 10^{14} cm^{-3} dopants in a depletion layer of thickness 400 nm, over 1 mm^2 sample area means detecting and quantifying 4×10^7 atoms. A determination of the depth distribution of (delta-doped) Be dopants in GaAs by SIMS is shown in figure 8.5. Note that deductions about the effects of annealing temperature and 'knock-on' effects during implantation have been made (reliably) from profiles containing a maximum of only 200 counts/channel (Schubert 1994).

Although STM and related spectroscopies (STS) have revolutionized surface imaging since the early 1980s, it is perhaps less clear whether similar advances in

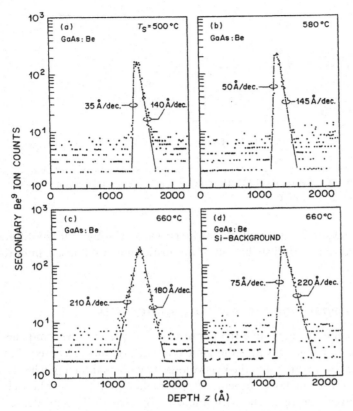

Figure 8.5. SIMS profile of Be delta-doped GaAs grown by MBE at: (a) 500, (b) 580 and (c) 660 °C; (d) inclusion of Si background doping reduces the segregation of Be to the surface (after Schubert 1994, reproduced with permission).

imaging buried interfaces, such as delta-doped layers, can be made. However, *ex situ* TEM/HREM and diffraction are very powerful, and have been widely applied to semiconductor interfaces. There are several related techniques including Fresnel (out of focus) imaging, which can image layers in profile with ML precision (Shih & Stobbs 1991). Convergent beam electron diffraction (CBED, see figure 3.1(d)) and convergent beam imaging (CBIM) are powerful means of measuring small strains and distortions in multilayers (Humphreys *et al.* 1988). *In situ* studies of reactions in UHV are very demanding research projects, which have been pursued by relatively few groups; but semiconductor interfaces have been prime targets for the application of these techniques (Yagi 1993, Ross *et al.* 1994, Gibson *et al.* 1997, Collazo-Davila *et al.* 1998, Marks *et al.* 1998).

One new STM-based technique has enabled some relevant device oriented studies to be performed on a microscopic scale, as described below. This variant, known as ballistic energy emission microscopy (BEEM) was invented in the late 1980s (Kaiser & Bell 1988, Bell & Kaiser 1988). Reviews of this rapidly developing area have been given by Prietsch (1995), by Bell & Kaiser (1996) and von Känel *et al.* (1997). The schematic

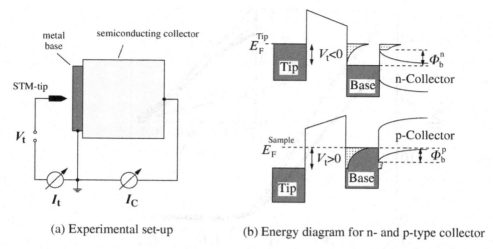

(a) Experimental set-up   (b) Energy diagram for n- and p-type collector

Figure 8.6. (a) Schematic set-up of a BEEM experiment, indicating tunneling and collector currents I_t and I_c and the tip voltage V_t; (b) energy level diagrams for forward BEEM at an n-type (top) and p-type collector (bottom) (after von Känel *et al.* 1995, reproduced with permission).

arrangement is shown in figure 8.6. BEEM can be understood as a hot electron triode, in which the tunneling current (I_t) into the metal serves as the source for the collected current (I_c) in the semiconductor. BEEM images can then be obtained by scanning the tip, and compared with STM images of the same area, as shown for a thin metallic CoSi$_2$ layer on Si(111) by von Känel *et al.* (1995). The application of STM and BEEM to study silicides is reviewed by Bennett & von Känel (1999).

The corresponding spectroscopy (BEES) is possible, also spatially resolved at the nanometer scale, and has become a very powerful means of determining local values of the Schottky barrier height, as indicated in figure 8.7. Here I_c is plotted against the tip voltage V_t for constant I_t, and goes to zero as $- V_t$ approaches ϕ_B. These ϕ_B values are at least as good as those produced by large area electrical methods: a quite remarkable achievement (Meyer & von Känel (1997).

It is a current research topic to understand the contrast mechanisms in both the microscopy and spectroscopy. If the metal base electrode is featureless on the lateral scale of interest, then BEEM contrast is thought to arise largely through effects occurring at the metal–semiconductor interface, though differences in experimental arrangements may have lead to inconsistencies between different groups. It is notable that transmission through the base layer as a function of thickness offers a direct measurement of the inelastic mean free path (imfp) for low energy electrons, well below the minimum in the imfp curves discussed earlier in section 3.3.4. For example, Ventrice *et al.* (1996) show that for Au films on Si(100) had $\lambda_i \sim 13$ nm at room temperature and 15 nm at 77 K, for an injected energy of around 1 eV above the Fermi level. It is thus reasonable to use base electrodes with thickness up to tens of nanometers in these techniques.

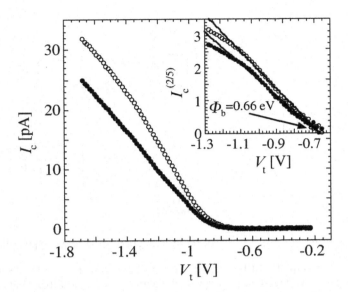

Figure 8.7. Ballistic electron emission spectra $I_c(V_t)$ normalized to the tunneling current, taken on top of (open circles) or next to (filled circles) an interfacial point defect. I_c is higher on the defect for V_t close to the barrier height ϕ_B. The inset shows the 2/5 power of I_c, yielding a value of $\phi_B = 0.66 \pm 0.01$ eV (after Meyer & von Känel 1997, reproduced with permisison).

One of the possibilities of BEEM is to identify the defects responsible for surface states. The images of misfit dislocations in the CoSi$_2$/Si(111) interface can be seen at high resolution to consist of a 'string of pearls' as shown in figure 8.8(b), rather than a continuous line image which one might expect from a dislocation line. In the same images there are also isolated point defects in the vicinity of the interface, and a sensitivity of below 10^{12} cm^{-2} is claimed (Meyer & von Känel, 1997); the hypothesis is that it is point defects trapped in the vicinity of dislocations, rather than the dislocations themselves, which are electrically active.

This sensitivity level is particularly inviting in relation to the SiO$_2$/Si and related interfaces. One of the main reasons why the various MOS and CMOS silicon device technologies work is the low density of surface states at the Si-SiO$_2$ interface, where values below 10^{12} cm^{-2} (i.e. $\sim 10^{-3}$ ML, or 10^{10} on 1 mm^2, or 10^4 on 1 μm^2), can be consistently achieved. However, it is extremely difficult to deduce what these defects actually are; the technologist is primarily interested in getting rid of them, and often the only means of assessing them are the same electrical properties which one is trying to optimize: a sure recipe for a black art.

Recently, it has been shown that BEEM can address such problems. Using a base electrode of thin granular Pt, electrons can be injected into, and can be trapped in, a 25 nm thick insulating film of SiO$_2$ on Si. Trapping of very few (~ 10) electrons results in a decrease of the BEEM current which can then be readily measured (Kaczer et al. 1996). A complementary (broad area) tool is electron spin resonance (ESR), which is

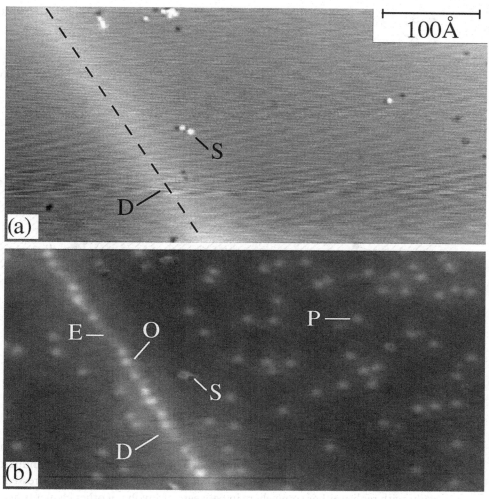

Figure 8.8. (a) STM topography image of a 2.8 nm thick $CoSi_2$ film on Si(111), showing a 0.06 nm high line due to the strain field of an interface dislocation (dashed line); (b) corresponding BEEM image showing interfacial point defects, S and P, and those trapped in the core of the interface dislocation D, which comprises empty (E) and occupied (O) regions (after Meyer & von Känel 1997, reproduced with permisison).

very good for detecting unpaired electrons; these are typically the centers which give rise to electron traps in SiO_2, some of which are indicated in figure 8.9. These centers, for example the P_b center, which is an electron trapped on a Si dangling bond, are different in detail on surfaces of different orientation (Helms & Poindexter, 1994). Another sensitive wide beam technique is called Total reflection X-ray Fluorescence (TXRF), which, by using glancing incidence X-rays, can detect below $10^{10}(cm^{-2})$ metal atoms on flat surfaces. It is now highly valued for examining the cleanliness of silicon wafers in production plant environments (Schroder 1998, section 10.4).

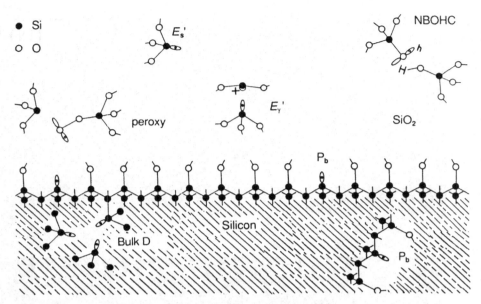

Figure 8.9. Paramagnetic point defects which have been observed in SiO_2/Si structures by electron spin resonance. The NBOHC configuration is the non-bridging-oxygen hole center (after Helms & Poindexter 1994, reproduced with permission).

8.2   Semiconductor heterojunctions and devices

8.2.1   *Origins of Schottky barrier heights*

There has been much discussion of the origin of Schottky barrier heights, and other related phenomena at metal–semiconductor and semiconductor–semiconductor interfaces. As often in physics, there are two limiting cases which can be addressed analytically, with reality either somewhere in between, or involving other elements not present in either. The story starts with a half-page paper by Schottky (1938), and continues with the opposing model of Bardeen (1947). The question to be answered is: what determines the energy levels in the semiconductor?

In the Schottky model, we bring together the metal and the semiconductor, and assume there is no electric field in the space between them. This means that we can form the barrier as the difference between the work function of the metal ϕ_M, and the electron affinity of the semiconductor χ_S: i.e

$$\phi_B = \phi_M - \chi_S. \tag{8.7}$$

Equation (8.7) can be simply tested: pick any semiconductor, deposit a series of metals on it, and measure the barrier height ϕ_B. This should scale directly with the metal work function ϕ_M. The test has been done many times (see e.g. Brillson 1982, Rhoderick & Williams, 1988 chapter 2, Lüth 1993/5 chapter 8) and the variation with metal workfunction is usually much weaker than this model implies. In the case of Mönch's work on Si(111)2×1 cited by Lüth, changing metals to give ϕ_M varying from 2 to 5.5 eV increases ϕ_B modestly from 0.3 to 0.9 eV.

The opposite Bardeen model assumes that surface states are sufficient to pin the Fermi level in the semiconductor, and notes that this energy level is placed at ϕ_0 above the valence band edge. The top of the conduction band, which forms the barrier, is at $(E_g - \phi_0)$ above the Fermi level; thus

$$\phi_B = E_g - \phi_0, \tag{8.8}$$

and the barrier height shouldn't vary at all with the work function of the metal. This is also rarely satisfied in experiment, and we must consider that these two models continue to be discussed because they are simple limiting cases. Once one begins to think in terms of the detailed mechanisms of what happens when two surfaces are put together to form the interface, then the basis of both models falls apart. For example, the two surfaces in vacuum may well be reconstructed, and this reconstruction will change, and may be eliminated in the resulting metal–semiconductor interface. Also the interfaces may well react chemically, and/or form a complex microstructure: do such 'metallurgical' effects have no influence on the result?

For many years these types of uncertainty lead to a whole series of tabulations of data, and empirical models which were all more or less specific to particular systems. This discussion was often played out at conferences, such as PCSI, *Physics and Chemistry of Semiconductor Interfaces*, or ICFSI, *International Conference on the Formation of Semiconductor Interfaces*, both still going at number 25 (January 1998, published in *J. Vac. Sci. Tech.*) and number 6 (June 1997, published in *Applied Surface Science*) respectively. Short of absorbing in detail a historical survey, such as those written by Brillson (1982, 1992, 1994) or Henisch (1984), and to a lesser extent by Rhoderick & Williams (1988) or Sutton & Balluffi (1995), the question for the 'interested reader' is: what can one extract of reasonable generality from this field?

The model which has most appeal for me is that introduced in 1965 by Heine, and developed by Flores & Tejedor (1979) and by Tersoff (1984, 1985, 1986). There is also an interestingly simple free electron model introduced by Jaros (1988). This topic is reviewed by Tersoff (1987) in the volume by Capasso & Margaritondo (1987), and by Mönch (1993, 1994). Termed MIGS, this refers not to a Russian fighter plane, but to metal-induced gap states: i.e. to states which are present in the band gap of the semiconductor, and are populated due to the proximity of the metal. This leads to the result that the Fermi level is pinned at an energy close to the middle of the gap, a similar result to the Bardeen model, but for different reasons. It further emphasizes the role of the 'interface dipole' and seeks to minimize this quantity. As such this becomes a (more or less) quantitative statement of the underlying point that *nature doesn't like long range fields*, which I have been stressing from section 1.5 onwards. The bones of this argument are summarized without attribution in a useful introductory text by Jaros (1989).

The ingredients of this model can be seen in figure 8.10. We know that there are forbidden energy regions in a bulk semiconductor, with an energy gap of width $E_g = E_C - E_V$. However, solution of the Schrödinger equation in a periodic potential does not say that these gap states cannot exist, it merely says that they can't propagate in an

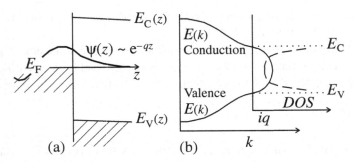

Figure 8.10. Elements of the MIGS model: (a) energy levels and wavefunction $\psi(z)$ of states in the gap close to a metal–semiconductor interface; (b) band diagram including the density of states (DOS, dashed line) of the MIGS, which peak near the band edges. Note that the length scale along z in panel (a) is much shorter than the scale d or L in figure 8.4, so that the bands are shown to be only gently sloping (adapted from Lüth 1993/5).

infinite medium. Mathematically this means that the **k**-vector has to have an imaginary component iq, which ensures decay of the wavefunction; we have seen this as a condition for a surface state in section 1.5. This decay is slower nearer to the band edges, and it is most rapid close to mid-gap.

Although this argument does depend in detail on the 3D reciprocal space geometry of the particular crystal, the 1D model illustrated here, and worked through by Lüth and by Mönch, gives the essential result. Thus the wavefunctions at these energies decay into the semiconductor, and must be matched to the traveling-wave solutions in the metal; this spill-over of charge creates an interface dipole, which is minimized if the Fermi level is around mid-gap. We have already seen, in section 6.1, that an ML array of relatively tiny electric dipoles can create quite large changes in electrostatic potential across the dipole sheet.

8.2.2 Semiconductor heterostructures and band offsets

The above points can be brought out even more forcefully by considering semiconductor heterostructures, as shown in figure 8.11. We bring together two semiconductors and ask how the bands will line up. If the semiconductors are similar, but the alignment is as in figure 8.11(a), electrons will spill over from right to left; this creates, or is the result of creating, a substantial interface dipole. However, in the more symmetric alignment of different semiconductors shown in figure 8.11(b), the electrons in the conduction band spill from right to left, whereas the sense is reversed in the valence band. The resulting charge distribution is much more compensated, i.e. the interface dipole is a lot smaller, and may even disappear. Simple, that's the answer!

Considered in terms of the Fermi energy, this problem is rather difficult to pose: we want the Fermi levels to line up, but at low temperature there are no states at E_F, so the problem appears to be undefined. In terms of the interface dipole, however, the problem appears concrete, even if it is still just a bit elusive. As Tersoff points out, if a reference level can be found, then the problem is trivial, it has already been solved: this

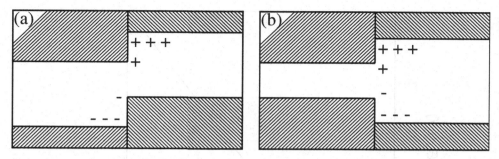

Figure 8.11. Band lineups at semiconductor–semiconductor interfaces, which result in (a) a strong interface dipole, (b) almost no interface dipole (after Tersoff 1987, redrawn with permission).

is really what was going on in the Bardeen and Schottky models, but the reference levels were *assumed*. Comparable models exist for semiconductor heterostructures. For example the ionization potential model (often called the electron affinity rule) is the exact analogue of the Schottky model, where the vacuum potential is the reference level, whereas the reference level in the case of metals is the Fermi energy. In Tersoff's model, the reference point is the branching point energy E_B, often called the charge neutrality level, E_n. It is difficult to pin down the exact definition of this quantity, but it corresponds to the energy where these states change over (branch) from being valence band-like to conduction band-like, as in figure 8.10(b), and so usually the energy lies near to mid-gap.

The electrostatic linear response model presented by Tersoff is instructive, as it shows why semiconductors are close to the metallic limit. In terms of a polarizability α, where $\alpha = \varepsilon - 1$, he finds that the valence band offset (VBO) at the interface, ΔE_V, is given by

$$\Delta E_V = [\alpha/(1+\alpha)]\Delta E_V^n + [1/(1+\alpha)]\Delta E_V^0, \tag{8.9}$$

where the first quantity ΔE_V^n is the difference between the charge neutrality levels, and the second ΔE_V^0 is the difference between the ionization energy levels of the materials in contact. Since for representative semiconductors, $\alpha \sim 10$, we can see that the first term on the right hand side of (8.9) dominates, unless $\Delta E_V^0 \gg \Delta E_V^n$. The response (to the lack of highly accurate *ab initio* calculations) has often been to make correlations which are expected to be true if the basic model is on the right track. If metal–semiconductor and semiconductor–semiconductor band alignments are similar in origin, then $(E_F - E_V)$ for the metal should parallel $(E_B - E_V)$ for the semiconductor, and this equals the (negative of the) barrier height for a p-type metal-semiconductor contact ϕ_{BP}, since the relevant acceptor states are close to E_V. This correlation, shown in figure 8.12, is perhaps the most successful prediction of the MIGS model.

The challenge for quantum mechanical calculations is that barrier heights and band offsets are of order 0.3–1.0 eV, and can be measured to ± 0.02 eV precision. As can be seen in figure 8.12 and in table 8.1, predictions in the late 1980s were good to ~ ± 0.1–0.15 eV. Some calculations have improved to a secure ± 0.1 eV, but claims to be much better are suspect. In particular, it is not easy to ensure that charge neutrality

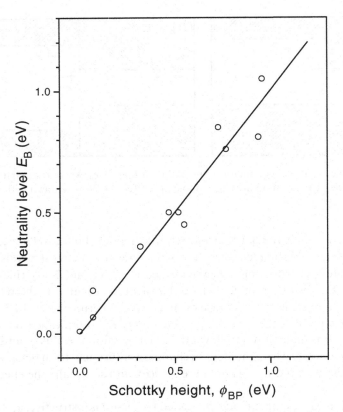

Figure 8.12. Correlation of the p-type Schottky barrier height for Au contacts ϕ_{BP} with the calculated position of the charge neutrality level $(E_B - E_V)$ for several 3–5 and 2–6 semiconductors (after Tersoff 1987, replotted with permission).

is maintained across the interface to the required accuracy. One important model of semiconductor interfaces which builds in this requirement is the 'model solid' approach originated by Van de Walle & Martin (1986, 1987), further developed by Van de Walle (1989), and reviewed by Franciosi & Van de Walle (1996) and Peressi *et al.* (1998) as discussed in the next section.

8.2.3 *Opto-electronic devices and 'band-gap engineering'*

Several books (e.g. Butcher *et al.* 1994, Kelly 1995, Davies 1998) and many conference articles make it clear that artificially tailored semiconducting heterostructures now form the leading edge of device technology, and indeed have done so for the past 20 years. By alternating thin layers of, for example, GaAs with (AlGa)As, one can produce structures with remarkable opto-electronic properties, such as the multiple quantum well (MQW) laser, and many others. They are fabricated by techniques such as MBE or MOCVD (see section 2.5) and can be patterned using optical or electron-based lithography techniques to form real devices (see e.g Kelly 1995, chapters 2 and 3).

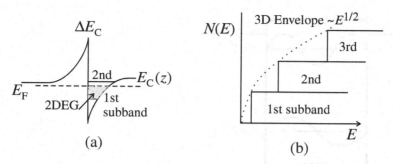

Figure 8.13. A semiconductor heterostructure giving rise to a 2D-electron gas at zero bias in n-type material. The energy levels shown in (a) are the onset energies of subbands whose density of states, $N(E)$, is indicated in (b).

These devices come in various geometries. Many devices use a material structured in one dimension, perpendicular to the layers, so that the electrons are confined in this direction and move (freely) in the other direction; this is the basis of the 2D-electron gas (2DEG) which is used for studying the quantum Hall effect (QHE) and many other effects. As can be seen in the n-type case illustrated in figure 8.13, the 2DEG exists in a thin layer where the conduction band of the narrower gap material dips below E_F; there are equivalent cases for p-type material. There are also 1D *wires* (1DEG) and zero-D *dots* (0DEG).

In a QW heterostructure, the electrons and holes are confined in a thin film of the narrower band gap material, e.g. GaAs, with $E_g = 1.42$ eV at room temperature or 1.52 eV at 0 K, surrounded by an alloy of (AlGa)As, as illustrated in figure 8.14 (the band gap of AlAs is 2.15 eV at room temperature, see table 8.1). A key quantity is the valence band offset, ΔE_V, or the conduction band offset ΔE_C, i.e. ($\Delta E_g - \Delta E_V$). The energy levels in the well are determined by how the band gap difference ΔE_g is partitioned between ΔE_V and ΔE_C. The quantity $Q_C = \Delta E_C / \Delta E_g$, the proportion of the gap difference which appears in the conduction band, is often quoted in data tables, but DFT and other theoretical models typically give ΔE_V with best accuracy.

Although the quantization of the energy levels in the z-direction leads to the 2DEG, electrons and holes can move in the x and y directions parallel to the interface; so what is shown in diagrams such as figures 8.13(a) or 8.14 is not a unique level, but the onset energy of a subband. Within a subband, the 2D density of states is a step function as shown in figure 8.13(b). For GaAs and similar materials, there are also light and heavy holes, related to spin-orbit splitting in the valence band, and the material, in contrast to Si, has a direct band gap. Thus the optical properties of the well are now determined by the electron and hole masses (three parameters), the well width L_z, and Q_C. Duggan (1987) has given a useful introduction to the determination of optical properties, typically pursued via optical absorption or photoluminescence (PL) experiments; another useful starting point is Kelly (1995, chapter 10). As shown in figure 8.15, the optical absorption (transmission) spectrum shows peaks which can be identified with light and heavy hole transitions. Note, in passing, that PL experiments only work at low temperatures, as the transitions are too broad at room temperature, and the

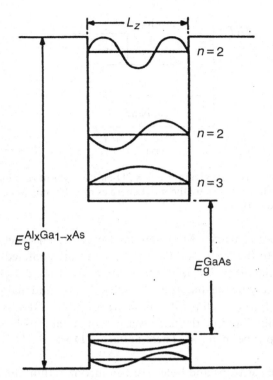

Figure 8.14. Simple energy levels in a quantum well, consisting of a narrower band gap material (GaAs) surrounded by a higher band gap material ($Al_xGa_{1-x}As$). Note, however, that the well wavefunctions must in practice spread into the surrounding layers, and the real band structure modifies this picture (from Duggan 1987, reproduced with permission).

number of parameters involved causes quite a bit of difficulty for data analysis. Nevertheless, several parameters can be determined (or assumed in order to get better values of other parameters), the early work leading to Q_C values typically in the range 0.6–0.8 for GaAs/AlGaAs heterostructures, and subsequently refined towards the lower end of this range (Yu *et al.* 1992, Davies 1998).

In the same volume (Capasso and Margaritondo 1987) there are extensive tabulations of early experimental valence band offsets ΔE_V, obtained largely by photoemission, but also by other techniques including extensions of C–V profiling and DLTS. Raman scattering has also found a useful niche (Menéndez *et al.* 1986, Menéndez & Pinczuk 1988) to measure inter-subband transitions for electrons, and hence ΔE_C. Later compilations and comments are given by Yu *et al.* (1992), Butcher *et al.* (1994) and Franciosi & Van de Walle (1996). Technology moved ahead in the 1990s, e.g. via the infrared devices based on resonant tunneling via minibands (Capasso & Cho 1994), but the science had more or less stabilized by the late 1980s. Table 8.1 gives some representative ΔE_V values for low strain interfaces.

However, it has become clear that pictures such as figures 8.10 and 8.11 are only a first step, and that *too simple* pictures may give a misleading impression. Valence band

Table 8.1. *Some calculated valence band offsets across low misfit (001) interfaces in comparison with experimental ΔE_V values*

Interface	Unstrained misfit at 300 K (%)[f]	Band gaps at 300 K (eV)[f]	Calculations ΔE_V (eV)	Experiment ΔE_V (eV)
Ge/GaAs	+0.09	0.66/1.42	0.63[a], 0.32[b] 0.58–0.62[e*]	0.53[b] 0.51±0.09[c]
GaAs/AlAs	−0.12	1.42/2.15	0.37[a], 0.55[b] 0.58±0.06[e]	0.50[b] 0.50±0.05[c]
InAs/AlSb	−1.27	0.35/1.62	−0.05[b] −0.17±0.1[c]	−0.13[f*] −0.18±0.05[c]
InAs/GaSb	−0.62	0.35/0.75	−0.38[a],−0.40[b] −0.54±0.02[d]	−0.51[b] −0.55±0.05[c]

References: (a) Van de Walle & Martin 1986; (b) Tersoff 1987; (c) range of values from Yu *et al.* 1992, mostly excluding measurements without error bars; (d) Montanari *et al.* 1996; (e) Peressi *et al.* 1998, (e*) for the two-layer mixed interfaces discussed in the text; (f) Davies 1998, Appendix 2, (f*) section 3.3.

offsets are often divided into three classes: type I, or *straddling* alignment, is as illustrated in figure 8.11(b) is appropriate for Ge/GaAs or GaAs/AlAs; type II, or *staggered* alignment, is as shown in figure 8.11(a), which is observed in the InAs/AlSb system. There is also a type III, or *broken-gap* alignment, observed for InAs/GaSb in which the bands do not overlap at all[1] (Davies 1998, section 3.3). There are two types of problem, one apparent, one real; let us get appearances out of the way first. Figure 8.11 shows schematically the energy positions E_V and E_C, but it does not show the dependence of these energies on the **k**-vector, i.e. the detailed band structure, which is shown simplified, in a rather unrealistically symmetric alignment, in figure 8.10.

Band structures are rather different for the various bulk semiconductors, as can be explored via the problems and projects given for chapter 7. When calculating energies such as E_B, a detailed integration is done for both the valence and conductions bands over the entire 3D Brillouin zone. In the integration, the Γ-point ($\mathbf{k}=0$) contributes negligibly; high densities of states are typically concentrated at the zone boundaries and near various maxima and minima, such as the valleys associated with indirect gaps. The calculation reflects the 'center of gravity' of the two bands. Zone boundary states correspond to standing waves, and in a bulk compound semiconductor such states may be preferentially located over one type of atom. The shift in E_B is associated with (partial) electron transfer from cations to anions which differs across the interface, also associated with different band structures (curvatures) in the two materials.

[1] Note that the language here can be confusing: Yu *et al.* (1992) have two subsets of type II to cover these two materials, and a different type III; there is also a sign convention which is unevenly applied. Here we use a negative sign for ΔE_V if the valence band edge of the narrower gap material is lower in energy than that of the wider gap material. This has the advantage of not having to remember which type is involved when interpreting ΔE_V and Q_C. There does not seem to be an accepted standard convention.

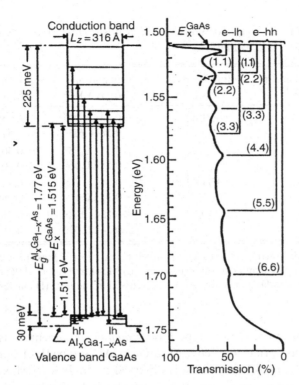

Figure 8.15. Effective mass analysis of a particular GaAs quantum well, surrounded by Al$_x$Ga$_{1-x}$As with $x = 0.21$. The left hand panel shows the predicted heavy and light hole transitions and band offsets, and the well thickness, all of which were deduced from the absorption spectrum shown in the right hand panel (from Duggan 1987, reproduced with permission).

So what real problems remain? Although the MIGS and related models are relatively satisfying, professionals in this field clearly do not believe that they contain the whole story, and can demonstrate that interface chemistry/ segregation plays an important role in addition (Brillson 1994, Mönch 1994). They can then consider tailoring the interfacial layers to produce particular desired offsets (Franciosi & Van de Walle 1996). For example, we can show that the composition of the interface layer *does* play a role. As argued by Peressi *et al.* (1998), the Tersoff model for the differences in ΔE_V between different heterostructures with the same substrate, or between different Schottky barrier heights $\Delta \phi_{BP}$ with the same metal as shown in figure 8.12, indicates that these differences are largely due to bulk properties. On the other hand, the absolute values are not merely bulk quantities, and the models discussed so far only work for non-polar interfaces, or for polar interfaces between homovalent materials, such as Ge/Si(001), which is of course strongly strained. The supposedly simple Ge/GaAs (001) junction would have two extreme ways of forming a sharp interface, i.e. termination with Ga or As, but as such an interface would be charged it must either reconstruct and/or intermix. There are two possibilities for an isolated neutral mixed plane, $(AsGe)_{1/2}$ or $(GaGe)_{1/2}$, which were calculated to have valence band offsets differing by 0.6 eV; this *difference* is much greater than the uncertainty in experimental values, but the *average* is closely the same as for the non-polar (110) interface.

The real material has many possible ways to intermix and thereby minimize charge imbalance across the interface, and a better option was thought to be mixing over two planes, so that the interface dipole could be reduced to zero. Again there are two options involving $(Ga_3Ge)_{1/4}$ followed by $(GaGe_3)_{1/4}$, or vice versa. Now the calculations shown in table 8.1 give 0.62 and 0.58 eV, much closer to each other and to experiment (Biasiol *et al.* 1992). To find the actual structure and the VBO at the same time is a challenge, since there are several structures worthy of attention which have similar energies. Nonetheles, Peressi *et al.* (1998) conclude that MIGS-related (or better termed, linear response theory) models form a very good starting point, provided one discusses the interface that is actually present. For the 'model solid' approach (Van de Walle 1989), the effects of strain can be incorporated in a natural way without further approximation; this method is therefore favored for calculations on strained layer interfaces.

The examples where these models clearly don't work correlate with strong chemical/metallurgical reactions and/or steps or other defects at the interface, with associated trapped charges and/or dipoles. One could counter by saying that in these cases, the interface is simply not what was initially postulated. If one adds the evidence now being obtained from BEEM about large lateral variations in barrier heights, and in transmission coefficients across such interfaces, then variability is not surprising. Technology in one sense is all about processing: in that context variability which one cannot control is the real disaster. But in making the transition to scientifically based industry, understanding is also very important. Without it, any small change in processing conditions forces a return to trial and error, with typically a huge parameter space to explore – preferably by yesterday, or you are out of business!

8.2.4 *Modulation and δ-doping, strained layers, quantum wires and dots*

In heterostructures, we also have to have provide carriers via doping. But if the layers are narrow enough, we may be able to put the dopants at different positions and thereby increase carrier mobilities by strongly reducing charged impurity scattering. This is one of the key points behind modulation doping, and is a factor in δ-doping, i.e. doping on a sub-ML scale, which can be used to change the shape of quantum wells (Schubert 1994). A limit to such techniques is the fact that dipoles are set up between the layers, which will also bend the bands. Depending on the doping level, the various length scales may or may not be comparable, and the models used will be different in detail. Understanding the effect of different length scales in models of condensed matter has a long history (Anderson 1972, Kelly 1995 chapter 3) and the topic continues to attract comment (Jensen 1998). Transitions in dimensionality, from 3D to 2D and so on down to 0D, are also of interest in the same sense. For example, a layered heterostructure which behaves as a 2DEG in zero magnetic field, becomes a 0DEG dot-like structure in a strong magnetic field, as the size of the cyclotron orbit becomes less than the lateral device dimensions. This is a key aspect of various devices based on 'quantum conduction': often the leading edge devices only work at low temperatures and/or in a high **B** field. Such devices are competitive in applications where ultimate sensitivity is required (e.g. telephone/TV satellite transmission and reception, or in astronomy), but not in domestic receivers whose emphasis is on optimum room temperature performance, where high electron mobility transistors (HEMTs) based on GaAs/(AlGa)As are a success story (Kelly 1995 chapter 5).

Some of these dimensional transitions are exemplified in strained layers, of which the archetype consists of GeSi alloys of various compositions on Si(100). The compressive strain due to 4.2% larger lattice constant of Ge means that the layers have a tetragonal distortion, which lifts the degeneracy of the four valence band minima with **k** parallel to the layer from the two with **k** perpendicular to the layer. This, and the switch of the position of the conduction band minimum at high Ge composition, influences both the band gap and offsets as a function of both composition and strain (see e.g. Kelly 1995 chapter 14, or Davies 1998 chapter 3). A realistic feel for the amount of work done on this one system may be obtained from the reprint collection made by Stoneham & Jain (1995), and the other references cited in section 7.3.3.

Quantum wires can be formed on a linear surface structures, the most obvious of which are vicinal surfaces consisting of arrays of steps. Experiments on a variety of configurations based on GaAs and AlGaAs are described by Petroff (1994). The problem is that individual steps are typically too rough to make this approach work, unless regular *multi*-atomic height steps can be reliably fabricated; this is a current research effort. It is not yet clear that such approaches can supplant lithography techniques (Prokes & Wang 1999). Similarly, quantum dots, in Ge/Si for example, need to be rather uniform in size to be useful; the question of whether one can persaude them to do this during growth of their own accord (i.e. via *self-organization*), or whether one uses lithographically patterned substrates is also a hot topic, which is discussed further in section 8.4. A discussion of early results using patterned layers is given by Kapon (1994); these efforts overlap with the topics discussed here in sections 5.3 and 7.3.

8.3 Conduction processes in thin film devices

The conductivity and the resistivity ρ are among the simplest material parameters to measure, one only needs a voltmeter and an ammeter. They are also some of the most useful properties, especially when they are non-linear and can thereby be used to amplify or store currents or voltages, as in essentially all active electronic devices. Yet it is an irony that what is easiest to measure and experience can also be the hardest to set on a firm scientific foundation, or to describe quantitatively with few unknown parameters. Here we explore in the simplest terms why this is the case, and indicate the role that surface and interface processes play in electrical and magnetic properties of thin films.

8.3.1　Conductivity, resistivity and the relaxation time

All electrical properties of thin film devices depend essentially on the number of charge carriers and the scattering processes to which these charges are subjected. To study such topics we need access to a book describing conduction processes in the relevant type of material. For normal metals and alloys, Rossiter (1987) is excellent; however, this book does not discuss superconductors or semiconductors even in outline. The reason is simply that these three topics are enormous fields, each with its own appropriate starting point. For superconductors, much effort has to be expended to describe the thermodynamic and quantum mechanical nature of the superconducting state, before one can consider the effects of bulk scattering and then thin film and surface effects (Tinkham 1996). The fabrication of useful *high-T_c* ceramic wires and tapes is a major technical challenge.

In particular, we can note that in semiconductors, most attention is paid to the number and type of carriers, and then we consider scattering processes which determine carrier mobility. For example, we write the conductivity σ as

$$\sigma = q(n\mu_e \text{ or } p\mu_h), \tag{8.10a}$$

in terms of the electron or hole mobilities μ_e and μ_h, given by

$$\mu = q\tau/m^*, \tag{8.10b}$$

where both the densities, n or p, and the scattering processes which lead to the effective scattering time τ, are determined by the defect density. Note that the word *or* is used here in the same sense as in section 8.1.2, to avoid too detailed a discussion of what happens if both n- and p-type dopants are present simultaneously, when extra care is always required. The effective mass is inversely proportional to the band curvature, and is in general a tensor quantity.

In metals, the number of carriers is essentially fixed, and the spotlight is on scattering processes. Elementary considerations start with the Drude model, and show that the conductivity is proportional to the density n of conduction electrons and the relaxation, or scattering, time τ between collisions as in equations (8.10) and (8.11). Moving to the quantum model with the correct Fermi–Dirac distribution function f_k, we realize that only those electrons $n(\mathbf{k})$ close to the Fermi energy participate in the scattering, and that the scattering time is now quite a complex average of all scattering processes. In the regime where the linearized Boltzmann transport equation (Rossiter 1987, chapter 1) is appropriate and we consider only elastic scattering,

$$\tau_k^{-1} = (2\pi)^{-3} \int d\mathbf{k}' Q_{kk'}(1 - f_{k'}), \tag{8.11}$$

where $Q_{kk'}$ is the matrix element for scattering processes from $\mathbf{k}$ to $\mathbf{k}'$ which conserve energy.

In general, the conductivity, relating the current density $\mathbf{J}$ to the electric field $\mathbf{E}$ is a second rank tensor $J_i = \sigma_{ij}E_j$ and the conductivity can be expressed as an integral over the Fermi surface $\mathbf{S}$ as

$$\sigma_{ij} = \{e^2/(2\pi^2 h)\} \int \tau_k v_i(\mathbf{k})v_j(\mathbf{k}) dS/ |v(\mathbf{k})|, \tag{8.12a}$$

where $v(\mathbf{k})$ is the Fermi velocity. For a cubic or amorphous metal, where the relaxation time is a function of the magnitude of k, but not of direction, (8.12a) simplifies to

$$\sigma = \{e^2/(6\pi^2 h)\} \int \tau_\mathbf{k} v(\mathbf{k}) d\mathbf{S}, \tag{8.12b}$$

so that σ and ρ are both scalar quantities and $\sigma = \rho^{-1}$. But, even so, we note that if a new scattering mechanism is added in (8.12), this will contribute to $\tau_\mathbf{k}^{-1}$, where the shortest time will dominate, but that the contribution to σ (in (8.12) as in (8.10)) is proportional to $\tau_\mathbf{k}$. Thus the contribution of a new scattering process to the resistivity ρ is as *an inverse of an inverse*.

8.3.2 *Scattering at surfaces and interfaces in nanostructures*

Given that many materials in general, and surface scattering in particular, are not intrinsically isotropic, the contribution of surface or 2D interface processes to the *resistivity* can be quite complicated. The first model applicable to thin films, which assumed an isotropic Fermi surface, was summarized by Sondheimer (1952) in terms of a single parameter p which characterized whether the scattering at the surface was specular ($p = 1$) versus diffuse ($p = 0$). Different formulae are applicable to wires or grain boundaries (Sambles & Preist 1982, Rossiter 1987 chapter 5), but for the thin film case, the Fuchs–Sondheimer formula is

$$\rho_\infty/\rho = 1 - \left[\frac{3(1-p)}{2\kappa} \int_1^\infty (t^{-3} - t^{-5}) \left(\frac{1 - \exp(-\kappa t)}{1 - p \exp(-\kappa t)} \right) dt \right], \tag{8.13}$$

where κ is d/Λ_∞, and t is the integration variable corresponding to the direction of electron travel. This formula clearly shows that as $p \to 1$ there is no extra scattering due to the surface, but if the scattering has a diffuse component, then the resistivity ratio ρ/ρ_∞ rises rapidly as the film thickness d becomes less than the scattering mean free path Λ_∞ in the corresponding bulk material, i.e. as $\kappa \to 0$. This formula is not the most accurate, but it is the simplest which has analytic limits.

The earliest comparisons with experiment typically showed agreement with (8.13) with p close to zero, the diffuse scattering limit. However, these early experiments were largely limited to thin polycrystalline films in which grain-boundary scattering played a crucial role. In several experiments, Al films were used which has a rough oxide on the surface. A particularly careful study of the resistivity of Au films grown on mica and KBr was made by Sambles *et al.* (1982), as illustrated in figure 8.16. Here Au(111) samples of different thicknesses were grown on mica, and the resistivity measured as a function of temperature. We can see in figure 8.16(a) that at low temperatures there is a substantial residual resistivity which increases with decreasing film thickness. The same data plotted on a logarithmic scale of ρ_∞/ρ versus κ shown in figure 8.16(b) gives a good fit to the more detailed model of Soffer (1967), but the fit may not be uniquely due to surface scattering. Internal surfaces such as grain boundaries can mimic surface scattering if the grain size is larger in thicker films (Sambles 1983). He concluded that 'p' in such films can be as high as 0.65, and that grain and twin boundary scattering

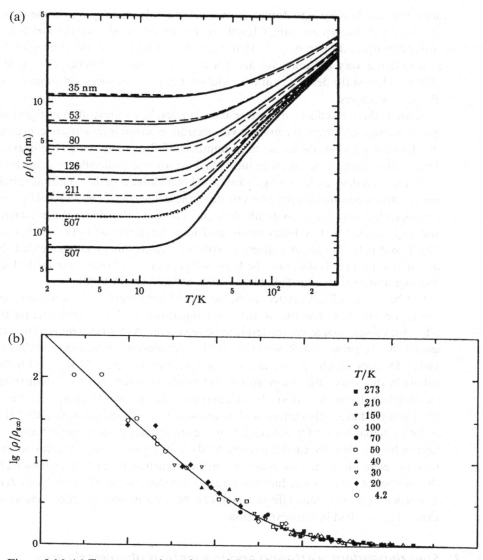

Figure 8.16. (a) Temperature dependence of the resistivity of Au(111) films of different thicknesses grown on mica, showing strongly thickness dependent residual resistivity; (b) the same data on a log–log plot of ρ/ρ_∞ versus $\kappa = d/\Lambda_\infty$ (after Sambles *et al.* 1982, reproduced with permission).

can also be strong. Thus from here on we consider both internal interfaces and external surfaces as sources of scattering.

Electron transport in semiconductor structures with dimensions down to a few nanometers have many interesting (and sometimes disturbing) properties for future devices. There is a limit due to the statistical distribution of donors, acceptors and scattering centers. A simple calculation shows that a device based on 10^{18} cm^{-3} donors or acceptors, a relatively high doping level, will have only one impurity somewhere in the width of a 10 nm square wire. Chaotic effects are expected once the number of electrons/ impurities drop to below ~ 10 per device cross section, so the size region < 30 nm is considered very dubious. We are not there yet: the next generation of 0.13 μm (130 nm) linewidth devices is in the pipeline, but it is not too soon to start worrying about these topics.

Second, the size effect on electronic energy levels, as illustrated in figure 8.14 for quantum well structures, means that via electron–electron interactions, energy levels in the device depend on the occupation numbers of the different electron energy levels. This is of course the same as in individual atoms and molecules, but in the case of > 10 nm sized devices, the energy levels are much closer together. Thus at suitably low temperatures, conduction processes through these devices will be affected by the effects known collectively as the Coulomb blockade. This is a rich field for theoretical research and experiments at low temperature and high magnetic fields (Ferry & Goodnick 1997), and is the origin of collective aspects of electronic behavior behind the fractional quantum Hall effect and the 1998 Nobel prize for Physics awarded to Luttinger, Störmer and Tsui (Mellor & Benedict 1998, Schwarzschild 1998).

But before we all get carried away, we should note that the reason these subtleties can be observed at all is due to the low temperature and high magnetic fields which allow the closely spaced energy levels to be separated. As Störmer himself noted on the award of the prize 'No, it won't revolutionize telecommunications' (Schwarzschild 1998). Devices which operate at room temperature (or even 77 K) need to be more robust in this sense. But many groups are involved in the race to demonstrate single electron transistors (SETs) which work at room temperature. The active regions of such devices must have a characteristic dimension < 10 nm (Devoret & Glattli 1998). This is the principal reason for believing that if even smaller devices are to become important in future, then the carrier density needs to be higher than in typical semiconductors. For example, such devices are increasingly sensitive to random radiation effects as the number of carriers is reduced. Interest is therefore turning to metallic systems, and in particular to magnetic effects which can be used in non-volatile devices; some of these are described in the next section.

8.3.3   Spin dependent scattering and magnetic multilayer devices

The giant magneto-resistance (GMR) effect is the reduction of (longitudinal) resistance in the presence of a (parallel) magnetic field, typically in a magnetic multilayer. In a large field, where the magnetization in all the layers are lined up, the 'spin-flip' scattering of the conduction electrons is minimized, whereas when some of the layers are aligned antiparallel it is greater. The biggest effects observed are changes of up to 80% of the resistance at high fields at low temperatures. Note that the *sign* of the magnetoresistance in ferromagnetic materials, decreasing as the field is increased, is opposite to that in normal metals, where the helical paths followed by electrons in an external field yield more opportunities for scattering. This distinction is spelled out in an influential report (Falicov *et al.* 1990), which was largely responsible for setting the agenda for research on magnetic materials and thin film devices during the 1990s. The push is now on to integrate magnetic superlattices with semiconductors, with the goal of high density non-volatile memory a realistic prospect in the not too distant future (DeBoeck & Borghs 1999, Daughton *et al.* 1999).

In magnetic materials we have to consider that there are two resistivity channels[2], $\rho\uparrow$ and $\rho\downarrow$. At low temperatures in ferromagnetic materials, spin-flip scattering is frozen out, meaning that the spin-up and spin-down channels behave independently and that conductivities add as $\sigma = \sigma\uparrow + \sigma\downarrow$, or equivalently $\rho^{-1} = (\rho\uparrow)^{-1} + (\rho\downarrow)^{-1}$. Within each resistivity channel, we have the different scattering mechanisms contributing in proportion to inverse relaxation times as discussed in section 8.3.1, and at finite temperatures magnon scattering, which tends to equalize the contributions of the two spin channels, is also possible. Thus it is not surprising that this topic can get quite complicated quite quickly; a useful introduction in the context of magnetic multilayer devices, where spin dependent scattering at interfaces is the most important effect, is given by Fert & Bruno (1994). Spin dependent scattering of the same type has also been demonstrated at aligned domain walls in pure Co and Ni films (Gregg *et al.* 1996).

Multilayer devices can be constructed in various different geometries, limiting cases being when the current is either in the plane (CIP) or perpendicular to the plane (CPP). The way in which one of the two resistivity channels can short circuit the other depends on the device geometry. For a sizable device, the CIP geometry tends to have the higher resistance, but the boundaries between the layers are less effective in producing spin flips than in the CPP geometry, which has higher *resistivity*. But the low *resistance* of the conventional CPP geometry means that the effects due to the thin device are very difficult to measure, because all the other resistances add in series. This CPP problem has lead some workers to go to considerable lengths to create structures in the CAP geometry, with the current at an *angle* to the plane, which can be done by using ridged substrates at an angle θ, is illustrated in figure 8.17(a).

The conductivity in the CAP geometry combines those of the other geometries as

$$\sigma_{CAP} = \sigma_{CIP}\cos^2\theta + \sigma_{CPP}\sin^2\theta, \tag{8.14}$$

and there are similar formulae for the combination magneto-resistance (Levy *et al.* 1995). Ono & Shinjo (1995) and Ono *et al.* (1997) have used Si(001) wafers, and have etched V-grooves with {111} facets into this substrate, on which the multilayers are grown. By varying the angle of the current direction ϕ, they could measure σ_{CAP} and σ_{CIP} on the same samples, and use interpolation formulae to estimate σ_{CPP}. The results of a particular sample are shown in figure 8.17(b); this corresponds to a 4 μm thick, 91-layer stack of four individual layers with composition Co(1.2)/Cu(11.6)/NiFe(1.2)/Cu(11.6), where the layer thickness in nm is given in brackets. Note that the MR ratio for this system in the CAP geometry reaches almost 50% at low temperature and is around 10% at room temperature. Although one could use higher values, these are

---

[2] This notation allows one to write an article entitled *The art of sp↑n electron↓cs* (Gregg *et al.* 1997), but the large number of authors on this paper suggests that they know they could only do this once and get away with it.

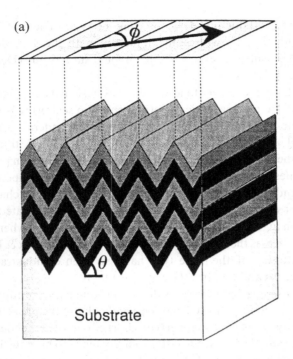

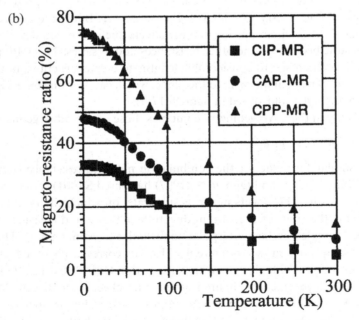

Figure 8.17. (a) Schematic illustration of a multilayer constructed on a substrate containing V-grooves at characteristic angle θ, with the current flowing in a direction given by the angle ϕ; (b) the measured magneto-resistance observed as a function of temperature in the CIP and CAP geometries, and the estimated CPP results, for a composite multilayer of Co/Cu/NiFe/Cu as described in the text (from Shinjo & Ono 1996, reproduced with permission).

good enough for magnetic field sensors, and read/record heads using GMR multilayers are already commercially available.

In the above example, the nonmagnetic spacer layers (of Cu) were sufficiently thick to ensure that the magnetism of the Co and NiFe layers are not coupled directly via the conduction electrons. This is in contrast to the much thinner spacer layers discussed in section 6.3.4 and illustrated in figure 6.24. At intermediate thicknesses, this spin-dependent coupling leads to complex magnetization curves, and the transfer of spin 'information' from one part of a device to another, which can be used in spin valves or spin transistors (Parkin *et al.* 1991, Johnson 1993, 1996, Parkin 1994, Monsma *et al.* 1995). The next stage may be to make use of spin polarized currents induced by magnetic elements into the substrate itself, and then to use these currents as the injector for a hot electron device (Gregg *et al.* 1996). The acronym for this UK-based development, SPICE (spin polarized injection current emitter), may have something to do with the existence of a popular all-female vocal group of the same name at around the same time. We shall see what becomes of both.

Another possible way forward, not involving magnetic coupling through the substrate, is to use magnetic wires grown on insulators; these wires have anisotropic magnetic properties, as well as being a more favorable geometry for CPP GMR measurements. NaCl(110) is a substrate with a high surface energy, and self-organizes into facets on (001) planes at 45° to the substrate plane. By deposition at a shallow angle, narrow wires will be produced at the tops of the ridges shown in figure 8.18(a), and these wires can then be capped to prevent oxidation, etc. In a series of experiments, Sugawara *et al.* (1997) first deposited a thin SiO layer on either NaCl(110) or (111), followed by Fe deposition at a shallow angle, followed by a further SiO layer.

This procedure allowed them to produce isolated islands aligned in one dimension, continuous parallel wires, or isolated Fe dots of various sizes. They could then remove the SiO/Fe/SiO assembly by dissolution of the substrate for TEM examination, and to make particle and wire density observations, exactly as described here in section 5.3, as shown in figure 8.18(b). The new feature is of course the ability to perform magnetic and magneto-optical (Kerr) measurements before this stage, similar to that shown here in figure 6.18, and hence explore magnetic anisotropy, dipole couplings and the paramagnetic to ferromagnetic transition as a function of particle size (Sugawara & Scheinfein 1997). There are a large range of parameters to explore, just within this one system, if anyone wants to take these results to the next stage of implementation as a working device.

Another system which clearly shows promise as a magneto-optical device is based on the nucleation of Co dots on Au(111), at the position of the surface vacancies which occur at the intersection of the (23×1) herringbone reconstruction, first observed by Voigtländer *et al.* (1991) and described here in section 5.5.3. A strong Kerr effect signal from ML deposits in these dots has been observed by Takashita *et al.* (1996), and Fruchart *et al.* (1999) have constructed well ordered Co pillars in Co/Au (111) multilayers, with improved magnetic properties. Whether or not this system will end up in a real device is not clear: are we ready to *use* surface reconstructions and surface point defects so directly in a manufacturing process? Whatever the

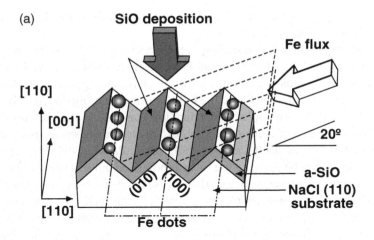

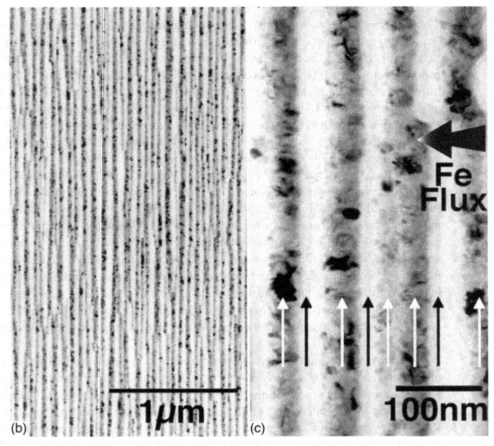

Figure 8.18. (a) Shadow deposition at glancing angle onto ridged substrates to produce Fe dots and, at higher coverage, nanowires; TEM pictures at lower (b) and higher (c) magnification, showing the length and width of the Fe wires in the SiO/Fe/SiO assembly (from Sugawara *et al.* 1997, reproduced with permission).

answer, these surface processes can be responsible for observed structures with lateral length scales in the 10 nm range.

8.4 Chemical routes to manufacturing

Although this book, including this chapter, has been written largely from the perspective of materials physics, materials-oriented chemistry plays an enormous role in device development. Chemists are also being drawn into the field of *combinatorial materials discovery*, a practically based mix of combinatorial chemistry and thin film deposition techniques, used to search for new compounds and compositions. Such developments are discussed briefly in this section.

8.4.1 *Synthetic chemistry and manufacturing: the case of Si–Ge–C*

The growth of Ge and SiGe alloys on Si(001) was considered in some detail in section 7.3.3. It is the simplest example of strained layer growth, and we described how the low mismatch alloys can be grown as layers suitable for heterostructures, or the strain can induce islands to form, which may (or may not) have potential as quantum dots. But do we want islands, or would layers be better? The answer, of course, depends on the application envisaged, but in general layers are preferable. So let's try mixing in another element, which relieves the strain and so promotes layer growth. This is the motivation behind research on the Si–Ge–C system; by mixing in the right amount of carbon one can hope to compensate for the strain introduced by Ge.

But how to do it? CVD is the most widely used manufacturing technique, starting from silane (SiH_4) and germane (GeH_4), so mix in some methane (CH_4) and see what happens. But some knowledge of thermochemistry is required, and this tells us that methane is much more stable both than silane and than germane, which means that CH_4 does not want to break up. Moreover, the solubility of C in Si or Ge is very low ($\sim 1 \times 10^{-6}\%$ in equilibrium at the Si melting point, and even lower in Ge), so the carbon which does form is likely to be in the form of SiC, which is not what we want; in practice up to 1–2% has been incorporated at SiGe CVD process temperatures $\sim 700\,°C$. Somehow we have to trick the system, thermodynamically and kinetically. This has been done by creating custom-built molecular precursors to use as the source material.

One group in Arizona (Kouvetakis *et al.* 1998a) has been developing compounds with an inbuilt tetrahedral arrangement involving one C atom surrounded by four SiH_3 or GeH_3 ligands, as illustrated in figure 8.19(a). These are sizable van der Waals molecules, which are liquids at room temperature and evaporate easily. They are also reactive, in that they lose hydrogen easily, and hence can incorporate the tetrahedral units Si_4C and Ge_4C into a growing film.

These molecules are crowded, and have bond lengths very different from normal, as indicated in the cluster calculation for Ge_nC shown in figure 8.19(b). This is because the small C atom is in the middle, and the much larger Si and especially Ge atoms are

(a)

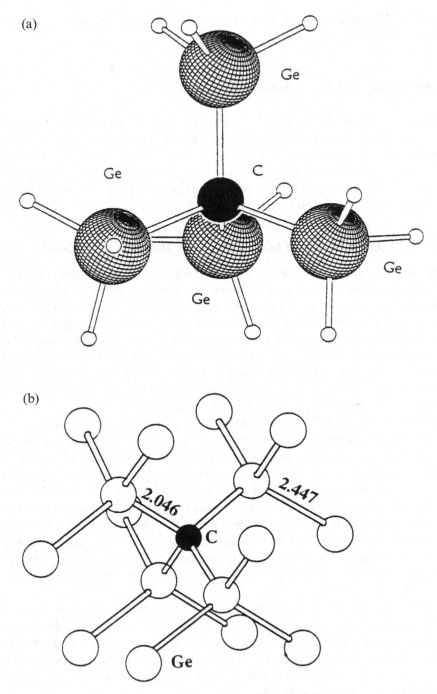

(b)

Figure 8.19. (a) The precursor molecule $(GeH_3)_3C$, with Ge–C and other bond lengths determined by gas phase electron diffraction; (b) structure of substitutional C in Ge and the associated Ge–C and Ge–Ge bond lengths in a calculated $(Ge)_nC$ cluster (after Kouvetakis *et al.* 1998a,b, reproduced with permission).

competing for space around it. However, this processing route suppresses the segregation of the C atoms, because they are always surrounded by Si or Ge, and Kouvetakis *et al.* (1998b) have been successful in incorporating 5–6%C into films grown at 470 °C, which have relatively few defects (and no SiC precipitates) as seen by TEM. Starting molecules with carbon on the outside would not be nearly as effective in this respect; in other words, by using these special molecules and relatively low processing temperatures, carbon has been tricked to remain in solution.

Whether this work represents a real breakthrough or just a very interesting development depends on the next stage – will the clever molecules be incorporated into actual manufacturing processes or not? The role of relatively small contract research firms in smoothing the path to manufacturing is interesting; such processes will certainly not be incorporated into large scale Fab lines overnight. Increasingly, it is the equipment manufacturers who incorporate new processes such as this one, in order to be able to persuade the large scale producer (of e.g. computer chips) to adopt such technologies when they reinvest the next few US$ billion.[3]

8.4.2　Chemical routes to opto-electronics and/or nano-magnetics

Optoelectronics, the integration of light sensitive devices with micro-electronics, is another huge field in which surfaces and thin films play a major role. One of the main interests in the growth of III–V and II–VI compounds is related to the ability to integrate such devices with silicon in the form of 'band gap engineering' as discussed in sections 7.3.4 and 8.2.3. Quantum dots are a hot topic touched on in section 8.2.4, where the intent is that uniform size and spacing can be achieved via self-organization. But, as always in technology, one has always to bear in mind that the most effective solution to the original challenge or problem, might come from a completely different route. One such possibility is the assembly and self-organization of ordered arrays of colloidal particles prepared by more or less traditional wet chemistry methods – a flask full of goo, a drying oven and a spray-gun may be all that is needed; I exaggerate, of course, but not by much. Such techniques are not specific to optoelectronics; magnetic particles can just as well be prepared and arranged in remarkably uniform arrays, as shown in figure 8.20.

Here the idea is to prepare II–VI materials such as CdSe in solutions containing additives which adsorb on the surface of colloidal crystals in the size range 5–10 nm, thus preventing further growth. These colloids, coated with self-assembled monolayers (SAMs), form a stable dispersion in solution, which can be made to crystallize out by gentle evaporation of the lighter component in the solution (e.g. octane from octanol at 80 °C). Further warming removes all but a few ML of the additives, to the extent that 3D colloidal superlattice crystals can be grown with sizes up to 50 μm (Murray *et al.* 1995, Heath 1995, Collier *et al.* 1998).

The size distribution is amazingly narrow, as shown in figure 8.20, and can be further controlled by manipulation of the supersaturation as a function of time during growth.

---

[3] Financiers should note that the US billion is used here which is *only* 10^9, rather than the UK 10^{12}.

(a)

(b)

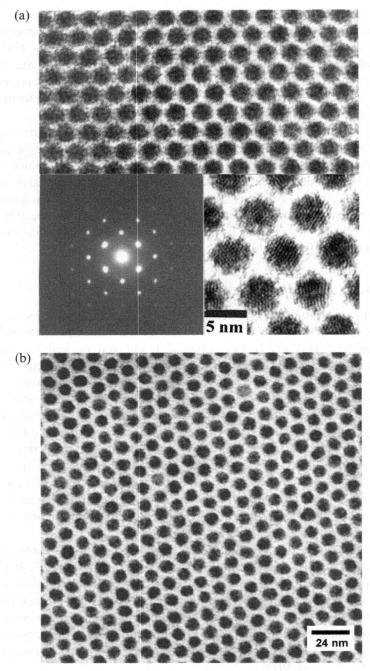

Figure 8.20. (a) 3D Optical superlattice formed from colloidal crystals of CdSe, spacing 6.5 nm, diameter 5 nm (similar to Murray *et al.* 1995); (b) 2D magnetic superlattice of Ag coated Co particles, spacing 13 nm, diameter 8 nm formed by colloidal techniques (after Murray 1999, both figures reproduced with permission). In both these cases the 2D hexagonal lattice extends over much larger distances than can be effectively portrayed here.

Other clever tricks are to cap these dots with a compatible material such as CdS having a higher refractive index or to introduce relevant dye sensitizers onto the surface of the dots. This latter technique has been a key ingredient in the color photographic processes using AgBr crystals for a long time. By the former means the confinement of the exciton wavefunction can be increased, and so produce stronger photoluminescence, quantum yields $>50\%$ having been demonstrated (Peng et al. 1997, Schlamp et al. 1997). However, until now the overall electro-luminescent energy efficiency has been low, and the devices are not yet stable enough for production (Peng et al. 1997, Alivisatos 1998, Collier et al. 1998).

A single electron transistor which functions at $T=4.2$ K has been demonstrated, simply by sprinkling individual colloidal dots of CdSe across the gate region of an FET (Klein et al. 1997). As a way forward, you might take this demonstration 'with a pinch of salt', but it is certainly spectacular. One of the arguments in favor of these colloidal crystals is their intrinsic cheapness, yet if there is no way of getting current in or out, or if they only work at low temperature, then we have just an intriguing demonstration, not yet an innovation. Another impressive demonstration is an FET which works at room temperature, made by dropping a single carbon nanotube across the source-drain gap (Tans et al. 1998a), and the observation of associated Coulomb blockade phenomena (Tans et al. 1998b).

8.4.3 Nanotubes and the future of flat panel TV

A further example centers around chemical routes to the production of field emission sources for computer or TV screens. As discussed in section 6.2.2, various carbides and nitrides have been researched over the years, but have never really been quite stable enough to make a serious impression on the market. Yet the conventional TV tube is the last remaining example of the vacuum triode in production; is it too destined for oblivion? If field emission could be made to work reliably in the planar geometry shown in figure 6.13(b), then maybe it could be saved!

These thoughts have gained impetus from the discovery of both multiple and single-walled carbon nanotubes, cylindrical intermediates between planar graphite and the closed cage fullerenes (Iijima 1991, Ebbesen & Ajayan 1992, Ajayan & Ebbesen 1997, Bernholc et al. 1997). The tubes grow as long filaments sticking out from the substrate, and emit electrons from the ends. Filaments can be produced in bundles or matrix arrays (Collins & Zettl 1997), which differ in detail depending on the production method. Using a plasma arc discharge, Terrones et al. (1998) have produced filaments containing a transition metal carbide particle near the end. The carbide particle catalytically converts C-containing compounds into tubes, almost as if it were knitting a sock, as shown in figure 8.21. Several different refractory carbides have been so encapsulated, and progress to date is reviewed by Terrones et al. (1999).

Nanotubes can join up in helical (chiral) as well as cylindrical geometries, p-n junctions along the length have been demonstrated where the chirality changes, and they also can be doped, or made, with various B–N–C mixtures. The case of BN has led to some exquisitely delicate analytical microscopy, combining pictures analogous to

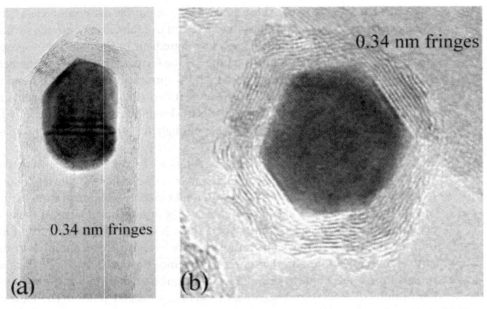

Figure 8.21. High resolution TEM images of TaC particles in nanotubes where (a) the TaC particles are encapsulated at tips, and (b) graphite fringes suggest epitaxial growth on the particles (from Terrones *et al.* 1998, reproduced with permission).

figure 8.21 with compositional analysis via EELS on the nanometer scale (Golberg *et al.* 1996, 1997). With this number of variables to explore, it seems only a matter of time before practical displays can be devised to rival those already demonstrated for the Spindt cathode and DLC geometries. But of course that does not mean that such a display will necessarily be successful in the marketplace, where it has to compete with all other flat display technologies as discussed earlier in section 6.2.2, and/or find its own particular niche. The advantage is the high intrinsic brightness, which may make it suitable for projection onto a large screen; the list of potential disadvantages will distinguish optimists from pessimists.

Note that in the above two examples the distinction between surfaces, thin films and molecules has all but disappeared: the molecule is *all surface* and *is* the thin film. What we are seeing here is an important stage in the development of single molecule electronics. The next stage is again the transition to manufacturing, which may well be a rocky road, and test to the limit who is serious. As they say on TV: 'don't go away', or equivalently 'watch this space'!

8.4.4 Combinatorial materials development and analysis

The combination of step-wise thin film deposition processes with analysis on a microscopic scale is coming to be known as combinatorial materials development. This series of techniques uses automated deposition and annealing sequences to produce a series of different compositions, thicknesses and/or doping levels in a matrix pattern of dots distributed in x and y. Different patterns can then be set up in several areas on the same chip, and then analyzed by a microscopic technique to screen for the 'best' properties. Typically, such techniques will be useful when one has to span a large range of parameters, and when theoretical models are of limited use. Stepper motors controlling stage movements and multiple deposition shutters, with a resolution of a few micrometers are now available more or less routinely (if expensively), which means that many thousands of different samples can be screened in a single experiment.

An example is the search for improved red phosphor materials to be used in flat panel displays (Danielson *et al.* 1997). The starting point materials were polycrystalline ternary or quaternary oxide layers acting as hosts for 'activator' rare earth ions such as Eu or Ce. Some 25 000 individual compositions deposited onto a 3-inch wafer were sampled, leading to a single best composition of $Y_{0.845}Al_{0.070}La_{0.060}Eu_{0.025}VO_4$. This composition was found to be as good as the existing commercial phosphors. It is clearly just a question of time before superior thin film materials are found using such techniques, which originated in the pharmaceutical industry as a means of accelerated drug discovery.

The next stage may well be to combine such approaches with patterned substrates, so that different areas explore different surface treatments or different growth regimes. A start in this direction, with emphasis on molecular recognition of metal and semi-conductor nanocrystals, has been made by Vossmeyer *et al.* (1998), and patterned arrays of bio-macromolecules have also been demonstrated. *Flexible* patterned substrates are also being produced via soft lithography, micro-contact printing and related techniques (Xia & Whitesides 1998). Such approaches involving microarrays of DNA are centrally involved in the future of the human genome project (DeRisi *et al.* 1997). The huge interest in microbiology means that such experiments, including seletively tagging the colloidal nanoparticles described in the section 8.4.2, are achieving widespread recognition in the materials community (Mirkin *et al.* 1996, 1999), even if, or perhaps because, most of the rest of us are very early on the learning curve. But we can all recognize the implied potential of the field.

All this suggests that we should consider creating patterned nests for microbes, so that we can then sit back and let them do our work for us. Indeed Richard Feynman was first with this suggestion in a famous lecture in 1959 entitled *'There's plenty of room at the bottom'*, republished as Feynman (1992) in a new *Journal of Microelectromechanical Systems*. I'll bet that H.G. Wells had the basic idea well before that – is there anything really new? But this is getting dangerously close to futurology, the proper business of the twenty-first century, not the twentieth. As this chapter is finalized, in December 1999, the decoding of chromosome 22 made headline news, special millennium issues of the journals arrived, and I started to read articles entitled 'The *Net* Century', etc. A few words on such topics and the educational/training implications in the short final chapter, and I'm done. The twenty-first century is essentially *yours*: good luck!

Further reading for chapter 8

Davies, J.H. (1998) *The Physics of Low-dimensional Semiconductors: an Introduction* (Cambridge University Press).

Jaros, M. (1989) *Physics and Applications of Semiconductor Microstructures* (Oxford University Press).

Ferry, D.K. & S.M. Goodnick (1997) *Transport in Nanostructures* (Cambridge University Press), chapters 1–4.

Kelly, M.J. (1995) *Low-dimensional Semiconductors* (Oxford University Press).

Lüth, H. (1993/5) *Surfaces and Interfaces of Solid Surfaces* (2nd/3rd Edns, Springer) chapters 7 and 8.

Rossiter, P.L. (1987) *The Electrical Resistivity of Metals and Alloys* (Cambridge University Press).

Sutton, A.P. & R.W. Balluffi (1995) *Interfaces in Crystalline Materials* (Oxford University Press) chapter 11.

Schroder, D.K. (1998) *Semiconductor Material and Device Characterization* (John Wiley).

Sze, S.M. (1981) *Physics of Semiconductor Devices* (2nd Edn, John Wiley) chapter 5.

Tinkham, M. (1996) *Introduction to Superconductivity* (2nd Edn, McGraw-Hill).

9 Postscript – where do we go from here?

This short postscript wraps up the book, and provides pointers to further reading and gathering information. The comments are personal impressions rather than cut and dried issues. Please take them suitably lightly – and then get on with the rest of your life which is, of course, all too short. *Time is the only real enemy* – someone must have said that before.

9.1 Electromigration and other degradation effects in nanostructures

Degradation over time is an important part of materials science, and is inherent in the metastabilty of all artificially tailored structures. It is a major reason why I will have to replace the laptop computer, on which I have been composing this book, in the not-too-distant future. There are many surface and thin-film related degradation processes involved, some of which have been discussed in the book, and some of which have been omitted for reasons of space and time. Others are being researched and could use a good review article. However, if I don't finish this book now, I never will (another example of degradation over time). I need to get on with other aspects of my life and so do you.

Polycrystalline metal wires forming interconnects can fail via necking of 'bamboo' structures, for the same basic reason as the tungsten filaments discussed in section 6.2.1. Such a wire has slight changes in thickness along its length, being slightly thinner where grain boundaries cross the wire, and slightly thicker either side (hence the name bamboo). When a current is passed, the resistance heating is greater in the grain boundary regions mostly because of the smaller cross section, and this leads to atom migration, principally by surface and grain boundary diffusion, towards the equilibrium structure.

Once we have a thin wire, the surface energy is important, and we know from the discussion in section 1.2 that the equilibrium shape is close to a sphere, and is given by the Wulff plot. Thus a thin wire tries to become shorter and more rounded, eliminating grain boundaries as it coarsens. This will happen more quickly for higher current densities, and at higher operating temperatures, both of which are implicated in smaller devices. Needless to say, eventually these effects sever the wire in two: bad news for me, and doubtless well calculated by the computer manufacturer who would love me to purchase a new one, subject to avoiding recalls on existing models and law suits. Under accelerated test conditions at high temperatures, this contraction, or expansion of a wire under load, forms the zero creep method of measuring surface energy, as mentioned in section 1.2.1.

Metal interconnects are very important components in integrated circuits (ICs), and have been mostly been made of aluminum, with copper being actively developed as a replacement for use at higher current densities. As wires and FET gate electrodes, these metals are laid down typically on the amorphous SiO_2 which is the insulating surface on Si-based MOS devices; they will grow as polycrystalline layers by nucleation and growth processes similar to those described in chapter 5. As contacts to silicon, they can be either ohmic (to get current in or out) or form control elements as in the Schottky barrier discussed in section 8.2.

Most metals are in fact very reactive to silicon itself. Metal–semiconductor interactions is an enormous topic, which has been visited at various points in this book, but not described systematically. This is an area where much work has been done, but it feels to me as if the definitive review has not yet been written; in any case it probably needs a whole book to itself. It is important to note here that substantial rearrangements of both metal and semiconductor at the interface results from this reactivity, as in the case of Ag/Ge or Si(111) shown in figure 1.20, and illustrated on the cover design. In many cases barrier layers consisting of more tightly bound metals, alloys and/or silicides, have been used to slow down interdiffusion across these interfaces and so prolong device lifetimes (Lloyd 1999).

Experiments on many metals deposited on semiconductors have investigated these interactions. As these are performed at elevated temperatures, some means of heating the semiconductor substrate is needed. Often simple direct current has been used, and as a result electromigration effects have been uncovered, that is, a directional movement of material due to a combination of electric potential and current. These effects have been reviewed by Yasunaga & Natori (1992), and have been found to occur in many systems since; it is also a topic which probably needs an update in review form.

As we look towards semiconductor devices during the twenty-first century, we can see that the issues aired in chapter 8 and in this section are capturing the imagination of relevant scientists and technologists (see, for example, Williams 1999). Although they all agree about the current economic dominance of CMOS Al–Si–SiO_2 based IC technology, there are increasing discussions of 'the end of the roadmap' or 'what happens below 0.1 μm [feature size]?' or 'what are the costs of the next generation of Fabs?'. These concerns center around what became known as Moore's law, named after one of the founders of Intel, who noted the exponential decrease of IC feature size as with time, corresponding roughly to a factor of two every three years over several decades.

Much of this information is being gleaned by science journalists, who can distill journal articles, and combine them with interviews of key industrial and academic *players*, in language which may be accessible to a wider public. One such article, published under the title *Failure analysis in the nanometer world*, quotes the head of failure analysis at Texas Instruments, Lawrence Wagner, to the effect that finding a fault in an

IC today is like 'looking for your lost keys from a helicopter hovering over Los Angeles'; by the end of the next decade, it is going to be like 'looking for your keys in the state of California' (Ouellette 1998). Thus we may well get the latest (and in this case some of the *best*) information from the news and comments sections of professional and trade journals such as *Physics Today, Physics World, The Industrial Physicist, Materials Research Bulletin*, or *Vacuum Solutions*, from special issues of magazines such as *New Scientist* and *Scientific American*, or even from major newspapers.

These articles tend not to be quoted in the archive literature; the thoughts expressed are not necessarily original, but they are *hot*. An example is the New Scientist special report of 7 November 1998: *Inventing the future in Silicon Valley* (Mullins *et al.* 1998). This issue, for example, discusses copper interconnects, and refers to the journal *Future Fab International*, which I have yet to locate. Who would have thought that such a title existed without skipping though these general interest articles? Even in the age of computerized information, such skipping is good exercise. However, in case you envisage a future weighed down with new subscriptions born out of desperation to keep up with the latest news, you can in fact relax. All these publications are (along with their advertisers) trying to get *your* attention, and similar stories will appear in many places. Another story on plans for copper interconnects and sub-0.2 μm technology in the major firms appeared in the March 1999 issue of *Semiconductor International*; it will not be the last such article. The advantages of copper, given its reactivity, are hotly disputed (Lloyd 1999); this is another reason to leave the topic out of this book.

9.2 What do the various disciplines bring to the table?

Readers who have stayed with me thus far will have realized that by now I am skating on thin ice (which is also good exercise until the ice cracks). I am coming to the end of the book, and also to the limits of my own professional expertise. My own training is in physics, and my professional interests lie mostly in the physics of materials, and within that in the subjects aired here. So my opinions outside this area may well not be professionally sound, and maybe I should keep them to myself in any case. But it is also interesting to share ideas and see how others react.

Several professions have contributed to the topics discussed in this book. As judged by the initial qualifications of the workers cited in the extensive reference list, this includes mathematics and computer science, physics, chemistry, materials science and engineering, and various branches of electrical, electronic, mechanical and chemical engineering. This list has not to date included biology, biochemical and medical sciences and biotechnology, but that is in the process of changing, and will change quickly in future. One of the most understated success stories over the past century is the impact that discoveries in physics have had on medical diagnosis and treatment; this achievement is simply vast, and should be more widely appreciated by the general public. One doesn't have to be clairvoyant to foresee a similar impact for micro-engineered surfaces and thin film devices, with chemical and/or biological selectivity and sensitivity, coupled to computer-based readout and analysis systems (Collins 1999). Does the range of expertise needed to function in these areas mean that subject specialisms will disappear in future? I don't think so. But it does mean that subject specialists are almost certain to have to work in teams, with a proper appreciation of the other person's/ subject's strengths and weaknesses.

The strength of the physicist's approach is that it looks for overarching themes, such as gravity, thermodynamics, quantum and statistical mechanics, and relativity, which have immense generality and thus condition our ways of thinking. Experimental physicists are adept at discovering and developing new instruments, at least in prototype form, and these can completely change our perception of what is observable and interesting. The most obvious example described here is the development of the various types of microscope (TEM, FIM, STM, etc., described in chapter 3) and their application to the study of kinetic processes described in chapter 5. Physicists are also good (but not uniquely so) at making simple models of such processes, provided they remember the quote from Einstein used by Pettifor (1995) 'as simple as possible, but not simpler'. Other disciplines are fond of quoting a joke against physicists, asked to make a model of a racehorse, who start off 'imagine a spherical lump of muscle' and go on to talk about symmetry breaking, which leads to the production of a head, a tail and four legs. There are many variants on this theme; such an approach can lead to a reputation for arrogance which is counterproductive.

Chemists see themselves as centrally positioned in science, the guardians of the periodic table, and the makers of new molecules *par excellence*. Historically this is indeed the case, and it remains so, though perhaps less celebrated, to the present. In my experience, a training in chemistry results in the ability to assimilate huge numbers of facts, often apparently unrelated, and then to try to make sense of them within whichever model is to hand. Quantum chemistry is a great success on the theoretical side, even if it must be annoying that sometimes the Nobel prizes for chemistry end up in the hands of people trained as mathematicians and physicists, as happened in 1998, noted here in chapters 6 and 7 and in the corresponding appendices J and K.

In parallel, experimental synthetic chemists try out large numbers of combinations of reactions to produce new molecules, which includes new synthetic materials and/or drugs, thus overlapping with materials scientists or biochemists, some of which was described in section 8.4. Coupled with this is an ability to spot the main chance: to decide an area is going nowhere and to move on to something more productive, which is an ability that more introspective scientists sometimes lack. This is of course not unique to chemists, but a chemists' training seems to encourage it. Thus it is perhaps not surprising that chemists, as much as engineers, are key to the production of electronic devices by the various forms of CVD, described here briefly in section 2.5, which is the main technology behind most of the semiconductor devices discussed in chapters 7 and 8. This technology is very largely empirically based, but functions continuously on a massive scale, and remains a challenge to analytical science and to process modeling. Chemists play similar roles in catalysis-based technologies in the petrochemical and related industry, indicated briefly in sections 2.4 and 4.5.

Surface and thin film processes is an area where the boundaries between physics, chemistry, materials science and engineering are particularly transparent, but even here there are different emphases in the choice of topic, and some mismatches in language which can take time and effort to sort out. The materials scientist and engineer also has a very broad training, one which has emerged from the traditional disciplines of mining and metallurgy. Many classes of crystalline and polymeric materials in bulk and thin film form are studied nowadays, of which this book has only discussed elemental adsorbed gases, crystalline metals and semiconductors, and some ionic compounds in any detail.

A training in materials science emphasizes materials properties and materials selection, and the role of microstructure in determining properties, involving the use of microscopy in all its analytical forms. Such instrumentation is needed to visualize, analyze and hence even to discuss the development of smaller devices which are now being developed. However, too much emphasis on microscopy alone can sometimes obscure the fact that much of our knowledge of the microscopic world has been achieved by techniques which average over relatively large areas, coupled with a well-founded model which interprets the data obtained at a microscopic level. This approach is exemplified by crystal and protein structure determination using X-rays, and is illustrated here in section 3.2 by surface structure determination using electrons, especially in conjunction with the corresponding microscopy.

The fields described in this book furnish good examples of serial versus parallel processing. Point by point techniques, such as STM, SEM and STEM are ideally suited to computer control and TV-like output, whereas the parallel processing diffraction-microscopy techniques (LEED-LEEM, RHEED-REM or THEED-TEM) are in general much faster, as emphasized in chapter 3. This serial-parallel choice is central in engineering terms, in manufacturing and inspection, and is a major factor in the rise of the 'throw-away' society. If finding individual faults in an IC will be like 'looking for your keys in California', then serial processing, and individual repair, techniques are at a great disadvantage compared to parallel processing of raw materials into new products.

9.3 What has been left out: future sources of information

It is of course a pity that the field of *Surface and thin film processes* will not stay still once I have finished this book. Such a thought condemns the writer to wanting to update the manuscript continually, to hope for further editions in which all mistakes will be corrected, and generally to wish that everything could be perfect in a manifestly imperfect world. In the present case I have heaps of reprints, half-formed ideas and vague thoughts on a wide range of topics from mechanical properties, indentation and wear, the related field of dislocations, quantum conduction in nano-wires, chemical and bio-sensors, microelectromechanical systems (MEMS, yet another acronym), you name it. Did I ever think I would get around to writing about these topics, or that I could master them sufficiently in a finite time or space? What a hope!

Sensors are an interesting case in point, since I mentioned them in the preface: it is therefore reasonable that you would expect me to write about them in the body of the text. These thin film, and maybe surface-related devices, have spawned a huge literature. But the near-impossibility of getting inside it in a finite time finally got to me, after reading a review of chemical sensors by the experts in the field (Janata *et al.* 1994). This review lists and says a few words about each of some 795 publications, notes the rapid year on year growth of this number, and ends with a pretty justified complaint about publishing practices in the field, with a quote adapted from Marvin Minsky, one of the founders of artificial intelligence and neural networks: "'Look', they say, 'it did this'. But they don't consider it equally wonderful to say, 'Look, it can't do that.'" As a result, there are lots of publications in this field, but there are not quite so many working devices.

Their more recent update on the same topic (Janata *et al.* 1998), gives the same treatment to a further 929 publications, and reaches a similar though more positive conclusion. Indeed, as I wrote this section, the June 1999 issue of the MRS Bulletin arrived in the mail, containing a series of articles on *Gas-Sensing Materials* (Watson & Ihokura 1999); this re-emphasizes the importance of belonging to the appropriate professional organization and scanning their journals. There is a lot of material in these reviews, but I know it would be next to impossible to find the full story simply by reading; I am a long way off writing anything useful.[1]

There have to be other, probably more attractive, visions of the future than adding yet more words; I should quit while I'm ahead. The web offers one such vision which is definitely worth exploring; and it is already an amazing resource. Several of the references cited in this book have been found from my office using the excellent WebofScience™. At the same time one can, allowing for errors and omissions, find out the apparent impact of papers, and whether there are more recent reviews on similar subjects. Just as putting my class notes up on the web originated this book, so the web offers a possible way of updating, or otherwise commenting on it, and of developing related teaching activities and projects.

Most of the appendices which follow this chapter are of the type you would normally expect in such a book. Appendix D, on the other hand, is an introduction and link to a series of web-based resources which will be updated, and can be used as a supplement to the book in whatever way seems reasonable to the reader. Some of the other appendices refer to web-based resources which can be accessed via Appendix D. In particular, I think it will be possible to use student projects, with their permission, to develop both background material, and to explore further topics both with students and colleagues, and present them in the form of web-pages for all to benefit. In an ideal world, this could be done collaboratively between co-workers in different fields, institutions and countries. A start has been made here, and I am grateful to colleagues for permission to include links to their web-based material; it will be interesting to see 'where we go from here'.

---

[1] Note added in proof: Dr Janata visited ASU in February 2000 to give a seminar, so we were able to discuss these issues in person. The current publication rate in the sensor field is about 1000 papers per year.

Appendix A
Bibliography

The following references are mainly textbooks, which can be used for obtaining background material and/or extended coverage of the material in this book. The sections where these sources are referred to are indicated in square brackets [*.*].

Adamson, A.W. (1990) *Physical Chemistry of Surfaces* (John Wiley, 5th Edn) [1.1, 1.2].
Ashcroft, N.W. & N.D. Mermin (1976) *Solid State Physics* (Saunders College) [6.1, 6.3, 7.1].
Blakely, J.W. (1973) *Introduction to the Properties of Crystal Surfaces* (Pergamon) [1.1].
Briggs, D. & M.P. Seah (1990) *Practical Surface Analysis, vols. I and II* (John Wiley) [3.3].
Bruch, L.W., M. W. Cole & E. Zaremba (1997) *Physical Adsorption: Forces and Phenomena* (Oxford University Press) [4.2, 4.4].
Buseck, P., J.M. Cowley & L. Eyring (Eds.) (1988) *High Resolution Transmission Electron Microscopy and Associated Techniques* (Oxford University Press) [3.1].
Chen, C.J. (1993) *Introduction to Scanning Tunneling Microscopy* (Oxford University Press) [3.1].
Clarke, L.J. (1985) *Surface Crystallography: an Introduction to Low Energy Electron Diffraction* (John Wiley) [1.4, 3.2].
Craik, D. (1995) *Magnetism: Principles and Applications* (John Wiley) [6.3].
Davies, J.H. (1998) *The Physics of Low-dimensional Semiconductors: an Introduction* (Cambridge University Press) [7.1, 8.2].
Desjonquères, M.C. & D. Spanjaard (1996) *Concepts in Surface Physics* (Springer) [1.1, 1.3, 4.5, 6.1, 7.1, Appendix K].
Dushman, S. & J. Lafferty (1992) *Scientific Foundations of Vacuum Technique* (John Wiley); this updates the 1962 book with the same title by S. Dushman [2.1, 2.3, Appendix G].
Feldman, L.C. & J.W. Mayer (1986) *Fundamentals of Surface and Thin Film Analysis* (North-Holland) [3.1, 3.4].
Ferry, D.K. & S.M. Goodnick (1997) *Transport in Nanostructures* (Cambridge University Press) [8.3].
Gibbs, J.W. (1928, 1948, 1957) *Collected Works, vol. 1* (Yale Univerity Press, New Haven); reproduced as (1961) *The Scientific Papers, vol. 1* (Dover Reprint Series, New York) [1.1].
Gil, B. (Ed.) (1998) *Group III Nitride Semiconductor Compounds* (Oxford University Press) [7.3].

Glocker, D.A. & S.I. Shah (Eds) (1995) *Handbook of Thin Film Process Technology* (Institute of Physics), parts A and B [2.3, 2.5].

Henrich, V.E. & P.A. Cox (1994, 1996) *The Surface Science of Metal Oxides* (Cambridge University Press) [1.4, 4.5].

Hill, T.L. (1960) *An Introduction to Statistical Thermodynamics* (Addison-Wesley, reprinted by Dover 1986), especially chapters 7–9, 15 and 16 [1.3, 4.2, Appendix E].

Hudson, J.B. (1992) *Surface Science: an Introduction* (Butterworth-Heinemann); reprinted in 1998 and published by John Wiley [1.1, 1.2, 2.1, 4.5].

Jaros, M. (1989) *Physics and Applications of Semiconductor Microstructures* (Oxford University Press) [8.2].

Jiles, D. (1991) *Magnetism and Magnetic Materials* (Chapman and Hall) [6.3].

Kelly, A. & G.W. Groves (1970) *Crystallography and Crystal Defects* (Longman) [1.3–4].

Kelly, M.J. (1995) *Low-dimensional Semiconductors* (Oxford University Press) [7.1, 8.2].

King, D.A. & D.P. Woodruff (Eds.) (1997) *Growth and Properties of Ultrathin Epitaxial Layers* (The Chemical Physics of Solid Surfaces and Heterogeneous Catalysis vol **8**, Elsevier) [5.1].

Kittel, C. (1976) *Introduction to Solid State Physics* (6th Edn, John Wiley) [6.3].

Liu, W.K. & M.B. Santos (Eds.) (1998) *Thin Films: Heteroepitaxial Systems* (World Scientific) [5.1, 7.3].

Lüth, H. (1993/5) *Surfaces and Interfaces of Solid Surfaces* (2nd/3rd Edns, Springer) [1.4, 2.2–3, 3.2–3, 7.2, 8.1–2].

Masel, R.I. (1996) *Principles of Adsorption and Reaction on Solid Surfaces* (John Wiley) [4.5].

Matthews, J.W. (Ed.) (1975) *Epitaxial Growth, part A* (Academic) [2.5]; *part B* (Academic) [5.1].

Mönch, W. (1993) *Semiconductor Surfaces and Interfaces* (Springer) [7.2, 8.2].

Moore, J.H., C.C. Davis & M.A. Coplan (1989) *Building Scientific Apparatus* (2nd Edn, Addison-Wesley) [2.3, 3.3].

O'Hanlon, J.F. (1989) *A Users Guide to Vacuum Technology* (John Wiley) [2.3, Appendix G].

Pettifor, D.G. (1995) *Bonding and Structure of Molecules and Solids* (Oxford University Press) [6.1, 7.1, 9.2, Appendix K].

Prutton, M. (1994) *Introduction to Surface Physics* (Oxford University Press) [1.4, 3.2–3, 3.5].

Rivière, J.C. (1990) *Surface Analytical Techniques* (Oxford University Press) [3.1, 3.3].

Rossiter, P.L. (1987) *The Electrical Resistivity of Metals and Alloys* (Cambridge University Press) [8.3].

Roth, A. (1990) *Vacuum Technology* (3rd Edn, North-Holland) [2.1–3, Appendix G].

Schroder, D.K. (1998) *Semiconductor Material and Device Characterization* (John Wiley) [8.1].

Smith, D.L. (1995) *Thin-Film Deposition: Principles and Practice* (McGraw-Hill) [2.5].

Smith, G.C. (1994) *Surface Analysis by Electron Spectroscopy* (Plenum) [3.3–4].

Stroscio, J. & E. Kaiser (Eds.) (1993) *Scanning Tunneling Microscopy* (Methods of Experimental Physics, Academic) vol. 27 [3.1].

Sutton, A.P. (1994) *Electronic Structure of Materials* (Oxford University Press) [6.1, 7.1].

Sutton, A.P. & R.W. Balluffi (1995) *Interfaces in Crystalline Materials* (Oxford University Press) [1.2, 6.1, 7.1, 8.2].

Sze, S.M. (1981) *Physics of Semiconductor Devices* (2nd Edn, John Wiley) [8.1].

Tinkham, M. (1996) *Introduction to Superconductivity* (2nd Edn, McGraw-Hill) [8.3].

Tringides, M.C. (Ed.) (1997) *Surface Diffusion: Atomistic and Collective Processes* (Plenum NATO ASI) **B360** [5.1].

Tsao, J.Y. (1993) *Materials Fundamentals of Molecular Beam Epitaxy* (Academic) [2.5, 7.3].

Walls, J.M. (Ed.) (1990) *Methods of Surface Analysis* (Cambridge University Press) [3.1].

Wiesendanger, R. (1994) *Scanning Probe Microscopy and Spectroscopy* (Cambridge University Press) [3.1, 7.2].

Woodruff, D.P. & T.A. Delchar (1986, 1994) *Modern Techniques of Surface Science* (Cambridge University Press) [3.2, 6.1].

Yu, P.Y. & M. Cardona (1996) *Fundamentals of Semiconductors: Physics and Materials Properties* (Springer) [7.1].

Zangwill, A. (1988) *Physics at Surfaces* (Cambridge University Press) [4.4–5, 6.1, 7.1–2].

Appendix B
List of acronyms

The following list gives the names of the acronyms used; the sections where they are first or most relevantly introduced are given in brackets.

0D, 1D, 2D, 3D: Zero, one, two or three (dimensions or dimensional) [throughout] as in 2DEG: two dimensional electron gas [8.2.3]

AES: Auger electron spectroscopy [3.3, 3.4]

AFM: atomic force microscopy [3.1.3]

ALE: atomic layer epitaxy [1.4.4]

APCVD: atmospheric pressure CVD [2.5.5]

APW: augmented plane waves (band structure method) [6.1.2]

AR: angular resolved, as in ARUPS [3.3.1]

BCF: Burton, Cabrera & Frank (see reference list for this 1951 paper) [1.3.2, 5.5.1]

BEEM, BEES: ballistic energy emission microscopy, spectroscopy [8.1.3]

CBED, CBIM: convergent beam electron diffraction [3.1.4, 8.1.3], imaging [8.1.3]

CERN: Centre Européenne pour la Recherche Nucléaire (accelerator laboratory) [2.1.3]

CFE: cold field emisison [6.2.3]

C–I: commensurate–incommensurate (phase or transition) [4.4.2]

CAP, CIP, and CPP: Current at an Angle, In or Perpendicular to the Plane [8.3.3]

CMOS: complementary metal–oxide–semiconductor (device) [8.1, 9.1]

C–V: current–voltage, as in C–V profiling [8.2.3]

CVD: chemical vapor deposition [2.5.5]

DAS: dimer–adatom–stacking fault model [1.4.5, 7.2.3]

DFT: density functional theory [4.5.2, 6.1.2, Appendix J]

DLC: diamond-like carbon [6.2.2, 8.4.3]

DLEED: diffuse LEED [3.2.2]

DLTS: deep level transient spectroscopy [8.1.3]

DNA: deoxyribo-nucleic acid [8.4.4]

EAM: embedded atom model [4.5.2, 6.1.2]

EELS: electron energy loss spectroscopy [3.3.1]

EMT: effective medium theory [4.5.2, 6.1.2]

ES: the Ehrlich–Schwoebel barrier [5.5.1]

ESCA: electron spectroscopy for chemical analysis [3.3.1]

ESR: electron spin resonance [8.1.3]
FEM, FES: field emission microscopy, spectroscopy [6.2.3]
FET: field effect transistor [8.4.2]
FIM: field ion microscopy [3.1.3, 5.4.2]
GGA: generalized gradient approximation [6.1.1]
GMR: giant magneto-resistance [8.3.3]
GSMBE: gas source MBE [2.5.3]
HAS: helium atom scattering [4.4.4, 5.4.3]
HEMT: high electron mobility transistor [8.2.4]
HREELS: high resolution electron energy loss spectroscopy [3.3.1]
HOLZ: higher-order Laue zone [3.2.2]
IA-IR: incommensurate aligned- incommensurate rotated (phase or transition) [4.4.2]
IBAD: ion beam assisted deposition [2.5.4]
IBSD: ion beam sputter deposition [2.5.4]
IC: integrated circuit [9.1]
ICB: ionized cluster beam [2.5.4]
ICISS: impact collision ion scattering spectroscopy [3.1.2]
I–V: current (I)–voltage (V), as in LEED I–V curves [3.2.1]
KKR: Korringa–Kohn–Rostoker (band structure method) [6.1.2]
KMC: kinetic Monte Carlo simulations [5.2]
KPZ: Kardar, Parisi & Zhang; an equation used in crystal growth modelling [7.3.4]
KTHNY: papers by Kosterlitz, Thouless, Halperin, Nelson & Young [4.4.2]
L: Langmuir (unit of gas dose) [4.5.4]
LDA: local density approximation [6.1.1]
LDOS: local density of states [4.5.2]
LEED: low energy electron diffraction [1.4, 3.2]
LEEM: low energy electron microscopy [3.1.3, 3.2]
LEP: large electron positron (ring) at CERN [2.3.2]
LPCVD: low pressure CVD [2.5.5]
LSD: local spin density [6.3.4]
MBE: molecular beam epitaxy [2.5.3]
MC: Monte Carlo (simulation method) [1.3]
MCD: magnetic circular dichroism [6.3.3]
MD: molecular dynamics (simulation method) [1.4.3, 4.5.2]
MFM: magnetic force microscopy [6.3.3]
MIDAS: a microscope for imaging, diffraction and analysis of surfaces [3.5.3, 6.3.3]
MIGS: metal induced gap states [8.2.1]
ML: monolayer [2.1.4] (do not confuse with mega (M) Langmuir (L))
MOKE: magneto-optical Kerr effect [6.3.3]; SMOKE is MOKE applied to surfaces.
MOMBE: metal-organic MBE [2.5.3]
MOS: metal-oxide-semiconductor (device) [8.1, 9.1]
MQW: multiple quantum well [7.3.2, 8.2.3]
MTP: multiply twinned particle [4.5.4]
NEA: negative electron affinity [1.5.5, 6.3.3]

OMVPE: organo-metallic vapor phase epitaxy [2.5.5]
OPW: Orthogonalized plane waves (band structure method) [6.1.2]
PBN: pyrolytic boron nitride [2.5.2]
PEEM: photo-electron emission microscopy [3.1.3]
PECVD: plasma enhanced CVD [2.5.5]
PLD: pulsed laser deposition [2.5.2]
PTFE: poly-tetrafluor ethylene [2.4.4, Appendix H]
QMS: quadrupole mass spectrometer [2.3.5]
RBS: Rutherford backscattering spectrometry [3.4.2]
REM: reflection electron microscopy [3.1.3]
RGA: residual gas analyzer [2.3.5]
RHEED: reflection high energy electron diffraction [2.5.3, 3.1.3, 3.2.2]
SAM: scanning Auger microscopy [3.5.1]; self-assembled monolayer [8.4.2]
SDW: static distortion wave [4.4.2]
SEM: scanning electron microscopy [3.1.3, 3.5.1]
SEMPA: scanning electron microscopy with polarization analysis [6.3.3]
SI: système internationale (of units) [2.1.1, Appendix C]
SIMS: secondary ion mass spectrometry [8.1.3]
SK: Stranski-Krastanov (growth mode) [5.1.3, 5.4.1]
SMOKE: surface magneto-optical Kerr effect [6.3.3]
SMP: small metal particle [2.4.4, 4.5.4]
SNOM: scanning near-field optical microscopy [3.1.3]
SNR: signal to noise ratio [3.1.3, problem 3.3]
SOS: solid on solid (model) [1.3.2]
SPA: spot profile analysis, as in SPA-LEED [3.2.1] or SPA-RHEED [3.2.2]
SPLEEM: spin-polarized low energy electron microscopy [6.3.3]
STEM: scanning transmission electron microscopy [3.1.3]
STM: scanning tunneling microscopy [1.4.5, 3.1.3]
TDS: thermal desorption spectroscopy [4.4.4]
TEA: triethylaluminum [2.5.3]
TEG: triethylgallium [2.5.3]
TEM: transmission electron microscopy [3.1.3]
TFE: thermal field emisison [6.2.3]
THEED: transmission high energy electron diffraction [1.4.5, 3.1.3]
TLK: terrace ledge kink (model) [1.2]
TSP: titanium sublimation pump [2.3.3]
TXRF: total reflection X-ray fluorescence [8.1.3]
UHV: ultra high vacuum [2]
UK: United Kingdom [2.1]
UPS: ultra-violet photoelectron spectroscopy [3.3.1]
VBO: valence band offset [8.2.3]
VdW: Van der Waals (interactions or forces) [4.1, 5.3.2]
XPS: X-ray photoelectron spectroscopy [3.3.1]
ZOLZ: zero-order Laue zone [3.2.2]

Appendix C
Units and conversion factors

To convert a quantity expressed in the units listed in the first column to those listed in the third column, multiply by the number given in the second column. The SI unit is given in brackets for each quantity. The values with ($\pm 1\sigma$) error bars are taken or deduced from the CODATA 1986 report (Cohen & Taylor 1987, 1998) as recommended by NIST in the USA and NPL in the UK. Note that it is not required to keep the full accuracy of these data to do the typical calculations encountered in this book, but it is helpful to have the extra decimal places were one to need them.

Mutiplication factors

The standard prefix is used for multiples and sub-multiples of units. These are shown in Table C1 below.

Table C1

Factor by which unit is multiplied	Prefix	Symbol
10^{12}	tera	T
10^{9}	giga	G
10^{6}	mega	M
10^{3}	kilo	k
10^{-2}	centi	c
10^{-3}	milli	m
10^{-6}	micro	μ
10^{-9}	nano	n
10^{-12}	pico	p

Length (m)

m	10^{10}	Å
Å	0.1	nm
kX (Cu Kα_1 X-ray unit)	1.00207789 ($\pm$70)	Å
a.u. (Bohr radius a_0)	0.529177249 ($\pm$24)	Å
inch	2.54	cm

Mass (kg)

m (electron mass)	$9.109\,389\,7\,(\pm54)\times10^{-31}$	kg
u (atomic mass unit)	$1.660\,540\,2\,(\pm10)\times10^{-27}$	kg
lb (pound)	$0.453\,592\,37$	kg

Time (s)

min	60	s
h	3600	s
day (mean solar)	86400	s
yr (sidereal)	3.1558×10^{7}	s

Angle (deg = 60 min = 3600 s of arc)

rad	57.295779	deg
rad	3437.7468	min
rad	206264.81	s

Force (N) and pressure (pascal (Pa) = N·m^{-2})

dyne	10^{-5}	N
mbar	100	Pa (N·m^{-2})
bar	10^{6}	dynes·cm^{-2}
Atm.	1.01325	bar
torr (mm Hg)	1.33322	mbar
pound weight (or force)	4.4822	N
ditto/square inch, i.e. (p.s.i)	68.9476	mbar

Energy (J)

erg	10^{-7}	J
cal	4.18400	J
litre-atm.	101.328	J
eV	$1.602\,177\,33\,(\pm49)\times10^{-19}$	J
a.u. (Hartrees)	$4.359\,748\,2\,(\pm26)\times10^{-18}$	J
a.u. (Hartrees)	$27.211\,396\,1\,(\pm81)$	eV

Energy related units

Energies are expressed in different units in different disciplines, and we also often need energies expressed in energies per unit area, per mole or per molecule.

Surface energy (J·m⁻²)

eV·nm⁻²	0.160217733 (±49)	J·m⁻²
eV·Å⁻²	100	eV·nm⁻²

Energy to temperature

Via $E = kT$, where Boltzmann's constant $k = 1.380658\ (\pm 12) \times 10^{-23}\ \text{J·K}^{-1}$. This conversion is useful for energy values in eV which occur in Arrhenius expressions $\exp(-E/kT)$.

eV	1.160444 (±10) × 10⁴	K
K	1.380658 (±12) × 10⁻²³	J

Energy to frequency

Via $E = h\nu$, where Planck's constant $h = 6.6260755\ (\pm 40) \times 10^{-34}\ \text{J·s}$. This conversion is needed in discussing (lattice) vibrations.

eV	2.4179884 (±7) × 10¹⁴	Hz
meV	0.24179884 (±7)	THz

Energy to wavenumber

Via $E = hc/\lambda$, where the velocity of light $c = 2.99792458 \times 10^8\ \text{m·s}^{-1}$ exactly. This conversion is often needed in (infrared) spectroscopy.

eV	0.80655411 (±24) × 10⁶	m⁻¹
meV	8.0655411 (±24)	cm⁻¹

Energy per mole, or per molecule (kJ·kmol⁻¹)

Energies of molecules (or atoms) can be expressed in units such as kJ·kmol^{-1}, J·mol^{-1}, kcal·mol^{-1}, K/molecule, ergs/molecule, eV/molecule etc. These are all related via the energy conversion table above and $N_A k = R = 8.314510\ (\pm 70)\ \text{J·mol}^{-1}\text{·K}^{-1}$, where Avogadro's number $N_A = 6.0221367\ (\pm 36) \times 10^{23}\ \text{mol}^{-1}$. Note: using kmol⁻¹ means the exponent is 26 rather than 23 in this expression, but the value of R stays the same if we use both kJ and kmol⁻¹.

E (kJ·kmol⁻¹)	1	E (J·mol⁻¹)
E/R (K·mol⁻¹)	1	E (K/molecule)
E/k (K/molecule)	8.314510 (±70)	E (J·mol⁻¹)
E/k (K/molecule)	1.98722 (±2)	E (cal·mol⁻¹)
E (eV/molecule)	23.060	E (kcal·mol⁻¹)
E (eV/molecule)	96.485	E (kJ·mol⁻¹)

Appendix D
Resources on the web or CD-ROM

Many resources are available on the World Wide Web (web for short) and some selected resources are available on CD-ROM. The advantages of the latter are immediate local availability and hence speed of access, but they lack the flexibility of the web, and typically cost real money. However, the disadvantage of referring to materials on the web is that they are not under one's own control; hence information may be out of date, and there can be no guarantee of accuracy. A partial solution to this problem is to construct one's own web page in the form of a 'portal' to relevant web sites, and only to link to those sites which have been created by colleagues whose work one trusts/admires. I plan to keep my site active for the immediate future, at the address http://venables.asu.edu/grad/appweb1.html which is accessible from my home page http://venables.asu.edu/index.html as *Web-based resources*. The initial appearance of this page is shown in figure D.1 below. The web-teaching experiences which lead to the construction of such pages are described by Venables (1998).

Potentially relevant resources on CD-ROM include the following:

(1) The NIST Surface structure database. Details are accessible via my page on data bases at http://venables.asu.edu/grad/appdat1.html

(2) The Matter project. This CD-ROM consists of a series of modules for teaching Materials Science to undergraduates, produced by P.J. Goodhew and co-workers, and available from Chapman and Hall as version 2.1 (1998). Many of these modules are useful as self-study materials for graduate students in other disciplines who need to know what materials science is about. The crystallography modules are described in print by Goodhew & Fretwell (1998).

(3) The VIMS CD-ROM. This CD-ROM (Visualization in Materials Science) has been produced by J.C. Russ and co-workers since 1995, and is published by PWS Publishing. It is widely used for undergraduate teaching in materials science.

(4) Advanced Computing in Electron Microscopy. CD-ROMs are marketed with specialist books on occasion, as in this text by Kirkland (1998), giving actual programs for simulating high resolution electron microscope images.

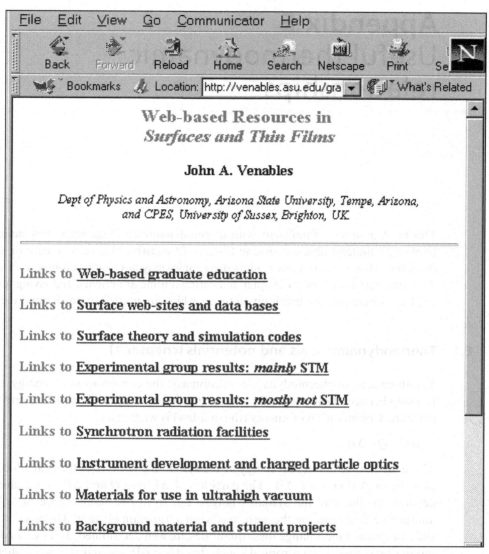

Figure D.1. Web-based resources available from my home page at
http://venables.asu.edu/grad/appweb1.html. Web-based comments, corrections and updates to
this book are at http://venables.asu.edu/book/contents.html.

Appendix E
Useful thermodynamic relationships

This book assumes a familiarity with thermodynamics and statistical mechanics at the level of an undergraduate course in Physics, Chemistry, Materials Science or a related discipline. However, for many readers these topics may not reside in foreground memory. This appendix is divided by chapter, indicating outline arguments, and giving references to places where one can brush up on topics which are needed to understand the text.

E.1 Thermodynamic laws and potentials (chapter 1)

The three laws of thermodynamics encompass: the conservation of energy, the relationship between heat and work, and the zero of entropy at the absolute zero of temperature. Conservation of energy (the first law) is written as

$$\Delta U = Q + \Delta W, \tag{E.1}$$

where the change in internal energy of the system ΔU comprises the heat added to it Q plus the work done on it ΔW. The partition of ΔU into Q and ΔW is not unique, but depends on the way the system changes, e.g. from one temperature or volume to another, i.e. it depends on the path, and is not a 'function of state'. However, in a reversible, or quasi-static, change these quantities are well determined, and we can define an infinitesimal change in entropy, dS such that $dU = TdS$ and $dW = -pdV$, where p and T are the pressure and temperature respectively.

Inserting these quantities into (E.1) leads to the equation for dU, which incorporates the second law into the first as

$$dU = TdS - pdV. \tag{E.2}$$

This is the first equation involving a thermodynamic potential U, and can be used to derive thermodynamic quantities at constant volume or constant entropy. For example, the specific heat at constant volume C_V is by definition

$$C_V = (\partial U/\partial T)_V = T(\partial S/\partial T)_V \tag{E.3}$$

or the bulk modulus at constant entropy B_S is

$$B_S = -V(\partial p/\partial V)_S = +V(\partial^2 U/\partial V^2)_S. \tag{E.4}$$

The relations obtained via different second partial derivatives involve consistency checks known as the four Maxwell relations; the one which follows from (E.2) is $(\partial T/\partial V)_S = -(\partial p/\partial S)_V$.

The above considerations apply to constant numbers of particles, N, but if N is itself a variable, then there is an extra term in (E.2) $+ \mu dN$, where μ is known as the chemical potential. This is the internal energy at which the particle is added to the system. When we are considering isolated systems at constant volume V and particle number N, the thermodynamic potential U is constant, and

$$dU = TdS - pdV + \mu dN. \tag{E.5}$$

This is what is meant by the entropy being maximized ($dS = 0$) in an isolated system at equilibrium, the second law of thermodynamics having shown that $dS \geq dQ/T$. The corresponding microscopic distribution is the micro-canonical ensemble of statistical mechanics. Consideration of constant pressure processes leads to definition of the enthalpy, $H = U + pV$, so that

$$dH = TdS + Vdp + \mu dN; \tag{E.6}$$

the enthalpy is useful in describing thermally isolated constant pressure processes. Often these correspond to (irreversible) flow processes, such as the flow of air over a wing or through a nozzle valve. Analogous to (E.3) we can now define the specific heat at constant pressure C_p as

$$C_p = (\partial H/\partial T)_p = T(\partial S/\partial T)_p. \tag{E.7}$$

The Maxwell relation which follows from (E.6) is $(\partial T/\partial p)_S = -(\partial V/\partial S)_p$.

Mostly we are concerned with systems at a given temperature, or in more technical terms 'in contact with a heat bath at temperature T'. For these the Helmholtz free energy $F = U - TS$ and the Gibbs free energy $G = U - TS + pV$ have been devised. These are particularly useful for discussing processes at constant volume and constant pressure respectively. Correspondingly we have

$$dF = -SdT - pdV + \mu dN, \tag{E.8}$$

and $dG = -SdT + Vdp + \mu dN.$ $\tag{E.9}$

In particular, F is minimum at constant (T, V and N) and G is minimum at constant (T, p and N), and the corresponding microscopic distribution is the Canonical ensemble. Section 1.1 is concerned with discussing the additional terms which arise when the area A of a surface is an additional thermodynamic variable.

The third law is concerned with establishing the zero of entropy for systems in equilibrium at the absolute zero of temperature (0 K). Since physical effects result only from differences in entropy and absolute zero cannot be reached, this may appear a bit academic. However, it has a certain fascination in the context of phase changes, especially since kinetics compete with thermodynamics, and can become very sluggish at low temperature. An example would be the entropy difference between solid N_2 and solid CO, which have the same crystal structure, Pa3. At absolute zero, CO should undergo head to tail ordering, whereas N_2 molecules, with identical ends, should not. However, the barriers to rotational ordering become completely unsurmountable at low temperatures, which leaves solid CO with non-zero entropy at 0 K. This argument can get quite complex if we consider nuclear ordering as well, e.g. in a system of $^{14}N-^{15}N$ molecules.

There are many, many books on thermodynamics and statistical mechanics, and it is perhaps unwise to single out particular ones. All of them will contain the above material, but may have slightly different notation. For example E may be used for internal energy instead of U, and upper (P) or lower (p) case for pressure. Books I have found useful include Hill (1960) and Mandl (1988); for a modern introduction, see Baierlein (1999). For a detailed list of thermodynamic quantities, and evaluation of properties which use the statistical mechanics of lattice vibrations in various models, see Klein & Venables (1976, 1977) especially chapters 6 (M.L. Klein & T.R. Koehler), 11 (R.K. Crawford), 12 (P. Korpiun & E. Lüscher) and 13 (C.A. Swenson).

E.2   Phase equilibria and phase transitions (chapter 4)

Two phases (subscripts 1 and 2) in equilibrium are described by the condition $\mu_1 = \mu_2$. In addition, adsorbed layers in equilibrium with the vapor phase have $\mu_1 = \mu_2 = \mu_v$; the spreading pressures also have to be equal: $\Phi_1 = \Phi_2$. Thus a phase transition line on a diagram such as figure 4.10 is defined thermodynamically by $d\mu_1 = d\mu_2 = d\mu_v$. Using $\mu_x = h_x - Ts_x$ for the phases $x = 1, 2$ and $\mu_v = h_v - T\sigma + kT\ln(p)$, we can obtain via (4.8)

$$d(\mu_x/T) = h_x\, d(1/T) - (1/Tn_x)\,d\Phi \tag{E.10a}$$

and

$$d(\mu_v/T) = h_v\, d(1/T) + k\,d(\ln p), \tag{E.10b}$$

in terms of the enthalpy per atom or molecule h, and the areal density of phase x, $n_x = N_x/A$. It follows that we can express $\ln p$ for transition lines as a function of $1/T$ involving differences as

$$\ln p = \left[\frac{-n_2(h_v - h_2) + n_1(h_v - h_1)}{(n_2 - n_1)kT} \right] + \left[\frac{n_2(\sigma - s_2) + n_1(\sigma - s_1)}{(n_2 - n_1)k} \right] \tag{E.11}$$

and hence

$$k\,d(\ln p)/d(1/T) = -h_v + \frac{n_2 u_2 - n_1 u_1}{(n_2 - n_1)}. \tag{E.12}$$

The isosteric heat of adsorption, which is the slope of the lines of constant coverage within a given phase, was discussed in section 4.3.1. It can be cast in a similar form to (E.12) as

$$-q_{st} = k[d(\ln p)/d(1/T)]_N = -h_v + [\partial(nu)/\partial n]_{A,T,V} \tag{E.13}$$

i.e. the finite difference has been replaced by a partial derivative. In the last two equations we have replaced the enthalpy h with the internal energy u, since the spreading pressures in the two phases are the same in (E.12) and the area is constant in (E.13). When one makes a specific model of an adsorbed layer, the internal energy includes lateral interactions. If these vary rapidly with the areal density n, as in a solid phase, the term in $n \partial u / \partial n$ can be an important component in q_{st} (Price & Venables 1976).

Appendix F
Conductances and pumping speeds, C and S

Useful formulae: (C, S are measured in liter/s)

Conductances C_i in series $\quad C^{-1} = \Sigma C_i^{-1}$, so that $S^{-1} = C^{-1} + S_0^{-1}$. $\qquad$ (F.1)

Conductances C_i in parallel $C = \Sigma C_i$; $S = \Sigma S_i$. $\qquad$ (F.2)

In the following conductance formulae, T = temperature (K), M = molecular weight, tube diameter D and length L are in cm. (Formulae in brackets are for air at 20 °C, and all formulae are for the molecular flow regime described in sections 2.2 and 2.3.)

C of an aperture = $2.86 \, (T/M)^{1/2} \cdot D^2$; ($C = 9.16 \, D^2$). $\qquad$ (F.3)

C of tube into large system = $3.81 \, (T/M)^{1/2} \cdot D^3 \, /(L + 1.33D)$;

$\quad (C = 12.1 \, D^3 \, /(L + 1.33D))$. $\qquad$ (F.4)

Table F1. *Conductances (liter/s) in typical situations for standard tube sizes*

D (inch) nominal size	D (mm) inside diameter	Aperture conductance	$L = 10$ cm (port conductance)	$L = 1$ m (no end, per meter)
1.5	36.9	125	40.7	6.0
2.5	61.2	343	153	27.7
4	99.4	905	512	119
6	149.7	2053	1357	406
8	200.4	3679	2657	974

Notes: Columns 3–5 use formulae (F.3) and (F.4) to provide conductance estimates for air at 20 °C.

Table F2. *Typical pump speeds for different pump types and sizes*

D (inch) nominal size	D (mm) nominal size	Turbomolecular pump	Diffusion pump + trap + valve	Ion or sputter ion pump
1.5	38	—	—	35
2.5	64	50	—	45
4	100	170–300	32	60
6	150	500	200	110–500
8	200	1400	415	>800

Notes: Columns 3–5 use data from several manufacturers to provide estimates for N_2 at 20 °C; for particular pumps check performance against specification.

Appendix G
Materials for use in ultra-high vacuum

This appendix gives a few indicators of suitable materials for use in a UHV environment. The main point is simply to emphasize that the materials need to have low outgassing rates per unit area exposed, and that they need to be stable at the temperatures not only of use, but also during bakeout. Some of the obvious candidates in the different categories are as follows. Much of this information can be gleaned from talking to practitioners, from vacuum technology books such as Dushman & Lafferty (1992), O'Hanlon (1989), Roth (1990) or from reading between the lines in design handbooks such as Yates (1997), or increasingly from the web. A page giving properties and some sources for materials is at http://venables.asu.edu/grad/appmat1.html

Structural materials

The most widely used structural material is 304 stainless steel, which is used to make chambers, flanges, etc., and can also be used for stages and other parts of the experiment itself. At very low temperatures this austenitic (largely f.c.c.) 18–20%Cr, 8–10%Ni Fe-based alloy transforms in part to the b.c.c. (martensitic) structure, and thereby becomes magnetic. If this could be important, then more technical details are needed, such as would be obtained from ASSDA – the Australian Stainless Steel Development Association – or AVS – the American Vacuum Society. Aluminum alloys are used for experimental pieces inside the vacuum system, and have also been used for whole chambers on occasion. This is the only material which has successfully achieved UHV pressures without baking, but it needs a special surface treatment developed in Japan for this to be effective, and the technique has not yet become widespread.

Titanium alloys are expensive, but have been successfully used for small parts, especially where fine bearing surfaces are required. Copper bronzes (zinc and lead free) are also useful for stages and related parts. Note that opposite bearing surfaces *must* be made of dissimilar materials; in UHV, removal of the oxide or interfacial layers between similar materials results in cold welding. Once this has happened it is too late to argue that you didn't know they would.

Electrical conductors

Oxygen-free high-conductivity (OFHC) copper is the standard material for electrical leads, provided they don't get too hot; silver (usually as a plating on copper) can be used to increase the conductivity of surface layers, which can be useful for high frequency applications. The normal 304 stainless steel can be used, but only for low current applications, possibly at high voltage, as in electron beam heating. The normal refractory materials are tungsten and tantalum as heater wire and electron source filaments. Often Ta supports are used to hold W filaments, which prevents the filaments welding to the support after use; Ta has the advantage of being less brittle than W.

Thermal conductors

Again OFHC copper is good, and silver, with a special surface treatment developed in Japan, can be used for low emissivity applications. The combination of high thermal conductivity with high electrical resistance is provided by sapphire, crystalline Al_2O_3. This is useful at low temperatures, and provides electrical isolation; this is also a possible role for diamond in thin films. Tantalum has good thermal conductivity to very high temperatures.

Electrical and thermal insulators

The following materials are discussed in decreasing order of thermal stability. Sapphire and alumina are useful up to high temperatures. These materials and fused quartz glass have very low expansion coefficients; fused quartz is an excellent insulator at low temperatures, so it can be used for supporting stable low temperature stages. Borosilicate glass (Pyrex®) is generally used for UHV windows, whereas fused quartz windows can be advantageous for special applications, including transmission into the ultra-violet, and if low birefringence is needed.

Machinable ceramic/glass (with trade names Macor®, Micalex® etc.) outgas more than the above glasses and so should be used rather sparingly, but they can readily be made into complex shapes. Various plastics can be used very sparingly, such as Kapton®, a polyimide polymer with very high dielectric strength. Teflon®, which is a high density PTFE (polytetrafluorethylene) can also be used, but tends to degas fluorine and other fluorine compounds. Wires coated with these polymers can be used, but detailed attention must be paid to bakeout temperatures, as emphasized in section 2.3.4.

Heat and electrical shields

The refractory metals molybdenum, tantalum and tungsten are used as heat shields, especially in several layers, or in a spiral wrap, because this cuts down substantially on heat transmission by radiation. Of the three metals, Ta is the most expensive, and Mo is best for reducing electrical (patch) fields in the vacuum, especially for low energy charged particles, as discussed in section 6.1.3.

Appendix H
UHV component cleaning procedures

General precautions

Take care not to damage knife edges, and don't transport grease and dirt (e.g. from bolt-holes) from the outside to the inside of the assembly. Clean polyethylene or *unpowdered* PVC gloves should be worn both as safety and contamination protection. More efficient cleaning results from doing small pieces in batches. Use new solvents for each piece or set of pieces. Handle parts with cleaned tweezers or tongs as much as is practicable. Large flanges or assemblies should be held by their outside surfaces only.

Cleaning procedures

The initial decision which has to be made is how dirty the pieces really are and whether you want to perform what I refer to informally as an *ultimate* clean or a *routine* clean. This difference can affect both the cost of the operation and the need for special solvent disposal procedures, with health and safety implications. For these reasons at least, it is strongly advisable to work closely with your laboratory manager, and to establish written procedures which are then followed carefully. An inexpensive routine clean with little or no disposal problems is described first, followed by an ultimate clean which is more expensive, and does have disposal implications.

Routine clean

(1) Prepare a solution of a standard laboratory cleaner in distilled water. Adjust the strength of the solution depending on how dirty the part is, and the type of contaminant. Heating the solution will decrease the necessary soak time, but don't exceed 95 °C.
(2) Put small parts in a beaker of the cleaning solution and clean ultrasonically for up to 15 min. Rinse in a beaker of distilled water. Large parts should be swabbed down with lint-free cloth, and soaked in the cleaning solution and then rinsed under

running distilled water to remove most of the loosely attached dirt and oil. If the parts will fit, they can be put directly into an ultrasonic cleaner tank filled with cleaning solution. After this step they should again be rinsed under running distilled water.

(3) Repeat steps (1) and (2) until the rinse water continues to wet the surface of the piece as the last trace drains off. Rinse several times after the last cleaning in the best quality deionized water you can find.

(4) Air dry on a clean surface. Large parts can be dried gently with a *clean* heat gun or in a *clean* oven.

Ultimate clean

If the contaminant on your parts doesn't come off with the procedure described above you may have to resort to organic solvents described below, or even chemical cleaning. In outline the same process is followed but solvent and rinse chemicals are different. Flammable liquids should not be used in an ultrasonic cleaner and large quantities should only be used in a high volume fume hood.

A typical procedure is to soak first in trichlorethylene, followed by up to 10 min in an ultrasonic cleaner. This is then followed by repeating the process with acetone, and then methanol. Finally the parts are dried using a clean air fan, for example a heat gun or hair dryer with the heat off.

Storage of UHV parts

Small parts are best stored in clean glassware. Large parts can have their ports covered with aluminum foil, stretched across them but not touching the sealing surface, and stored in polythene bags if necessary. All aluminum foil and plastic bags have some organic or silicone oil on them, which could recontaminate your clean parts, so beware!

Sample cleaning

The objective of sample cleaning is to prepare a surface with a well-defined chemical and structural state in a *reproducible* manner. The particular process used inside the vacuum system tends to be very material specific, as discussed in the text. However, some general types of technique used for preparation of samples are as follows. The first three apply to procedures before the samples are introduced to the vacuum system.

(1) Various forms of mechanical, chemical or electrochemical polishing are used to make the sample surface as parallel as possible to the desired crystallographic plane.

(2) Solvent cleaning is used to remove as much of the polishing contaminants as possible.

(3) Chemical cleaning and passivation of the surface is sometimes used with reactive materials to reduce atmospheric surface contamination.

(4) Heating in the vacuum by resistive means, electron bombardment, or laser annealing can be used both to evaporate contaminants and to remove crystallographic imperfections from the surface layers. Vapor pressure versus temperature curves are very useful in this process.

(5) Ion bombardment or sputtering is used to remove surface material. Heating is used to anneal out the subsequent damage caused to the structure.

(6) Oxidation is used both to remove contaminants and damaged surface material, especially in refractory elements.

More details for selected elements can be found in the research literature, including Musket *et al.* (1982). I have checked, using the WebofScience™ citation index, that this reference is still used by workers researching surface and thin film processes. We can therefore use the web, in conjunction with older references, to track down what has happened more recently, and which papers are thought to be valuable.

Appendix J
An outline of local density methods

Although we introduced density functional theory (DFT) in section 6.1 in the context of Lang and Kohn's work on metal surfaces, the concept itself is much broader. It consists of setting up a general *single* particle method to solve the Schrödinger equation for the ground state of a *many* electron system by: (1) showing that the equation can be solved variationally to give an upper bound to the energy of the system expressed in terms of the electron density $n(\mathbf{r})$, sometimes written $\rho(\mathbf{r})$; this theorem was introduced by Hohenberg & Kohn (1964); and (2) proposing practical schemes whereby this theorem can be implemented as an iterative computational method, starting from a set of approximate wave functions describing the ground state of the electron system. The main non-relativistic scheme in use is due to Kohn & Sham (1965). The pervasiveness of these methods was recognized in 1998 by the award of the Nobel prize for *chemistry* to Walter Kohn (Levi 1998).

Writing down too many equations specifically here will take too much space, and may encourage the reader to believe that the method is simpler than it actually is. Some of the key review articles have been cited in sections 6.1.2 and 7.1.3. So many words have already be spilt on the topic, the methods are so widespread, and yet no-one can give a measure of just how good an approximation DFT represents, or say categorically whether further developments such as GGA necessarily improve matters, that there is no sense in which I should try to confuse you further. Nonetheless, this is a good topic for an (ongoing) student project, and figure 6.1 was produced by Ben Saubi from the original Lang and Kohn output data tables.

The DFT method is based on three coupled equations, and auxiliary 'orbitals' ψ_i, which are othogonal to each other, but should not be confused with the real (many body) wavefunctions of the system. The energy E is expressed as the sum of the electrostatic energy due to the *external* potential and a *functional* $F[n(\mathbf{r})]$ of the electron density $n(\mathbf{r})$ at vector position in the material $\mathbf{r}$. This functional is written as a sum of the kinetic energy $T[n(\mathbf{r})]$, the Coulomb self-energy of the electrons, expressed as the product $0.5n(\mathbf{r})\varphi(\mathbf{r})$, which[1] is integrated over the space $d^3\mathbf{r}$, and the exchange-correlation term

---

[1] Here $\varphi(\mathbf{r})$ is the electrostatic potential provided by all the *other* electrons. This potential is subject to the consistency check provided by Poisson's equation $\nabla^2\varphi(\mathbf{r}) + 4\pi n(\mathbf{r}) = 0$. Don't try to check the units, since most theoretical papers use 'atomic' units, in which $\hbar$, e and m have all been set equal to 1.

E_{xc}. The Coulomb energy is non-local, but is explicit if one knows $n(\mathbf{r})$; E_{xc} is in principle non-local also. However, the *local density* approximation (LDA) consists in providing an expression for E_{xc} in terms of $n(\mathbf{r})$, and thereby also $\mu_{xc} = \mathrm{d}E_{xc}/\mathrm{d}n$. Kohn & Sham (1965) were the first to do this explicitly, but work has continued to research accurate forms valid over a wide density range (e.g. Callaway & March 1984, Jones & Gunnarson 1989).

The above terms constitute the first equation, for which we need an explicit expression for the kinetic energy term. This is obtained by treating the $\psi_i(\mathbf{r})$ as if they were *real* orbitals, and expressing $T[n(\mathbf{r})]$ as $\psi_i^* \nabla^2 \psi_i$ integrated over the space $\mathrm{d}^3\mathbf{r}$ and summed over all the orbitals i. The cycle is then closed by expressing $n(\mathbf{r})$ as the sum of $|\psi_i(\mathbf{r})|^2$, and iterating to find the minimum energy E_0 and, at the same time, the correct $n_0(\mathbf{r})$. We're done!

However, as one can appreciate from this outline description, actually doing a real calculation is computationally very intensive. Individual orbitals are typically expanded in plane waves with a large number M of independent coefficients $c_{\mathbf{k}}$ of the various $\mathbf{k}$ vectors which have to be computed as the calculation proceeds. For N electrons, the number of computing operations scales as $(MN)^3$. However, many of the operations required for each $\mathbf{k}$ are identical, so the code can be written for implemention on parallel (super)computers. By 1999, systems with N in the region of several hundreds and typical $M \sim 1000$ can be tackled. The virtue of pseudopotential calculations is that they reduce the number of electrons per atomic site, at the cost of increased complexity of the ionic (external) potential. There is a strong impetus to reduce the cubic power law to something lower, which is what I meant by the exclamation in section 6.1.2: $O(N)$ methods are *in!* However, no actual methods are that good, the best perhaps scaling as $N\ln(N)$, and one also needs to look critically at the multiplying constants. Many careers have been spent trying to crack these highly technical conceptual and computational problems.

One further conceptual aid is to discuss how we can visualize the exchange-correlation term in real space. In an electron gas, we have one electron per Wigner–Seitz sphere of radius r_s, i.e. of approximately atomic dimensions. Thus the 'electron' or quasi-particle, when it moves, carries around a sphere of about this size which is deficient in electrons, due to their mutual interaction, i.e. the electron position and motions are *correlated*. Moreover, this 'sphere of influence' has Friedel oscillations associated with it, and depends on the electron spin, like spins repelling each other via the Pauli exclusion principle, and unlike spins ignoring each other. This *exchange* effect does not have to be added in separately, it is already there: on average in LDA, and explicitly in the LSD models of magnetic materials discussed in section 6.3.4.

Appendix K
An outline of tight binding models

In lecturing on semiconductors, I have typically started from a two-page handout, reproduced from Pettifor (1995, chapter 3, p. 54, and chapter 7, p. 198–201). This short description is adapted with permission, and extended in the direction of further reading.

The basic idea of bonding and anti-bonding orbitals which arise when two atoms overlap is sketched in figure K.1(a). The initially degenerate energy levels E_A of the atoms are split by overlap of the wavefunctions, into an bonding level $|h|$ below E_A and an anti-bonding level $|h|$ above E_A, and so the level splitting $w = 2|h|$. The lower level can contain two electrons, one of each spin, as in the hydrogen molecule; there is also a shift for both levels via the Pauli exclusion principle. These interactions are characterized by h, the *bond integral*, and S, the *overlap integral*, given for two atoms A and B by

$$h = \int \psi_A^* \bar{V} \psi_B d\mathbf{r}, \text{ and } S = \int \psi_A^* \psi_B d\mathbf{r}, \tag{K.1}$$

where the relevant potential $\bar{V} = (V_A + V_B)/2$. For dissimilar atoms, the corresponding equations lead to diagonalizing a 2×2 matrix, from which we arrive at both the asymmetric distribution of charge on the atoms A and B, as illustrated in figure K.1(b), and the quadratic relation 7.1, which can be obtained from the approximate solution for the energies $E^\pm$ for the bonding($+$) and antibonding($-$) orbitals,

$$E^\pm = \bar{E} + |h|S \mp 0.5[4h^2 + (\Delta E)^2]^{1/2}, \tag{K.2}$$

where $\bar{E} = (E_A + E_B)/2$ and $\Delta E = (E_A - E_B)$; this equation is valid to second order in the quantities h and S (Pettifor 1995, chapter 3, pp. 50–54).

In the case of diamond-like tetrahedral structures of semiconductors, there are two atoms per basis in the f.c.c. unit cell, and four valence electrons per atom, leading to the need to diagonalize an 8×8 matrix (project 7.2). The construction of this matrix may be simplified by considering the formation of sp^3 hybrids, and the corresponding energy levels illustrated in figure 7.1(b). The hybrid energy is

$$E_0 = (E_S + 3E_P)/4, \tag{K.3}$$

and the hybrid bond integral is

$$h = (ss\sigma - 2\sqrt{3}sp\sigma - 3pp\sigma)/4; \tag{K.4}$$

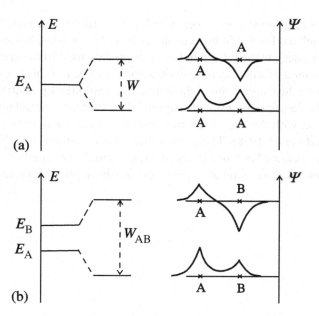

Figure K.1. Bonding and antibonding states for (a) homonuclear and (b) heteronuclear diatomic molecules. The shift in energy levels due to overlap repulsion has not been shown (after Pettifor 1995, redrawn with permission).

here the notation spσ means, for example, the bond integral between an s-orbital on one atom and a p-orbital on the other, arranged to form a σ-bond. As all these terms are 3D integrals which are functions of the internuclear separation, one can see that there is a strong incentive to use symmetry arguments to the maximum extent, and to neglect terms which are not essential. Using the hybrid orbitals detailed in (7.2), the energies at the top and bottom of the valence band can be evaluated as illustrated in figure 7.1(b), namely

$$E_V^{t,b} = E_0 + h + c_{t,b}\Delta E_{sp},\qquad\qquad\text{(K.5)}$$

where the constants $c_t = +1/4$ and $c_b = -3/4$. Similar formulae exist for the conduction band, so that, from figure 7.1(b), we can see that the band gap in this model is

$$E_g = 2|h| - \Delta E_{sp}.\qquad\qquad\text{(K.6)}$$

Pettifor (1995, pp. 202–206) then proceeds to discuss the ideas of bond-order, bond-order potentials, and the second moment of the electron energy distribution μ_2 in terms of the ratio ($\Delta E_{sp}/2|h|$), and derives the angular dependent terms which are an essential part of empirical potentials widely used to describe the diamond structure semiconductors.

A much more detailed discussion of tight-binding methods for bulk semiconductors is given by Yu & Cardona (1996, chapter 2) and, in the context of surfaces, given by Desjonquères & Spanjaard (1996, chapter 5); these features can be explored further using project 7.2. The shapes and magnitudes of the pseudopotentials for the different s-, p- and d-orbitals can be gleaned from papers which parametrize the behavior of the individual materials, such as that for Si by Lenosky *et al.* (1997). One is, however, left with the impression that, despite the large number of input data used from experiment to determine the fit, the models still only have limited applicability. For this reason, the many computations in the literature using tight-binding and empirical potentials have been down-played in the descriptions given in chapter 7. However, one should note that this situation can change, with the rise of schemes described as *ab initio* or *first principles* tight binding (Turchi *et al.* 1998). These and other related methods, in which the parameters needed are calculated independently of experimental data, imply that tight binding methods do not have to remain at the most obviously empirical level for ever.

References

Note: the square brackets [*.*.*] indicate the sections in which the reference is used; the textbook references with two numbers only [*.*], except for chapter 9, are also listed in Appendix A, and in the further reading at the end of the corresponding chapters.

Abernathy, C.R. (1995) in *Handbook of Thin Film Process Technology* (Eds. D.A. Glocker and S.I. Shah, Institute of Physics) section A2.1 [2.5.3].

Abraham, F.F (1982) *Rep. Prog. Phys.* **45** 1113–1161 [4.4.3].

Adamson, A.W. (1990) *Physical Chemistry of Surfaces* (John Wiley, 5th Edn) [1.1–2].

Ajayan, P.M. & T.W. Ebbesen (1997) *Rep. Prog. Phys.* **60** 1025–1062 [8.4.3].

Akhter, P. & J.A. Venables (1981) *Surface Sci.* **102** L41–45; *Surface Sci.* **103** 301–314 [both 6.2.4].

Albrecht, T.R., S. Akamine, T.E. Carver & C.F. Quate (1990) *J. Vac. Sci. Tech.* **A8** 3386–3396 [3.1.3].

Alerhand, O.L., A.N. Berker, J.D. Joannopoulos, D. Vanderbilt, R.J. Hamers & J.E. Demuth (1990) *Phys. Rev. Lett.* **64** 2406–2409 [7.3.1].

Alivisatos, A.P. (1998) *Materials Research Bulletin*, February issue, pp. 18–23 [8.4.2].

Alldredge, G.P. & L. Kleinman (1974) *Phys. Rev.* **B10** 559–573 [6.1.2].

Allen, R.E. & F.W. deWette (1969) *Phys. Rev.* **179** 873–886; *Phys. Rev.* **188** 1320–1323 [1.4.3].

Amar, J.G. & F. Family (1995) *Phys. Rev. Lett.* **74** 2066–2069 [5.2.5, 5.4.3, 5.5.3].

Ambacher, O. (1998) *J. Phys. D: Appl. Phys.* **31** 2653–2710 [7.3.4].

Anderson, P.W. (1972) *Science* **177** 393–396 [8.2.4].

Anton, R., A. Schmidt & V. Schunemann (1990) *Vacuum* **41** 1099–1101 [5.3.2].

Appelbaum, J.A. & D.R. Hamann (1976) *Rev. Mod. Phys.* **48** 479–496 [6.1.2].

Arthur, J.R. (1994) *Surface Sci.* **299/300** 818–823 [7.3.4].

Ashcroft, N.W. & N.D. Mermin (1976) *Solid State Physics* (Saunders College) [6.1, 6.3, 7.1].

Bader, S.D. & V.L. Erskine (1994) in *Ultrathin Magnetic Structures* (Eds. B. Heinrich and J.A.C. Bland, Springer) **2** chapter 4, pp. 297–325 [6.3.3].

Baierlein, R. (1999) *Thermal Physics* (Cambridge University Press) [1.3.1, Appendix E].

Bak, P., P. Kleban, W.N. Unertl, J. Ochab, G. Akinci, N.C. Bartelt & T.L. Einstein (1985) *Phys. Rev. Lett.* **54** 1539–1542 [4.5.1].

Bales, G.S. (1996) *Surface Sci.* **356** L439-L444 [5.4.3]; (1996) *Mater. Res. Soc. Symp.* **399** 37–44 [5.5.3].

Bales, G.S. & D.C. Chrzan (1994) *Phys. Rev.* **B50** 6057–6067 [5.2.2, 5.2.5].

Barabási, A.L. & H.E. Stanley (1995) *Fractal Concepts in Surface Growth* (Cambridge University Press) [7.3.4].

Bardeen, J. (1947) *Phys. Rev.* **71** 717–727 [8.2.1].

Barker, J.A., D. Henderson & F.F. Abraham (1981) *Physica A***106** 226–238 [4.2.4].

Bartelt, M.C. & J.W. Evans (1992) *Phys. Rev. B***46** 12675–12687; (1994) *J. Vac. Sci. Tech. A***12** 1800–1808 [both 5.2.5].

Barth, J.V., H. Brune, G. Ertl & R.J. Behm (1990) *Phys. Rev. B***42** 9307–9318 [6.1.2].

Baski, A.A., S.C. Erwin & L.J. Whitman (1997) *Surface Sci.* **392** 69–85 [7.3.1].

Bassett, D.W. (1983) in *Surface Mobilities on Solid Materials* (Ed. V.T. Binh, Plenum) NATO ASI **B86** 63–108 [3.1.3, 5.4.2].

Bassett, D.W. & P.R. Webber (1978) *Surface Sci.* **70** 520–531 [5.4.2].

Bassett, G.A. (1958) *Phil. Mag.* **3** 1042–1045, plus 4 figures following page 1056 [5.5.1].

Batchelor, D.R., P. Rez, D.J. Fathers & J.A. Venables (1988) *Surf. Interface Anal.* **13** 193–201 [3.4.1].

Batchelor, D.R., H.E. Bishop & J.A. Venables (1989) *Surf. Interface Anal.* **14** 700–716 [3.4.1].

Batson, P.E. & J.F. Morar (1993) *Phys. Rev. Lett.* **71** 609–612 [6.2.2].

Batson, P.E., D.W. Johnson & J.C. H. Spence (1992) *Ultramicroscopy* **41** 137–145 [6.2.2].

Bauer, E. (1958) *Z. Kristallogr.* **110** 372–394 in German [5.1.3]; (1975) in *Interactions on Metal Surfaces* (Ed. R. Gomer, Topics in Applied Physics **4**, Springer) 225–274 [3.3.1–2]; (1984) in *The Chemical Physics of Solid Surfaces and Heterogeneous Catalysis* (Eds. D.A. King and D.P. Woodruff, Elsevier) **3B** 1–57 [5.4.1]; (1994) *Surface Sci.* **299/300** 102–115; (1994) *Rep. Prog. Phys.* **57** 895–938 [both 3.1.3, 3.2.3, 6.3.3]; (1997) in *The Chemical Physics of Solid Surfaces and Heterogeneous Catalysis* (Eds. D.A. King and D.P. Woodruff, Elsevier) **8**, chapter 2, pp. 46–65 [5.4.1].

Bauer, E. & H. Poppa (1972) *Thin Solid Films* **12** 167–185 [5.1.1].

Bauer, E., H. Poppa, G. Todd & P.R. Davis (1977) *J. Appl. Phys.* **48** 3773–3787 [3.4.2, 5.4.1].

Bauer, E., T. Duden, H. Pinkvos, H. Poppa & K. Wurm (1996), *J. Mag. Mag. Mat.* **156** 1–6 [6.3.3].

Bayard, R.T. & D. Alpert (1950) *Rev. Sci. Inst.* **21** 571–572 [2.3.5].

Beaume, R., J. Suzanne & J.G. Dash (1980) *Surface Sci.* **92** 453–466 [4.5.4].

Becker, W. & K.H. Bernhardt (1983) *Proc. 9th Int. Vacuum Cong.* (Madrid) pp. 212–216 [2.3.2].

Bell, A.T. (Ed.) (1992) *Catalysis Looks to the Future* (National Academy Press, Washington D.C.) [2.4.4].

Bell, L.D. & W.J. Kaiser (1988) *Phys. Rev. Lett.* **61** 2368–2371; (1996) *Ann. Rev. Mater. Sci.* **26** 189–222 [both 8.1.3].

Bennett, P.A. & H. von Känel (1999) *J. Phys. D: Appl. Phys.* **32** R71–R87 [8.1.3].

Bennett, P.A., J.C. Fuggle, F.U. Hillebrecht, A. Lenselink & G.A. Sawatsky (1983) *Phys. Rev. B***27** 2194–2209 [3.3.4].

Bentz, J.A., R.V. Thompson & S.K. Loyalka (1997) *Vacuum* **48** 817–824 [2.3.5].

Berman, A. (1996) *Vacuum* **47** 327–332 [2.1.3].

Bermond, J.M. & J.A. Venables (1983) *J. Cryst. Growth* **64** 239–256 [4.5.4].

Bermond, J.M., J.J. Métois, X. Egéa & F. Floret (1995) *Surface Sci.* **330** 48–60 [7.3.1–2, project 7.4].

Bernhardt, K.H. (1983) *J. Vac. Sci. Tech.* **A1** 136–139 [2.3.2].

Bernholc, J., C. Roland & B.I. Yakobson (1997) *Current Opinion in Solid State and Materials Sci.* **2** 706–715 [8.4.3].

Besenbacher, F. (1996) *Rep. Prog. Phys.* **59** 1737–1802 [3.5.4].

Besenbacher, F. & J.K. Nørskov (1993) *Prog. Surface Sci.* **44** 5–66 [4.5.3].

Besocke, K. (1987) *Surface Sci.* **181** 145–153 [3.1.3].

Besocke, K., B. Krahl-Urban & H. Wagner (1977) *Surface Sci.* **68** 39–46 [6.1.3].

Bethge, H. (1962) *Phys. Stat. Sol.* **2** 3–27, 775–820, both in German; (1990) in: *Kinetics of Ordering and Growth at Surfaces* (Ed. M.G. Lagally, Plenum) NATO ASI **B239** 125–144 [all 5.5.1].

Biasiol, G., L. Sorba, G. Bratina, R. Nicolini, A. Franciosi, M. Peressi, S. Baroni, R. Resta & A. Baldereschi (1992) *Phys. Rev. Lett.* **69** 1283–1286 [8.2.3].

Biegelsen, D.K., R.D. Bringans, J.E. Northrup & L.E. Swartz (1990) *Phys. Rev.* **B41** 5701–5706 [7.3.4].

Binnig, G., H. Rohrer, C. Gerber & E. Weibel (1982) *Phys. Rev. Lett.* **49** 57–61 [3.1.3, 7.2.3].

Bishop, H.E. & J.C. Rivière (1969) *J. Appl. Phys.* **40** 1740–1744 [3.4.1].

Blakely, J.W. (1973) *Introduction to the Properties of Crystal Surfaces* (Pergamon) [1.1].

Boguslawski, P. & J. Bernholc (1999) *Phys. Rev.* **B59** 1567–1570 [7.3.2].

Bonzel, H.P. (1983) in *Surface Mobilities on Solid Materials* (Ed. V. T. Binh, Plenum) NATO ASI **B86** 195–241 [5.5.2].

Bonzel, H.P. & C.H. Kleint (1995) *Prog. Surface Sci.* **49** 107–153 [3.3.1–2].

Booth, J.G. (Ed.) (1996) *Second International Symposium on Metallic Multilayers (MML'95)*, *J. Mag. Mag. Mat.* **156** 1–453 [6.3.4].

Borovsky, B., M. Krueger & E. Ganz (1997) *Phys. Rev. Lett.* **78** 4229–4232; (1999) *Phys. Rev.* **B59** 1598–1601 [both 7.3.2].

Bott, M., M. Hohage, T. Michely & G. Comsa (1993) *Phys. Rev. Lett.* **70** 1489–1492 [6.1.2].

Bott, M., M. Hohage, M. Morgenstern, T. Michely & G. Comsa (1996) *Phys. Rev. Lett.* **76** 1304–1307 [5.4.3].

Brack, M. (1993) *Rev. Mod. Phys.* **65** 677–732 [6.1.1].

Briggs, D. & M.P. Seah (1990) *Practical Surface Analysis, vols. I and II* (John Wiley) [3.3].

Brillson, L.J. (1982) *Surface Sci. Rep.* **2** 123–326; (1992) in *Handbook of Semiconductors* (Ed. P.T. Landsberg, Elsevier) **1** 281–417; (1994) *Surface Sci.* **299/300** 909–927 [all 8.2.1, 8.2.3].

Brock, J.D., R.J. Birgenau, J.D. Lister & A.Aharony (1989) *Contemp. Phys.* **30** 321–335 [4.4.3].

Brodie, I. (1995) *Phys. Rev.* **B51** 13660–13668 [6.1.3].

Brodie, I. & C.A. Spindt (1992) *Adv. Electronic and Electron Phys.* (Academic) **83** 1–106 [6.2.2].

Bromann, K., H. Brune, H. Röder & K. Kern (1995) *Phys. Rev. Lett.* **75** 677–680 [5.5.2].

Bromann, K., H. Röder, K. Kern, J. Jacobsen, P. Stoltze, K.W. Jacobsen & J. Nørskov, *Phys. Rev.* **B52** (1995) R14380–R14383 [5.5.2].

Brommer, K.D., M. Needels, B.E. Larson & J.D. Joannopoulos (1992) *Phys. Rev. Lett.* **68** 1355–1358; see also *Phys. Rev. Lett.* **71** (1993) 3612–3613 for ensuing correspondence [7.2.3].

Brown, W.A., R. Kose & D.A. King (1998) *Chem. Rev.* **98** 797–831 [4.4.4].

Browning, R. (1984) *J. Vac. Sci. Tech.* **A2** 1453–1456; (1985) *J. Vac. Sci. Tech.* **3** 1959–1964 [3.5.2].

Bruch, L.W. (1991) in *Phase Transitions in Surface Films 2* (Eds. H. Taub *et al.*, Plenum) pp. 67–82 [chap. 4, prob. 4.2].

Bruch, L.W., M.W. Cole and E. Zaremba (1997) *Physical Adsorption: Forces and Phenomena* (Oxford University Press) [4.2, 4.4].

Brune, H. (1998) *Surface Sci. Rep.* **31** 121–229 [5.4.3, 5.5.2–3].

Brune, H. & K. Kern (1997) in *The Chemical Physics of Solid Surfaces and Heterogeneous Catalysis* (Eds. D.A. King and D.P. Woodruff, Elsevier) **8** chapter 5 pp. 149–206 [5.4.3, 5.5.2].

Brune, H., G.S. Bales, J. Jacobsen, C. Boragno & K. Kern (1999) *Phys. Rev.* **B60** 5991–6006 [5.2.2, 5.2.5].

Brune, H., H. Röder, C. Boragno & K. Kern (1994) *Phys. Rev. Lett.* **73** 1955–1958 [5.4.3].

Brune, H., K. Bromann, H. Röder, K. Kern, J. Jacobsen, P. Stoltze, K. Jacobsen & J.K. Nørskov (1995) *Phys. Rev.* **B52** 14380–14383 [5.4.3, 5.5.2].

Brune, H., J. Wintterlin, R.J. Behm & G. Ertl (1992) *Phys. Rev. Lett.* **68** 624–626 [4.5.3].

Brune, H., J. Wintterlin, J. Trost, G. Ertl, J. Weichers & R.J. Behm (1993) *J. Chem. Phys.* **99** 2128–2148 [4.5.3].

Brune, H., M. Giovannini, K. Bromann & K. Kern (1998) *Nature* **394** 451–453 [5.4.3].

Brush, S.G. (1967) *Rev. Mod. Phys.* **39** 883–893 [4.5.1].

Bunshah, R.F. (Ed.) (1991) *Handbook of Deposition Technologies for Films and Coatings* (2nd Edn, Noyes Publications) [2.5.4].

Burhop, E.H.S. (1952) *The Auger Effect and Other Radiationless Transitions* (Cambridge University Press) [3.3.3].

Burton, W.K., N. Cabrera and F.C. Frank (1951) *Phil. Trans. R. Soc. Lond.* **A243** 299–358 [1.3.2].

Buseck, P., J.M. Cowley & L. Eyring (Eds.) (1988) *High Resolution Transmission Electron Microscopy and Associated Techniques* (Oxford University Press) [3.1].

Butcher, P., N.H. March & M.P. Tosi (Eds.) (1993) *Physics of Low-dimensional Semiconductor Structures* (Plenum) [8.2.3].

Calisti, S., J. Suzanne & J.A. Venables (1982) *Surface Sci.* **115** 455–468 [4.4.2, 4.4.4].

Callaway, J. & N.H. March (1984) *Solid State Physics* **38** 135–221 [Appendix J].

Cammarata, R.C. (1994) *Prog. Surface Sci.* **46** 1–38 [7.3.1].

Campbell, C.T. (1997) *Surface Sci. Rep.* **27** 1–112 [4.5.4].

Capasso, F. & A.Y. Cho (1994) *Surface Sci.* **299/300** 878–891 [8.2.3].

Capasso, F. & G. Margaritondo (Eds.) (1987) *Heterostructure Band Discontinuities: Physics and Device Applications* (North Holland) [8.2.1, 8.2.3].

Car, R. & M. Parinello (1985) *Phys. Rev. Lett.* **55** 2471–2474 [7.1.3].

Cardona, M. & L. Ley (Eds.) (1978) *Photoemission in Solids* (Springer) **1** [3.3.1–2].

Carlsson, J.O. (1991) in *Handbook of Deposition Technologies for Films and Coatings* (Ed. R.F. Bunshah, 2nd Edn, Noyes Publications) chapter 7, pp. 374–433 [2.5.5].

Cerny, S. (1983) in *The Chemical Physics of Solid Surfaces and Heterogeneous Catalysis* (Eds. D.A. King and D.P. Woodruff, Elsevier) **2** chapter 1, pp. 1–57 [4.4.1].

Chadi, D.J. (1979) *Phys. Rev. Lett.* **43** 43–47 [1.4.4, 7.2.4]; (1987) *Phys. Rev Lett.* **59** 1691–1694 [7.3.2]; (1989) *Ultramicroscopy* **31** 1–9 [1.4.4, 7.1.3–4, 7.2.1–2, 7.2.4, 7.3.4]; (1994) *Surface Sci.* **299/300** 311–318 [1.4.4].

Chambers, A., R.K. Fitch, & B.S. Halliday (1998) *Basic Vacuum Technology* (2nd Edn, Institute of Physics) [2.3.1].

Chambliss, D.D. & K.E. Johnson (1994) *Phys. Rev. B***50**, 5012–5015 [5.5.3].

Chambliss, D.D., R.J. Wilson & S. Chiang (1991) *Phys. Rev. Lett.* **66** 1721–1724 [5.5.3].

Chan, E.M., M.J. Buckingham & J.L. Robins (1977) *Surface Sci.* **67** 285–298 [5.3.2].

Chang, C.C. (1974) *Surface Sci.* **25** 53–74 [3.3.3].

Chaparro, S.A., J.S. Drucker, Y. Chang, D. Chandrasekhar, M.R. McCartney & D.J. Smith (1999) *Phys. Rev. Lett.* **83** 1199–1202 [7.3.3].

Chapman, J.N. & K.J. Kirk (1997) in *Magnetic Hysterisis in Novel Magnetic Materials* (Ed. G.C. Hadjipanayis, Kluwer) NATO ASI **E338**195–206 [6.3.2].

Chapman, J.N., A.B. Johnston, L.J. Heyderman, S. McVitie & W.A.P. Nicholson (1994) *IEEE Trans. Magnetics* **30** 4479–4484 [6.3.2].

Chattarji, J. (1976) *The Theory of Auger Transitions* (Academic) [3.3.2–3].

Chen, C.J. (1993) *Introduction to Scanning Tunneling Microscopy* (Oxford University Press) [3.1].

Chen, C. & T.T. Tsong (1990) *Phys. Rev. Lett.* **64** 3147–3150 [5.4.2].

Chen, X., F. Wu, Z. Zhang & M.G. Lagally (1994) *Phys. Rev. Lett.* **73** 850–853 [7.3.3].

Chino, K. (1973) *Solid State Electron.* **16** 119–121 [8.1.1].

Cho, K., J.D. Joannopoulos & A.N. Berker (1996) *Phys. Rev. B***53** 1002–1005 [7.3.2].

Chui, S.T. (1983) *Phys. Rev. B***28** 178–194 [4.4.3].

Chung, M.F. & L.H. Jenkins (1970) *Surface Sci.* **22** 479–485 [3.3.3].

Clarke, L.J., (1985) *Surface Crystallography: an Introduction to Low Energy Electron Diffraction* (John Wiley) [1.4, 3.2].

Cohen, E.R. & B.N. Taylor (1987) *Rev. Mod. Phys.* **59** 1121–1148; (1998) *Physics Today,* August Buyers Guide BG7–BG14 [Appendix C].

Cohen, M.L. (1984) *Phys. Rep.* **110** 293–309 [7.1.3].

Collazo-Davilla, C., E. Bengu & L.D. Marks (1998) *Phys. Rev. Lett.* **80** 1678–1681 [8.1.3].

Collier, C.P., T. Vossmeyer & J.R. Heath (1998) *Ann. Rev. Phys. Chem.* **49** 371–404 [8.4.2].

Collins, F.S. (1999) *Microarrays and Macroconsequences*: see several review articles in *Nature Genetics* **21** 2–60 [9.2].

Collins, P.G. & A. Zettl (1997) *Phys. Rev. B***55** 9391–9399 [8.4.3].

Copel, M., M.C. Reuter, E. Kaxiras & R.M. Tromp (1989) *Phys. Rev. Lett.* **63** 632–635 [7.3.3].

Craik, D. (1995) *Magnetism: Principles and Applications* (John Wiley) [6.3].

Crawford, R.K. (1977) in *Rare Gas Solids* (Eds. M.L. Klein and J.A. Venables, Academic) **2** chapter 11 pp. 663–728 [1.3.1].

Crommie, M.F., C.P.Lutz, D.M. Eigler & E.J. Heller (1995) *Physica D***83** 98–108 [6.1.1].

Crowell, C.R., J.C. Sarace & S.M. Sze (1965) *Trans. Met. Soc. AIME* **233** 478–481 [8.1.1].

Cullis, A.G., D.J. Robbins, A.J. Pidduck & P.W. Smith (1992) *J. Cryst. Growth* **123** 333–343 [7.3.3].

Cumpson, P.J. & M.P. Seah (1997) *Surf. Interface Anal.* **25** 430–446 [3.4.2].

Daimon, H. (1988) *Rev. Sci. Inst.* **59** 545–549 [3.3.1].

Daimon, H., T. Nakatani, S. Imada, S. Suga, Y. Kagoshima & T. Miyahara (1995) *Rev. Sci. Inst.* **66** 1510–1512 [3.3.1–2, 6.3.3].

Danielson, E., J.H. Golden, E.W. McFarland, C.M. Reaves, W.H. Weinberg & X.D. Wu (1997) *Nature* **389** 944–948 [8.4.4].

Darling, G.R. & S. Holloway (1995) *Rep. Prog. Phys.* **58** 1595–1672 [4.5.2].

Daruka, I. & A.L. Barabási (1997) *Phys. Rev. Lett.* **79** 3708–3711 [7.3.3].

Daughton, J.M., A.V. Pohm, R.T. Fayfield & C.H. Smith (1999) *J. Phys. D: Appl. Phys.* **32** R169–R177 [8.3.3].

Davies, G.J. & D. Williams (1985) in *The Technology and Physics of Molecular Beam Epitaxy* (Ed. E.H.C. Parker, Plenum) chapter 2 pp. 15–46 [2.5.2].

Davies, J.H. (1998) *The Physics of Low-dimensional Semiconductors: an Introduction* (Cambridge University Press) [7.1, 8.2].

Debe, M.K. & D.A. King (1977) *J. Phys. C***10** L303-L308; *Phys. Rev. Lett.* **39** 708–711 [1.4.3].

DeBoeck, J. & G. Borghs (1999) *Physics World*, April issue pp. 27–32 [8.3.3].

DeBoer, F.R., R. Boom, W.C.M. Mattens, A.R. Miedema & A.K. Niessen (1988) *Cohesion in Metals: Transition Metal Alloys* (North-Holland) [6.1.4].

DeRisi, J.L., V.R. Iyer & P.O. Brown (1997) *Science* **278** 680–686 [8.4.4].

Delchar, T.A. (1993) *Vacuum Physics and Techniques* (Chapman and Hall) chapter 2, pp. 26–46 [2.2.3, 2.3.1].

Deng, X. & M. Krishnamurthy (1998) *Phys. Rev. Lett.* **81** 1473–1476 [7.3.3].

Denier van der Gon, A.W., J.M. Gay, J.W.M. Frenken & J.F. van der Veen (1991) *Surface Sci.* **241** 335–345 [7.2.3].

Desjonquères, M.C. & D. Spanjaard (1996) *Concepts in Surface Physics* (Springer) [1.1, 1.3, 4.5, 6.1, 7.1, project 7.4, Apendix K].

Deutsch, P.W., L.A. Curtiss & J.P. Blaudeau (1997) *Chem. Phys. Lett.* **270** 413–418 [7.1.3].

Devoret, M.H. & C. Glattli (1998) *Physics World,* September issue pp. 29–33 [8.3.2].

Diehl, R.D. & R. McGrath (1997) *J. Phys. Condens. Matter* **9** 951–968 [6.1.3].

Dijkamp, D., T. Venkatesan, X.D. Wu, S.A. Shabeen, N. Jiswari, Y.H. Min-Lee, W.L. McLean & M. Croft (1987) *Appl. Phys. Lett.* **51** 619–621 [2.5.2].

Donohoe, A.J. & J.L. Robins (1972) *J. Cryst. Growth* **17** 70–76; (1976) *Thin Solid Films* **33** 363–372 [both 5.3.1].

Drinkwine, M.J. & D. Lichtman (1979) *Partial Pressure Analyzers and Analysis* (American Vacuum Society) [2.3.5].

Drucker, J.S. (1993) *Mat. Res. Soc. Symp.* **280** 389–392; (1993) *Phys. Rev. B***48** 18203–18206 [7.3.3].

Duckworth, H.E., R.C. Barber & V.S. Venkasubramanian (1986) *Mass Spectroscopy* (2nd Edn, Cambridge University Press) [2.3.5].

Duggan, G. (1987) in *Heterostructure Band Discontinuities: Physics and Device Applications* (Eds. F. Capasso and G. Margaritondo, North Holland) chapter 5, pp. 207–262 [8.2.3].

Duke, C.B. (1992) *J. Vac. Sci. Tech. A***10** 2032–40; (1993) *Appl. Surf. Sci.* **65** 543–552; *Festkörper-probleme/Advances in Solid State Physics* **33** 1–36; (1994) *Scanning Microscopy* **8** 753–764; (1996) *Chemical Reviews* **96** 1237–1260 [all 7.1.3, 7.2.1]; Duke, C.B. (Ed.) (1994) *Surface Science: the First Thirty Years* (*Surface Sci.* **299/300** 1–1054) [preface].

Dürr, H., J.F. Wendelken & J.K. Zuo (1995) *Surface Sci.* **328** L527–L532 [5.4.3].

Dushman, S. & J. Lafferty (1992) *Scientific Foundations of Vacuum Technique* (John Wiley); this updates the 1962 book with the same title by S. Dushman [2.1, 2.3, Appendix G].

Dwyer, V.M. & J.A.D. Matthew (1983) *Vacuum* **33** 767–769; (1984) *Surface Sci.* **143** 57–83 [both 3.4.2].

Dylla, H.F. (1996) *Vacuum* **47** 647–651 [2.1.3].

Eaglesham, D.J. & M.Cerullo (1990) *Phys. Rev. Lett.* **64** 1943–1946 [7.3.3].

Eaglesham, D.J., A.E. White, L.C. Feldman, N. Moriya & D.C. Jacobson (1993) *Phys. Rev. Lett.* **70** 1643–1646 [7.3.1–2].

Eastman, D.E., J.J. Donelon, N.N. Hien & F.J. Himpsel (1980) *Nucl. Inst. Meth.* **172** 327–336 [3.3.2].

Ebbesen, T.W. & P.M. Ajayan (1992) *Nature* **358** 220–222 [8.4.3].

Ehrlich, G. (1991) *Surface Sci.* **246** 1–12; (1994) *Surface Sci.* **299/300** 628–642; (1995) *Surface Sci.* **331/333** 865–877; (1997) in *Surface Diffusion: Atomistic and Collective Processes* (Ed. M.C. Tringides, Plenum) NATO ASI **B360** 23–43 [all 3.1.3, 5.4.2].

Ehrlich, G. & F.G. Hudda (1966) *J. Chem. Phys.* **44** 1039–1055 [5.5.1].

Eigler, D. & Schweizer (1990) *Nature* **344** 524–526 [4.3.2].

Einstein, T.L. (1996) in *Handbook of Surface Science* (Ed. W.N. Unertl, Elsevier) **2** chapter 11 pp. 577–650 [4.5.1–2, 5.4.2].

Eiswirth, M., K. Krischer & G. Ertl (1990) *Appl. Phys. A***51** 79–90 [4.5.4].

Eiswirth, M., M. Bär & H.H. Rotermund (1995) *Physica D***84** 40–57 [4.5.4].

El-Gomati, M.M., A.P. Janssen, M. Prutton & J.A. Venables (1979) *Surface Sci.* **85** 309–316 [3.5.1].

El-Gomati, M.M., M. Prutton, B. Lamb & C.G. Tuppen (1988) *Surf. Interface Anal.* **11** 251–265 [3.5.2].

Elliott, A.G. (1974) *Surface Sci.* **44** 337–359 [5.3.3].

Ellis, T.H., G. Scoles, U. Valbusa, H. Jónsson & J.H. Weare (1985) *Surface Sci.* **155** 499–534 [4.4.1].

Engdahl, G. & G. Wahnström (1994) *Surface Sci.* **312** 429–440 [4.5.3].

Ertl, G. (1994) *Surface Sci.* **299/300** 742–754 [4.5.4].

Estrup, P.J. (1994) *Surface Sci.* **299/300** 722–730 [1.4.3, 6.1.2].

Eustathopoulos, N., J.-C. Joud & P. Desré (1973) *J. Chim. Phys.* **70** 42–48 [6.1.4].

Falicov, L.M., D.T. Pierce, S.D . Bader, R. Gronsky, K.B. Hathaway, H.J. Hopster, D.N. Lambeth, S.S.P. Parkin, G. Prinz, M. Salamon, I.K. Schuller & R.H. Victora (1990) *J. Materials Res.* **5** 1299–1340 [8.3.3].

Falkenberg, G. & R.L. Johnson (1999) unpublished drawing of experimental chamber used at HASYLab [2.4.3].

Falta, J. & M. Henzler (1992) *Surface Sci.* **269/270** 14–21 [7.3.2].

Feenstra, R.M. (1994) *Surface Sci.* **299/300** 965–979 [3.5.4, 7.2.1].

Feenstra, R.M., J.A. Stroscio, J. Tersoff & A.P. Fein (1987) *Phys. Rev. Lett.* **58** 1192–1195 [7.2.1].

Feibelman, P.J. (1990) *Phys. Rev. Lett.* **65** 729–732 [5.4.2]; (1997) *Phys. Rev.* **B56** 2175–2182 [4.5.4].

Feibelman, P.J., J.S. Nelson & G.L. Kellogg (1994) *Phys. Rev.* **B49** 10548–10556 [5.4.3].

Feidenhans'l, R. (1989) *Surface Sci. Rep.* **10** 105–188 [3.2.1, 7.2.3].

Feidenhans'l, R., J.S. Pedersen, J. Bohr, M. Nielsen, F. Grey & R.L. Johnson (1988) *Phys. Rev.* **B38** 9715–9720 [7.2.3].

Feldman, L.C. & J.W. Mayer (1986) *Fundamentals of Surface and Thin Film Analysis* (North-Holland) [3.1, 3.4].

Félix, C., G. Vandoni, W. Harbich, J. Buttet & R. Monot (1996) *Phys. Rev.* **B54** 17039–50 [5.4.3].

Felter, T.E., R.A. Barker & P.J. Estrup (1977) *Phys. Rev. Lett.* **38** 1138–1141 [1.4.3].

Ferrari, A.M. & G. Pacchioni (1996) *J. Phys. Chem.* **100** 9032–9037 [5.3.2–3].

Ferrario, B. (1996) *Vacuum* **47** 363–370 [2.3.2].

Ferry, D.K. & S.M. Goodnick (1997) *Transport in Nanostructures* (Cambridge University Press) [8.3].

Fert, A. & P. Bruno (1994) in *Ultrathin Magnetic Structures* (Eds. B. Heinrich and J.A.C. Bland, Springer) **2** chapter 2.2, pp. 82–118 [8.3.3].

Feynman, R.P. (1992) *J. Microelectromechanical Systems* **1** 60–66 [8.4.4].

Fink, H.W. (1988) *Physica Scripta* **38** 260–263 [6.2.2].

Flores, F. & C. Tejedor (1979) *J. Phys.* **C12** 731–749 [8.2.1].

Floro, J.A., E. Chason, R.D. Twesten, R.Q. Hwang & L.B. Freund (1997) *Phys. Rev. Lett.* **79** 3946–3949 [7.3.3].

Floro, J.A., G.A. Lucadamo, E. Chason, L.B. Freund, M. Sinclair, R.D. Twesten & R.Q. Hwang (1998) *Phys. Rev. Lett.* **80** 4717–4720 [7.3.3].

Follstaedt, D.M. (1993) *Appl. Phys. Lett.* **62** 1116–1118 [7.3.1].

Fournier, R., S.B. Sinnott & A.E. DePristo (1992) *J. Chem. Phys.* **97** 4149–4161 [7.1.3].

Franciosi, A. & C.G. Van de Walle (1996) *Surface Sci. Rep.* **25** 1–140 [8.2.2–3].

Frank, L. (1991) *Measurement Sci. Tech.* **2** 312–317 [3.5.1].

Frankl, D.R. & J.A. Venables (1970) *Adv. Phys.* **19** 409–456 [5.1.4, 5.3.3].

Freeman, A.J., C.L. Fu, S. Onishi & M. Weinert (1985) in *Polarized Electrons in Surface Physics* (Ed. R. Feder, World Scientific) chapter 1 pp. 3–66 [6.3.4].

Frenken, J.W.M. & P. Stoltze (1999) *Phys. Rev. Lett.* **82** 3500–3503 [6.1.4].

Friedel, J. (1969) in *The Physics of Metals* (Ed. J.M. Ziman, Cambridge) pp. 340–408 [6.3.4].

Frohn, J., J.F. Wolf, K. Besocke & M. Teske (1989) *Rev. Sci. Inst.* **60** 1200–1201 [3.1.3].

Fruchart, O., M. Klaua, J. Barthel & J. Kirschner (1999) *Phys. Rev. Lett.* **83** 2769–2772 [8.3.3].

Fuggle, J.C., F.U. Hillebrecht, R. Zeller, Z. Zolnierek, P.A. Bennett & C. Freiburg (1982) *Phys. Rev.* **B27** 2145–2178 [3.3.4].

Futamoto, M., M. Nakazawa & U. Kawabe (1980) *Surface Sci.* **100** 470–480; (1983) *Vacuum* **33** 727–732 [6.2.1].

Futamoto, M., M. Hanbücken, C.J. Harland, G.W. Jones & J.A. Venables (1985) *Surface Sci.* **150** 430–450 [3.5.1–2, 6.2.4].

Gangwar, R. & R.M. Suter (1990) *Phys. Rev.* **B42** 2711–2714 [4.3.1].

Gangwar, R., N.J. Collella & R.M. Suter (1989) *Phys. Rev.* **B39** 2459–2471 [4.3.1, 4.4.3].

Garber, E., S.G. Brush & C.W.F. Everitt (Eds.) (1986) *Maxwell on Molecules and Gases* (MIT Press) pp. 295–296 [2.1.3].

García, A. & J.E. Northrup (1993) *Phys. Rev.* **B48** 17350–17353 [7.3.1].

Gates, A.D. & J.L. Robins (1982) *Surface Sci.* **116** 188–204 [5.3.3, 5.5.1]; (1987a) *Thin Solid Films* **149** 113–128; [5.3.2–3]; (1987b) *Surface Sci.* **191** 492–517 [5.5.1]; (1988) *Surface Sci.* **194** 13–43 [5.3.2].

Genzken, O. & M. Brack (1991) *Phys. Rev. Lett.* **67** 3286–3289 [6.1.1].

Gibbs, J.W. (1928, 1948, 1957) *Collected Works, vol. 1* (Yale Univerity Press, New Haven); reproduced as (1961) *The Scientific Papers, vol. 1* (Dover Reprint Series, New York) [1.1].

Gibson, J.M. X. Chen & O. Pohland (1997) *Surface Rev. Lett.* **4** 559–566 [8.1.3].

Gibson, K.D. & S.J. Sibener (1985) *Phys. Rev. Lett.* **55** 1514–1517; *Faraday Disc. Chem. Soc.* **80** 203–215 [4.4.4].

Giesen, M. & H. Ibach (1999) *Surface Sci.* **431** 109–115 [5.4.3, 5.5.2].

Gil, B. (Ed.) (1998) *Group III Nitride Semiconductor Compounds* (Oxford University Press) [7.3].

Glocker, D.A. & S.I. Shah (Eds.) (1995) *Handbook of Thin Film Process Technology* (Institute of Physics), parts A and B [2.3, 2.5].

Glueckstein, J.C., M.M.R. Evans & J. Nogami (1996) *Phys. Rev.* **B54** R11066-R11069 [3.4.2].

Godbey, D.J. & M. Ancona (1992) *Appl. Phys. Lett.* **61** 2217–2219; (1993) *J. Vac. Sci. Tech.* **B11** 1392–1395; (1997) *J. Vac. Sci. Tech.* **A15** 976–980; (1998) *Surface Sci.* **395** 60–68 [all 7.3.3].

Golberg, D., Y. Bando, M. Eremets, K. Kurashima, T. Tamiya, K. Takemura & H. Yusa (1996) *Appl. Phys. Lett.* **69** 2045–2047; (1997) *J. Electron Microscopy* **46** 281–292 [both 8.4.3].

Goldys, E., Z.W. Gortel & H.J. Kreuzer (1982) *Surface Sci.* **116** 33–65 [4.4.4].

Gomer, R. (1961) *Field Emission and Field Ionization* (Harvard Press) [6.1.3, 6.2.2–3]; (1983) in *Surface Mobilities on Solid Materials* (Ed. V. T. Binh, Plenum NATO ASI) **B86** 1–25 [5.5.2]; (1990) *Rep. Prog. Phys.* **53** 917–1002 [5.5.2, 6.2.3]; (1994) *Surface Sci.* **299/300** 129–152 [6.2.2–3].

Goodhew, P.J. & T.A. Fretwell (1998) *J. Mater. Ed.* **20** 68–75 [Appendix D].

Gossard, A.C. (Ed.) (1994) *Epitaxial Microstructures* (Semiconductors and Semimetals **40**, Academic) [7.3.4].

Gotoh, Y. & S. Ino (1978) *Jap. J. Appl. Phys.* **17** 2097–2109 [3.2.2].

Graper, E.B. (1995) in *Handbook of Thin Film Process Technology* (Eds. D.A. Glocker and S.I. Shah, Institute of Physics) sections A1.0, A1.1, A1.2 [2.5.2].

Greene, J.E. (1991) in *Handbook of Deposition Technologies for Films and Coatings* (Ed. R.F. Bunshah, 2nd Edn, Noyes Publications) chapter 13 pp. 681–739 [2.5.4].

Gregg, J.F., W. Allen, K. Ounadjela, M. Viret, M. Hehn, S.M. Thompson & J.M.D. Coey (1996) *Phys. Rev. Lett.* **77** 1580–1583 [8.3.3].

Gregg, J.F., W. Allen, N. Viart, R. Kirschman, C. Sirisathikul, J.-P. Schille, M. Gester, S. Thompson, P. Sparks, V. Da Costa, K. Ounadjela & M. Skvarla (1997) *J. Mag. Mag. Mat.* **175** 1–9 [8.3.3].

Griffith, J.E. & G.P. Kochanski (1990) *Crit. Rev. Solid State and Mater. Sci.* **16** 255–289 [7.2.4].

Grimley, T.B. (1967) *Proc. Phys. Soc. (London)* **90** 751–764; (1967) *Proc. Phys. Soc. (London)* **92** 776–782 [both 4.5.2].

Grimley, T.B. (1983) in *The Chemical Physics of Solid Surfaces and Heterogeneous Catalysis* (Eds. D.A. King and D.P. Woodruff, Elsevier) **2** 333–376 [4.5.2].

Gringerich, K.A., I. Shim, S.K. Gupta & J.E. Kingcade, Jr. (1985) *Surface Sci.* **156** 495–503 [5.3.1].

Haas, G., A. Menck, H. Brune, J.V. Barth, J.A. Venables & K. Kern (2000), *Phys. Rev.* **B61** 11105–11108 [5.3.3].

Halperin, B.I. & D.R. Nelson (1978) *Phys. Rev. Lett.* **41** 121–124; *Phys. Rev. Lett.* **41** 519 [both 4.4.3].

Hamichi, M., A.Q.D. Faisal, J.A. Venables & R. Kariotis (1989) *Phys. Rev.* **B39** 415–425 [4.4.2].

Hamichi, M., R. Kariotis & J.A. Venables (1991) *Phys. Rev.* **B43** 3208–3214 [4.4.2].

Hammer, B & J.K. Nørskov (1997) in *Chemisorption and Reactivity on Supported Clusters and Thin Films* (Eds. R.M Lambert and G. Pacchioni, Kluwer) NATO ASI **E 331** 285–351 [4.5.2, project 4.4].

Han, J., R.L. Gunshor & A.V. Nurmiko (1999) in *Thin Films: Heteroepitaxial Systems* (Eds. W.K. Liu and M.B. Santos, World Scientific) chapter 12 pp. 520–620 [7.3.4].

Hanbücken, M., M. Futamoto & J.A. Venables (1984) *Surface Sci.* **147** 433–450 [3.4.2].

Hannon, J.B., C. Klünker, M. Giesen, H. Ibach, N.C. Bartelt & J.C. Hamilton (1997) *Phys. Rev. Lett.* **79** 2506–2509 [5.4.3].

Harding, J.H., A.M. Stoneham & J.A. Venables (1998) *Phys. Rev.* **B57** 6715–6719 [5.3.2, 5.5.1].

Harland, C.J. & J.A. Venables (1985) *Ultramicroscopy* **17** 9–20 [3.4.2].

Harland, C.J., G.W. Jones, T.N. Doust & J.A. Venables (1987) *Scanning Micr. Suppl.* **1** 109–114 [6.2.4].

Harrison, W.A. (1980) *Electronic Structures and the Properties of Solids* (Freeman) [7.1.1].

Harsdorff, M. (1982) *Thin Solid Films* **90** 1–14; (1984) *Thin Solid Films* **116** 55–74 [5.3.3].

Hartig, K., A.P. Janssen & J.A. Venables (1978) *Surface Sci.* **74** 69–78 [5.4.1].

Hauenstein, R.J. (1999) in *Thin Films: Heteroepitaxial Systems* (Eds. W.K. Liu and M.B. Santos, World Scientific) chapter 11 pp. 512–559 [7.3.4].

Healy, S.D., K.R. Heim, Z.J. Wang, G.G. Hembree, J.S. Drucker & M. Scheinfein (1994) *J. Appl. Phys.* **75**, 5592–5594 [5.5.3].

Heath, J.R. (1995) *Science* **270** 1315–1316 [8.4.2].

Heim, K.R., S.D. Healy, Z.J. Yang, J.S. Drucker, G.G. Hembree & M.R. Scheinfein (1993) *J. Appl. Phys.* **74** 7422–7430 [3.5.3, 6.3.3].

Heim, K.R., S.T.Coyle, G.G. Hembree, J.A. Venables & M.R. Scheinfein (1996) *J. Appl. Phys.* **80** 1161–1170 [5.3.3].

Heinrich, B. & J.A.C. Bland (Eds.) (1994) *Ultrathin Magnetic Structures* (Springer) **1** and **2** [6.3].

Heinrich, B. & J.F. Cochran (1993) *Adv. Phys.* **42** 523–639 [6.3].

Heinz, K. (1994) *Surface Sci.* **299/300** 433–446 [3.2.1]; (1995) *Rep. Prog. Phys.* **58** 637–704 [3.2.1, 3.2.3].

Helmer, J.C. & G. Levi (1995) *J. Vac. Sci. Tech.* **A13** 2592–2599 [2.3.2].

Helms, C.R. & E.H. Poindexter (1994) *Rep. Prog. Phys.* **57** 791–852 [8.1.3].

Hembree, G.G. & J.A. Venables (1992) *Ultramicroscopy* **47** 109–120 [3.4.2, 3.5.3].

Hembree, G.G., J. Unguris, R.J. Celotta & D.T. Pierce (1987) in *Physical Aspects of Microscopic Characterization of Materials* (*Scanning Micr. Suppl.*) **1** 229–240 [6.3.3].

Hembree, G.G., J. S. Drucker, S.D. Healey, K.R. Heim, Z.J. Yang & M.R. Scheinfein (1994) *Appl. Phys. Lett.* **64** 1036–1038 [6.3.3].

Henisch, H.K. (1984) *Semiconductor Contacts: an Approach to Ideas and Models* (Oxford University Press) [8.2.1].

Henrich, V.E. & P.A. Cox (1994, 1996) *The Surface Science of Metal Oxides* (Cambridge University Press) [1.4, 4.5].

Henry, C.R. (1998) *Surface Sci. Rep.* **31** 231–326 [4.5.4].

Henry, C.R., C. Chapon & B. Mutaftschiev (1985) *Surface Sci.* **163** 409–434 [5.5.1].

Henry, C.R., C. Chapon, C. Duriez & S. Gorgio (1991) *Surface Sci.* **253** 177–189 [4.5.4].

Henry, C.R., C. Chapon, C. Goyhenex & R. Monot (1992) *Surface Sci.* **272** 283–288 [4.5.4].

Henry, C.R., C. Chapon, S. Giorgio & C. Goyhenex (1997) in *Chemisorption and Reactivity on Supported Clusters and Thin Films* (Eds. R.M Lambert and G. Pacchioni, Kluwer) NATO ASI **E 331** 117–152 [4.5.4].

Henzler, M. (1977) in *Electron Spectroscopy for Surface Analysis* (Ed. H. Ibach, Springer) 117–149; (1997) *Surface Rev. Lett.* **4** 489–500 [both 3.2.1, 3.2.3].

Herbots, N., O.C. Hellman, P. Ye & X. Wang (1994) in *Low Energy Ion-Surface Interactions* (Ed. J. W. Rabalais, John Wiley) chapter 8 pp. 387–480 [2.5.4].

Herring, C. (1951) *Phys. Rev.* **82** 87–95 [1.2.1]; (1953) in *Structure and Properties of Solid Surfaces* (Eds. R. Gomer and C.S. Smith, Chicago University Press) pp. 5–72 [1.2.1, 1.2.3].

Heun, S., J. Falta & M. Henzler (1991) *Surface Sci.* **243** 132–140 [7.3.2].

Heyraud, J.C. & J.J. Métois (1980) *Acta Metall.* **28** 1789–1797 [6.1.4]; *J. Cryst. Growth* **50** 571–574 [6.1.4]; (1983) *Surface Sci.* **128** 334–342 [1.2.3, 6.1.4].

Higashi, G.S., E.A. Irene and T.Ohmi (Eds.) (1993) *Mater. Res. Soc. Symp.* **315** [2.4.1].

Higashi, G.S., M. Hirose, S. Raghavan and S. Verhaverbeke (Eds.) (1997) *Mater. Res. Soc. Symp.* **477** [2.4.1].

Hill, T.L. (1960) *An Introduction to Statistical Thermodynamics* (Addison-Wesley, reprinted by Dover 1986), [1.3, 4.2, Appendix E].

Himpsel, F.J. (1994) *Surface Sci.* **299/300** 525–540 [3.3.2, 3.3.4].

Himpsel, F.J., F.R. McFeely, A. Taleb-Ibrahimi, J.A. Yarnoff & G. Hollinger (1988) *Phys. Rev.* **B38** 6084–6096 [3.3.4].

Hobson, J.P. (1983) *Proc. 9th Int. Vacuum Cong. (Madrid)* 273–282; (1984) *J. Vac. Sci. Tech.* **A2** 144–149 [2.3.4].

Hobson, J.P. & E.V. Kornelsen (1979) *J. Vac. Sci. Tech.* **16** 701–707 [2.4.3].

Hohenberg, P. & W. Kohn (1964) *Phys. Rev.* **136B** 864–871 [Appendix J].

Holian, B.L. (1980) *Phys. Rev.* **B22** 1394–1404 [4.2.4].

Holloway, P.H. (1977) *Surface Sci.* **66** 479–494; (1980) *Adv. Electronics and Electron Physics* **54** 241–298 [both 3.4.2].

Hölzl, J. & F.K. Schulte (1979) in *Solid Surface Physics* (Ed. G. Höhler, Springer Tracts in Modern Physics **85**) 1–150 [6.1.3, 6.2.1].

Honig, R.E. & D.A. Kramer (1969) *RCA Review* **30** 285–304 [1.3.1, project 7.4].

Hopster, H. (1994) in *Ultrathin Magnetic Structures* (Eds. B. Heinrich and J.A.C. Bland, Springer) **1** chapter 4.1 pp. 123–152 [6.3.3].

Horch, S., P. Zeppenfeld & G. Comsa (1995) *Surface Sci.* **331/333** 908–912; (1995) *Appl. Phys.* **A60** 147–153 [both 4.3.2].

Horn-von Hogen, M., B.H. Müller, A. Al-Falou & M. Henzler (1993) *Phys. Rev. Lett.* **71** 3170–3173 [7.3.3].

Howes, P.B., C. Norris, M.S. Finney, E. Vlieg & R.G. van Silfhout (1993) *Phys. Rev.* **B48** 1633–1642 [1.4.6].

Howie, A. (1995) *J. Microscopy* **180** 192–203 [3.5.1].

Huang, M., C.J. Harland & J.A. Venables (1993) *Surf. Interface Anal.* **20** 666–674 [3.3.2].

Hudson, J.B. (1992) *Surface Science: an Introduction* (Butterworth-Heinemann); reprinted in 1998 and published by John Wiley [1.1–2, 2.1, 4.5].

Hüfner, S. (1996) *Photoelectron Spectroscopy* (2nd Edn, Springer) [3.2.2, 3.3.4].

Hull, R. & E.A. Stach (1999) in *Thin Films: Heteroepitaxial Systems* (Eds. W.K. Liu and M.B. Santos, World Scientific) chapter 7 pp. 299–367 [7.3.3].

Humphreys, C.J., D.M. Maher, H.L. Fraser & D.J. Eaglesham (1988) *Phil. Mag.* **A58** 787–798; (1998) *Ultramicroscopy* **26** 13–24 [both 8.1.3].

Hurle, D.T.J. (Ed.) (1993, 1994) *Handbook of Crystal Growth* (Elsevier): **1**-*Fundamentals*; **2**-*Bulk Crystal Growth*; **3**-*Thin Films and Epitaxy* [1.3.2].

Ibach, H. (Ed.) (1977) *Electron Spectroscopy for Surface Analysis* (Springer); (1994) *Surface Sci.* **299/300** 116–128 [both 3.3.1].

Ibach, H. & D.L. Mills (1982) *Electron Energy Loss Spectroscopy and Surface Vibrations* (Academic) [3.3.1].

M. Ichikawa, M. & T. Doi (1988) in *Reflection High Energy Electron Diffraction and Reflection Electron Imaging of Surfaces* (Eds. P.K. Larsen and P.J. Dobson, Plenum) NATO ASI **B188** 343–369 [3.2.2, 3.5.1].

Ichimaya, A., Y. Ohno & Y Horio (1997) *Surface Rev. Lett.* **4** 501–511 [3.2.3].

Idzerda, Y.U., L.H. Tjeng, H.J. Lin, C.J. Gutierrez, G. Meigs & C.T. Chen (1993) *Phys. Rev. B***48** 4144–4147 [6.3.3].

Iijima, S. (1991) *Nature* **354** 56–58 [8.4.3].

Inglesfield, J.E. & G.A. Benesh (1988) *Phys. Rev. B***37** 6682–6700 [6.1.3].

Inglesfield, J.E. (1985) *Prog. Surface Sci.* **20** 105–164 [6.1.2].

Ino, S. (1977) *Jap. J. Appl. Phys.* **16** 891–908; (1988) in *Reflection High Energy Electron Diffraction and Reflection Electron Imaging of Surfaces* (Eds. P.K. Larsen and P.J. Dobson, Plenum) NATO ASI **B188** 3–28 [both 3.2.2].

Inomata, K., S.N. Okuno, Y. Saito & K. Yusu (1996) *J. Mag. Mag. Mat.* **156** 219–223 [6.3.4].

Irving, A.C. & D.W. Palmer (1992) *Phys. Rev. Lett.* **68** 2168–2171 [8.1.3].

Isogai, H. (1997) *Vacuum* **48** 175–179 [2.3.5].

Itoh, M., G.R. Bell, A.R. Avery, T.S. Jones, B.A. Joyce & D.D. Vvedensky (1998) *Phys. Rev. Lett.* **81** 633–636 [7.3.4].

Itoh, T. (Ed.) (1989) *Ion Beam Assisted Film Growth* (Elsevier); (1995) in *Handbook of Thin Film Process Technology* (Eds. D.A. Glocker and S.I. Shah, Institute of Physics) section A3.3 [both 2.5.4].

Jablonski, A. (1990) *Surf. Interface Anal.* **15** 559–566 [3.4.2].

Jackson, K.A. (1958) in *Liquid Metals and Solidification* (ASM, Cleveland, Ohio) pp. 174–186; in *Growth and Perfection of Crystals* (Eds. R.H. Doremus, B.W. Roberts and D. Turnbull, John Wiley) pp. 319–324 [both 1.3.2].

Jackson, K.A., D.R. Uhlmann & J.D. Hunt (1967) *J. Cryst. Growth* **1** 1–36 [1.3.2].

Jacobsen, J., B. Hammer, K.W. Jacobsen & J.K. Nørskov (1995) *Phys. Rev. B***52** 14954–14962 [4.5.3].

Jacobsen, J., K.W. Jacobsen & J.K. Norskov (1996) *Surface Sci.* **359** 37–44 [6.1.4].

Jacobsen, J., K. W. Jacobsen & J. P. Sethna (1997) *Phys. Rev. Lett.* **79** 2843–2846 [5.4.2].

Jacobsen, K.W. (1988) *Comments on Cond. Matter Physics* **14** 129–161 [6.1.2].

Jacobsen, K.W., J.K. Nørskov & M.J. Puska (1987) *Phys. Rev. B***35** 7423–7442 [6.1.2].

Jakubith, S., H.H. Rotermund, W. Engel, A. von Oertzen & G.Ertl (1990) *Phys. Rev. Lett.* **65** 3013–3016 [4.5.4].

Janata, J., M. Josowicz & D.M. DeVaney (1994) *Analytical Chemistry* **66** 207R–228R [9.3].

Janata, J., M. Josowicz, P. Vanysek & D.M. DeVaney (1998) *Analytical Chemistry* **70** 179R–208R [9.3].

Jank, W. & J. Hafner (1990) *Phys. Rev. B***41** 1497–1515 [7.1.1].

Janssen, A.P., C.J. Harland & J.A. Venables (1977) *Surface Sci.* **62** 277–292 [3.3.1, 3.5.1].

Janssen, A.P., P. Akhter, C.J. Harland & J.A. Venables (1980) *Surface Sci.* **93** 453–470 [6.2.4].

Jaros, M. (1988) *Phys. Rev.* **B37** 7112–7114 [8.2.1]; (1989) *Physics and Applications of Semiconductor Microstructures* (Oxford University Press) [8.2].

Jensen, P. (1998) *Physics Today*, July issue pp. 58–59 [8.2.4].

Jeong, H.C. & E.D. Williams (1999) *Surface Sci. Rep.* **34** 171–294 [7.3.1].

Jesson, D.E., K.M. Chen, S.J. Pennycook, T. Thundat & R.J. Warmack (1996) *Phys. Rev. Lett.* **77** 1330–1333 [7.3.3].

Jiles, D. (1991) *Magnetism and Magnetic Materials* (Chapman and Hall) [6.3].

Jin, A.J., M.R. Bjurstrom & M.H.W. Chan (1989) *Phys. Rev. Lett.* **62** 1372–1375 [4.3.1, 4.4.3].

Johnson, M. (1993) *Science* **260** 320–323; (1996) *J. Mag. Mag. Mat.* **156** 321–324 [both 8.3.3].

Johnson, M.D., J. Sudijono, A.W. Hunt & B.G. Orr (1993) *Surface Sci.* **298** 392–398 [7.3.4].

Johnson, M.D., C. Orme, A.W. Hunt, D. Graff, J. Sudijono, L.M. Sander & B.G. Orr (1994) *Phys. Rev. Lett.* **72** 116–119 [7.3.4].

Johnson, R.L. (1991) *Festkörperprobleme – Advances in Solid State Physics* **31** 115–132 [2.4.3, 3.2.1].

Johnston, A.B., J.N. Chapman, B. Khamsehpour & C.D.W. Wilkinson (1996) *J. Phys.: Appl. Phys.* **D29** 1419–1427 [6.3.2].

Jona, F., H.D. Shih, A. Ignatiev, D.W. Jepsen & P.M. Marcus (1977) *J. Phys.* **C10** L67-L72 [1.4.4].

Jones, G.W. & J.A. Venables (1985) *Ultramicroscopy* **18** 439–444 [3.5.1, 5.4.1, 6.2.4].

Jones, G.W., J.M. Marcano, J.K. Nørskov & J.A. Venables (1990) *Phys. Rev. Lett.* **65** 3317–3320 [5.4.1, 5.4.3].

Jones, R.O. & O. Gunnarson (1989) *Rev. Mod. Phys.* **61** 689–746 [Appendix J].

Joós, B. & M.S. Duesbery (1985) *Phys. Rev. Lett.* **55** 1997–2000 [4.4.3].

Joyce, B.A., D.D. Vvedensky & C.T. Foxon (1994) in *Handbook of Semiconductors* (Ed. S. Mahajan, Elsevier) **3** 275–368 [7.3.4].

Kaczer, B., Z. Meng & J.B. Pelz (1996) *Phys. Rev. Lett* **77** 91–94 [8.1.3].

Kahn, A. (1994) *Surface Sci.* **299/300** 469–486; (1996) *Surface Rev. Lett.* **3** 1579–1595 [7.1.3, 7.2.1].

Kaiser, W.J. & L.D. Bell (1988) *Phys. Rev. Lett.* **60** 1406–1409 [8.1.3].

Kandel, D. & J.D. Weeks (1995) *Phys. Rev.* **B52** 2154–2164 [7.3.4].

Kandel, D. (1997) *Phys. Rev. Lett.* **78** 499–502 [5.2.3].

Kapon, E. (1994) in *Epitaxial Microstructures* (Ed. A.C. Gossard) *Semiconductors and Semimetals* **40** 259–336 [8.2.4].

Kardar, M., G. Parisi & Y.C. Zhang (1986) *Phys. Rev. Lett.* **56** 889–892 [7.3.4].

Kariotis, R., J.A. Venables, M. Hamichi & A.Q.D. Faisal (1987) *J. Phys.* **C19** 5717–5726 [4.4.2].

Kariotis, R., J.A. Venables & J.J. Prentis (1988) *J. Phys.* C21 3031–3046 [4.2.1, 4.4.2].

Keller, K.W. (1986) *J. Cryst. Growth* 74 161–171; *Cryst. Growth* 76 469–475; *Cryst. Growth* 78 509–518 [all 5.5.1].

Kellogg, G.L. (1994) *Surface Sci. Rep.* 21 1–88 [3.1.3, 5.4.2]; (1997) *Phys. Rev. Lett.* 79 4417–4420 [5.4.2].

Kellogg, G.L. & P.J. Feibelman (1990) *Phys. Rev. Lett.* 64 3143–3146 [5.4.2].

Kelly, A. & G.W. Groves (1970) *Crystallography and Crystal Defects* (Longman) [1.3–4, problem 4.1].

Kelly, M.J. (1995) *Low-dimensional Semiconductors* (Oxford University Press) [7.1, 8.2].

Kenny, P.G., I.R. Barkshire & M. Prutton (1994) *Ultramicroscopy* 56 289–301 [3.5.2].

Kern, K., R. David, R.L. Palmer & G. Comsa (1986) *Phys. Rev. Lett.* 56 620–623; *Surface Sci.* 175 L669-L674 [both 4.4.2, 4.4.4].

Kern, K. & G. Comsa (1988) in *Chemistry and Physics of Solid Surfaces* (Eds. R. Vanselow and R.F. Howe, Springer) 7 65–108 [4.4.2, 4.4.4]; (1989) *Adv. Chem. Phys.* 76 211–280 [4.4.4].

Kern, R. & M. Krohn (1989) *Phys. Stat. Sol. (a)* 116 23–38 [5.5.1].

Kern, R., G. LeLay & J.J Métois (1979) in *Current Topics in Materials Sci.* (Ed. E. Kaldis, North-Holland) 3 139–419 [5.1.1, 5.5.1, 7.2.3].

Kief, M.T & W.T. Egelhoff, Jr. (1993) *Phys. Rev.* B47 10785–10814 [5.5.3].

Kiejna, A. (1999) *Prog. Surf. Sci.* 61 85–125 [6.1.1, 6.1.4].

King, D.A. & D.P. Woodruff (Eds.) (1997) *Growth and Properties of Ultrathin Epitaxial Layers* (*The Chemical Physics of Solid Surfaces and Heterogeneous Catalysis*, Elsevier) 8 [1.4.8, 4.5, 5.1].

Kingcade, J.E., H.M. Nagarathna-Naik, I. Shim & K.A. Gringerich (1986) *J. Phys. Chem.* 90 2830–2834 [7.1.3].

Kirkland, E.J. (1998) *Advanced Computing in Electron Microscopy* (Plenum) [Appendix D].

Kirschner, J. (Ed.) (1985) *Polarized Electrons at Surfaces* (Springer Tracts in Modern Physics 106) [6.3.3].

Kittel, C. (1976) *Introduction to Solid State Physics* (6th Edn, John Wiley) [6.3].

Klaua, M. (1987) in *Electron Microscopy in Solid State Physics* (Eds. H. Bethge and J. Heydenreich, Elsevier) chapter 19 pp. 454–470 [5.4.3, 5.5.1].

Klein, D.L., R. Roth, A.K.L. Lim, A.P. McEuen (1997) *Nature* 389 699–701 [8.4.4].

Klein, M.L. & J.A. Venables (Eds.) (1976) *Rare Gas Solids* (Academic) 1 chapters 1, 4 and 6 [1.3.1, 1.4.3, Appendix E], 2, chapters 11, 12 and 13 (1.3.1, Appendix E).

Kley, A., P. Ruggerone & M. Scheffler (1997) *Phys. Rev. Lett.* 79 5278–5281 [7.3.4].

Knall, J. & J.B. Pethica (1992) *Surface Sci.* 265 156–167 [7.3.1].

Koch, S.W., W.E. Rudge & F.F. Abraham (1984) *Surface Sci.* 145 329–344 [4.4.2].

Kohn, W. & L.J. Sham (1965) *Phys. Rev.* 140 A 1133–1138 [Appendix J].

Köhler, U., O. Jusko, B. Müller, M. Horn-von Hogen & M. Pook (1992) *Ultramicroscopy* 42–44 832–837 [7.3.3].

Kohmoto, S. & A. Ichimiya (1989) *Surface Sci.* 223 400–412 [7.2.3].

Kolaczkiewicz, J. & E. Bauer (1984) *Phys. Rev. Lett.* 53 485–488 [4.3.2, 6.1.3].

Kosterlitz, J.M. & D.J. Thouless (1973) *J. Phys.* **C6** 1181–1203 [4.4.3].

Kouvetakis, J., A. Haaland, D.J. Shorokhov, H.V. Volden, G.V. Girichev, V.I. Skolov & P. Matsunaga (1998a) *J. Am. Chem. Soc.* **120** 6738–6744 [8.4.1].

Kouvetakis, J., D. Nesting & D.J. Smith (1998b) *Chem. Mater.* **10** 2935–2949 [8.4.1].

Krahl-Urban, B., E.A. Niekisch & H. Wagner (1977) *Surface Sci.* **64** 52–68 [6.1.3].

Kreuzer, H.J. (1982) *Appl. Surface Sci.* **11/12** 793–802 [4.4.4].

Krischer, K., M. Eiswirth & G. Ertl (1991) *Surface Sci.* **251/252** 900–904 [4.5.4].

Krishnamurthy, M., J.S. Drucker & J.A. Venables (1991) *J. Appl. Phys.* **69** 6461–6471; (1991) *Mater. Res. Soc. Symp.* **202** 77–82 [both 7.3.3].

Krishnamurthy, M., A. Lorke & P.M. Petroff (1994) *Surface Sci.* **304** L493-L499 [7.3.4].

Krug, J. & H. Spohn (1992) in C. Godrèche (Ed.) *Solids Far From Equilibrium* (Cambridge University Press) chapter 6 pp. 479–582 [7.3.4].

Krüger, P. & J. Pollmann (1994) *Appl. Phys.* **A59** 487–502; (1995) *Phys. Rev. Lett.* **74** 1155–1158 [both 7.1.3, 7.2.4].

Kubby, J.A. & J.J. Boland (1996) *Surface Sci. Rep.* **26** 61–204 [3.5.4].

Kubiyak, R.A., P. Driscoll & E.H.C. Parker (1982) *J. Vac. Sci. Tech.* **20** 252–253 [2.5.2].

Kyuno, K., A Gölzhäser & G. Ehrlich (1998) *Surface Sci.* **397** 191–196 [5.4.3].

Klein, D.L., R. Roth, A.K.L. Lim, A.P. Alivisatos & P.L. McEuen (1997) *Nature* **389** 699–701 [8.4.2].

Lagally, M.G. (1975) in *Surface Physics of Materials* (Ed. J.W. Blakely, Academic) **2** chapter 9 419–473 [1.4.3, 3.2.1, 3.2.3].

Lagally, M.G., T.-M. Lu & G.-C. Wang (1980) in *Ordering in Two Dimensions* (Ed S.K. Sinha, Elsevier- North Holland) pp. 113–121; (1980) *Surface Sci.* **92** 133–144 [4.5.1].

Lahrer, Y. Thèse, Orsay (1970), in French [4.4.1].

Landolt, M. (1985) in *Polarized Electrons in Surface Physics* (Ed. R. Feder, World Scientific) chapter 9 pp. 385–421 [6.3.3].

Lang, N.D. & W. Kohn (1970) *Phys. Rev.* **B1** 4555–4568; (1971) *Phys. Rev B3* 1215–1223 [6.1.1, 6.1.4].

Lang, N.D. (1973) *Solid State Physics* **28** 225–300 [6.1.1].

Langelaar, H. & D.O. Boerma (1996) *Surface Sci.* **352–4** 597–601 [5.4.3].

Leamy, H.J., G.H. Gilmer & K.A Jackson (1975) in *Surface Physics of Materials* (Ed. J.W. Blakely, Academic) **1** chapter 3 pp. 121–188 [1.3.1–2].

Leckey, R.C.G., J.D. Riley & A. Stampfl (1990) *J. Elect. Spectrosc.* **52** 855–866 [3.3.2].

Lee, G.D. C.Z. Wang, Z.Y. Lu and K.M. Ho (1999) *Surface Sci.* **426** L427-L432 [7.3.2].

Lennard-Jones, J.E. & A.F. Devonshire (1937) *Proc. R. Soc. Lond.* **A163** 53–70; (1938) *Proc. R. Soc. Lond.* **A165** 1–11 [both 4.2.4].

Lenosky, T.J., J.D. Kress, I. Kwon, A.F. Voter, B. Edwards, D.F. Richards, S. Yang & J.B. Adams (1997) *Phys. Rev.* **B55** 1528–1544 [7.1.3, Appendix K].

Levi, B.G. (1998) *Physics Today* December issue pp. 20–22 [Appendix J].

Levy, P.M., S. Shang, T.Ono & T. Shinjo (1995) *Phys. Rev.* **B52** 16049–16054 [8.3.3].

Lewis, B. & J.C. Anderson (1978) *Nucleation and Growth of Thin Films* (Academic) [5.2.2].

Ley, L. & M. Cardona (Eds.) (1979) *Photoemission in Solids* (Springer) **2** [3.3.2].

Li, Y., G.G. Hembree & J.A. Venables (1995) *Appl. Phys. Lett.* **67** 278–278 [3.4.2, 7.3.3].

Liu, F. & M.G. Lagally (1997) in *The Chemical Physics of Solid Surfaces and Heterogeneous Catalysis* (Eds. D.A. King and D.P. Woodruff, Elsevier) **8** chapter 7 pp. 258–296 [7.3.2].

Liu, J., G.G. Hembree, G.E. Spinnler & J.A. Venables (1993) *Ultramicroscopy* **52** 369–376 [3.5.3].

Liu, W.K. & M.B. Santos (Eds.) (1998) *Thin Films: Heteroepitaxial Systems* (World Scientific) [5.1, 7.3].

Lloyd, J.R. (1999) *J. Phys. D: Appl.* **32** R109–R118 [9.1].

Lorensen, H.T., J.K. Nørskov & K.W. Jacobsen (1999) *Phys. Rev.* **B60** R5149–R5152 [5.4.2].

Luo, F.C.H., G.G. Hembree & J.A. Venables (1991) *Mater. Res. Soc. Symp.* **202** 49–54 [3.4.2].

Lüth, H. (1993/5) *Surfaces and Interfaces of Solid Surfaces* (2nd/3rd Edns, Springer) [1.4, 2.2–3, 3.2–3, 7.2, 8.1–2].

MacKenzie, J.K., A.J.W. Moore & J.F. Nicholas (1962) *J. Phys. Chem. Solids* **23** 185–205 [1.2.3].

McGilp, J. & P. Weightman (1976) *J. Phys.* **C9** 3541–3556; (1978) *J. Phys.* **C11** 643–650 [both 3.3.3].

McVitie, S., J.N. Chapman, L. Zhou, L.J. Heyerman & W.A.P. Nicholson (1995) *J. Mag. Mag. Mat.* **148** 232–236 [6.3.2].

Madden, H.H. (1981) *J. Vac. Sci. Tech.* **18** 677–689 [3.3.4].

Madhukar, A. & S.V. Ghaisas (1988) *Crit. Rev. Solid State & Mater. Sci.* **14** 1–130 [7.3.4].

Maksym, P.A. (1997) *Surface Rev. Lett.* **4** 513–524; (1999) *Surface Rev. Lett.* **6** 451–460 [3.2.3].

Mandl, F. (1988) *Statistical Physics* (2nd Edn, Wiley) [1.3.1, Appendix E].

Mankos, M., J.M. Cowley & M.R. Scheinfein (1996) *Phys. Stat. Sol. (a)* **154** 469–504 [6.3.2].

Marisco, V., M. Blanc, K. Kuhnke & K. Kern (1997) *Phys. Rev. Lett.* **78** 94–97 [4.4.4].

Markov, I. (1995) *Crystal Growth for Beginners* (World Scientific) [1.2.2]; (1996) *Phys. Rev.* **B53** 4148–4155 [5.2.3].

Marks, L.D. (1985) *Surface Sci.* **150** 358–366; (1994) *Rep. Prog. Phys.* **57** 603–649 [both 1.2.3].

Marks, L.D., E. Bengu, C. Collazo-Davilla, D. Grozea, E. Landree, C. Leslie & W. Sinkler (1998) *Surface Rev. Lett.* **5** 1087–1106 [8.1.3].

Marten, H. & G. Meyer-Ehmsen (1988) in *Reflection High Energy Electron Diffraction and Reflection Electron Imaging of Surfaces* (Eds. P.K. Larsen and P.J. Dobson, Plenum) NATO ASI **B188** 99–115 [3.2.3].

Martin, J.W. & R.D. Doherty (1976) *Stability of Microstructure in Metallic Systems* (Cambridge University Press) chapter 4, especially pp. 154–193 [1.2.1, 1.2.3, 5.5.2].

Marton, D. (1994) in *Low Energy Ion–Surface Interactions* (Ed. J. W. Rabalais, John Wiley) chapter 9 pp. 481–534 [2.5.4].

Masel, R.I. (1996) *Principles of Adsorption and Reaction on Solid Surfaces* (John Wiley) [4.5].

Matthew, J.A.D., A.R. Jackson & M.M. El-Gomati (1997) *J. Elect. Spectrosc.* **85** 205–219 [3.4.2].

Matthews, J.W. (Ed.) (1975) *Epitaxial Growth, part A* (Academic) [2.5]; *part B* (Academic) [5.1].

Matthews, J.W. (1975) in *Epitaxial Growth, part B* (Ed. J. W. Mathews, Academic) chapter 8, pp. 559–609 [7.3.2].

Matthews, J.W. & A.E. Blakeslee (1974) *J. Cryst. Growth* **27** 118–125; (1975) *J. Cryst. Growth* **29** 273–280; (1976) *J. Cryst. Growth* **32** 265–273 [all 7.3.2].

Mavrikakis, M., Hammer, B & J.K. Nørskov (1998) *Phys. Rev. Lett.* **81** 2819–2822 [project 4.4].

Meade, R.D. & D. Vanderbilt (1989) *Phys. Rev.* **B40** 3905–3913 [7.2.3, 7.3.1].

Medeiros-Ribeiro, G., A. M. Bratovski, T.I. Kamins, D.A.A. Olberg & R.S. Williams (1998) *Science* **279** 353–355 [7.3.3].

Mejías, J.A. (1996) *Phys. Rev.* **B53** 10281–10288 [5.3.2].

Mellor, C. & K. Benedict (1998) *Physics World* November issue pp. 49–50 [8.3.2].

Men, F.K. (1994) *Phys. Rev.* **B50** 15469–15472 [7.3.2].

Men, F.K., W.E. Packard & M.B. Webb (1988) *Phys. Rev. Lett.* **61** 2469–2471 [7.3.1].

Menéndez, J. & A. Pinczuk (1988) *IEEE J. Quantum Electronics* **24** 1698–1711[8.2.3].

J. Menéndez, A. Pinczuk, D.J. Werder, A.C. Gossard & J.H. English (1986) *Phys. Rev.* **B33** 8863–8865 [8.2.3].

Menzel, D. (1994) *Surface Sci.* **299/300** 170–182 [3.3.4].

Metcalfe, F.L. & J.A.Venables (1996) *Surface Sci.* **369** 99–107 [3.4.2].

Methfessel, M., D. Hennig & M.M. Scheffler (1992) *Phys. Rev.* **B46** 4816–4829 [6.1.1, 6.1.3–4].

Métois, J.J & J.C. Heyraud (1989) *Ultramicroscopy* **31** 73–79 [1.2.3].

Métois, J.J, J.C. Heyraud & R. Kern (1978) *Thin Solid Films* **78** 191–208 [5.5.1].

Métois, J.J., G.D.T. Spiller & J.A. Venables (1982) *Phil. Mag. A***46** 1015–1022 [4.5.4].

Meyer, G. & N.M. Amer (1988) *Appl. Phys. Lett.* **53** 1045–1047; 2400–2402 [3.1.3].

Meyer, H.J. & B.J. Stein (1980) *J. Cryst. Growth* **49** 707–717 [5.5.1].

Meyer, J.A. & R.J. Behm (1995) *Surface Sci.* **322** L275–L280 [5.5.3].

Meyer, J.A., I.D. Baikie, E. Kopatzki & R.J. Behm (1995) *Surface Sci.* **365** L647–L651 [5.5.3].

Meyer, T. & H. von Känel (1997) *Phys. Rev. Lett.* **78** 3133–3136 [8.1.3].

Mezey, L.Z. & J. Giber (1982) *Surface Sci.* **117** 220–231 [6.1.4].

Michaelson, H.B. (1977) *J. Appl. Phys.* **48** 4729–4733 [6.1.1, 6.1.3].

Miguel, J.J. de, C.E. Aumann, R. Kariotis & M.G. Lagally (1991) *Phys. Rev. Lett.* **67** 2830–2833 [7.3.2].

Milman, V., D.E. Jesson, S.J. Pennycook, M.C. Payne, M.H. Lee & I. Stich (1994) *Phys. Rev.* **B50** 2663–2666 [7.3.2].

Milman, V., S.J. Pennycook & D.E. Jesson (1996) *Thin Solid Films* **272** 375–385 [7.3.2].

Milne, R.H., M. Azim, R. Persaud & J.A. Venables (1994) *Phys. Rev. Lett.* **73** 1396–1399; (1995) *Surface Sci.* **336** 63–75 [6.2.4].

Mirkin, C.A., R.L. Letsinger, R.C. Mucic & J.J. Storhoff (1996) *Nature* **382** 607–609; (1999) for a summary of work by this group see *Materials Research Bulletin*, March issue, p 80 [8.4.4].

Mo, Y.W. & M.G. Lagally (1991) *Surface Sci.* **248** 313–320 [7.3.2].

Mo, Y.W., J. Kleiner, M.B. Webb & M.G. Lagally (1991) *Phys. Rev. Lett.* **66** 1998–2001; (1992) *Surface Sci.* **268** 275–295 [both 7.3.2].

Modinos, A. (1984) *Field, Thermionic and Secondary Electron Spectroscopy* (Plenum) [6.2.2].

Mogren, S. & R. Reifenburger (1991) *Surface Sci.* **254** 169–181 [6.2.3].

Mönch, W. (1990) *Rep. Prog. Phys.* **53** 221–278 [8.1.1]; (1993) *Semiconductor Surfaces and Interfaces* (Springer) [7.2, 8.2]; (1994) *Surface Sci.* **299/300** 928–944 [8.1.1, 8.2.1, 8.2.3].

Monsma, D.J., J.C. Lodder, T.J.A. Popma & B. Dieny (1995) *Phys. Rev. Lett.* **74** 5260–5623 [8.3.3].

Montanari, B., M. Peressi, S. Baroni & E. Molinari (1996) *Appl. Phys. Lett.* **69** 3218–3220 [8.2.3].

Moore, J.H., C.C. Davis & M.A. Coplan (1989) *Building Scientific Apparatus* (2nd Edn, Addison-Wesley) [2.3, 3.3].

Morgenstern, K., G. Rosenfeld, E. Laegsgaard, F. Besenbacher & G. Comsa (1998) *Phys. Rev. Lett.* **80** 557–559 [5.4.3, 5.5.2].

Morimoto, A. & T. Shimizu (1995) in *Handbook of Thin Film Process Technology* (Eds. D.A. Glocker and S.I. Shah, Institute of Physics) section A1.5 [2.5.2].

Moruzzi, V.L., J.F. Janak & A.R. Williams (1978) *Calculated Electronic Properties of Metals* (Pergamon) [6.1.2, 6.3.4].

Mulheran, P.A. & J.A. Blackman (1995) *Phil. Mag. Lett.* **72** 55–60; (1996) *Phys. Rev.* **B53** 10261–10267 [both 5.2.5].

Müller, B. & M. Henzler (1995) *Rev. Sci. Inst.* **66** 5232–5235 [3.2.3].

Müller, B., L. Nedelmann, B. Fischer, H. Brune & K.Kern (1996) *Phys. Rev.* **B54** 17858–17865 [5.4.3].

Mullins, J., M. Walker, J. Webb & D. MacKenzie (1998) *New Scientist* **160**, issue 2159, 4, 32–34, 42–57 [9.1].

Mullins, W.W. (1957) *J. Appl. Phys.* **28** 333–339 [4.5.4].

Murray, C.B., C.R. Kagan & M.G. Bawendi (1995) *Science* **270** 1335–1338 [8.4.2].

Murray, C.B. (1999) private communication; but see e.g. Sun, S.H. & C.B. Murray (1999) *J. Appl. Phys.* **85** 4325–4330 [8.4.2].

Musket, R.G., W. McLean, C.A. Colmenares, D.M. Makowiecki & W.J. Siekhaus (1982) *Appl. Surf. Sci.* **10** 143–207 [Appendix H].

Mutaftschiev, B. (1980) in *Dislocations in Solids* (Ed. F.R.N. Nabarro, North-Holland) **5** chapter 19 [5.5.1].

Myers-Beaghton, A.K. & D.D. Vvedensky (1991) *Phys. Rev.* **A44** 2457–2468 [5.2.5].

Nakamura, S. (1998) *Ann. Rev. Mater. Sci.* **28** 125–152 [7.3.4].

Naumovets, A.G. & Y.S. Vedula (1985) *Surface Sci. Rep.* **4** 365–434 [5.5.2].

Naumovets, A.G. (1994) *Surface Sci.* **299/300** 706–721 [5.5.2].

Neddermeyer, H. (1996) *Rep. Prog. Phys.* **59** 701–769 [3.5.4].

Needs, R.J., M.J. Godfrey & M. Mansfield (1991) *Surface Sci.* **242** 215–221 [6.1.2, 7.3.1].

Nelson, D.R. & B.I. Halperin (1979) *Phys. Rev.* **B19** 2457–2484 [4.4.3].

Nemanich, R.J., C.R. Helms, M. Hirose & G.W. Rubloff (Eds.) (1992) *Mater. Res. Soc. Symp.* **259** [2.4.1].

Nettesheim, S., A. von Oertzen, H.H. Rotermund & G. Ertl (1993) *J. Chem. Phys.* **98** 9977–9985 [4.5.4].

Newns, D. (1969) *Phys. Rev.* **178** 1123–1135 [4.5.2].

Nicholas, J.F. (1965) *An Atlas of Models of Crystal Structures* (Gordon and Breach) [1.2.3].

Nichols, F.A. & W.W. Mullins (1965) *J. Appl. Phys.* **36** 1826–1835 [4.5.4].

Noro, H. (1994) D.Phil. thesis, University of Sussex [2.4.1].

Noro, H., R. Persaud & J.A. Venables (1995) *Vacuum* **46** 1173–1176 [2.4.1, 5.4.1, 5.5.3]; (1996) *Surface Sci.* **357/358** 879–884 [5.4.1, 5.4.3, 5.5.2, 6.2.4].

Nørskov, J.K. (1990) *Rep. Prog. Phys.* **53** 1253–1295 [4.5.2]; (1993) in *The Chemical Physics of Solid Surfaces and Heterogeneous Catalysis* (Eds. D.A. King and D.P. Woodruff, Elsevier) **6** 1–27 [4.5.2]; (1994) *Surface Sci.* **299/300** 690–705 [4.5.2].

Nørskov, J.K., K.W. Jacobsen, P. Stolze & L.B. Hansen (1993) *Surface Sci.* **283** 277–282 [6.1.2].

Northrup, J.E. (1989) *Phys Rev. Lett.* **62** 2487–2490 [7.3.4]; (1993) *Phys. Rev.* **B47** 10032–10025 [7.3.2].

Northrup, J.E. & M.L. Cohen (1983) *Chem. Phys. Lett.* **102** 440–441 [7.1.3].

Northrup, J.E., M.T. Yin & M.L. Cohen (1983) *Phys Rev.* **A28** 1945–1950 [7.1.3].

Novaco, A.D. & J.P. McTague (1977) *J. de Physique* **38-C4** 116–120; (1977) *Phys. Rev. Lett.* **38** 1286–1289 [both 4.4.2].

Nozières, P. (1992) in C. Godrèche (Ed.) *Solids Far From Equilibrium* (Cambridge University Press) chapter 1, pp. 1–154 [1.2.3, 1.3.1].

Ogawa, S. & S. Ino (1971) in *Advances in Epitaxy and Endotaxy* (Eds. H.G. Schneider and V. Roth, VEB, Leipzig) chapter 4 pp. 183–226; (1972) *J. Cryst. Growth* **13/14** 48–56 [both 4.5.4].

O'Hanlon, J.F. (1989) *A Users Guide to Vacuum Technology* (John Wiley) [2.3, Appendix G]; (1994) *J. Vac. Sci. Tech.* **A12** 921–927 [2.4.4, 2.5.5].

Ono, T. & T. Shinjo (1995) *J. Phys. Soc. Japan* **64** 363–366 [8.3.3].

Ono, T., Y. Sugita, K. Shigeto, K. Mibu, N. Hosoito & T. Shinjo (1997) *Phys. Rev.* **B55** 14457–14466 [8.3.3].

Orloff, J. (1984) in *Electron Optical Systems for Microscopy, Microanalysis and Microlithography* (SEM Pfefferkorn Conference Series) **3** 149–162 [6.2.1, 6.2.3].

Orme, C. & B.G. Orr (1997) *Surface Rev. Lett.* **4** 71–105 [7.3.4].

Ouellette, J. (1997) *The Industrial Physicist*, September issue, pp. 7–11 [2.4.4]; (1998) *The Industrial Physicist*, June issue, pp. 11–14 [9.1].

Panish, M.B. & S. Sumski (1984) *J. Appl. Phys.* **55** 3571–3576 [2.5.3].

Papaconstantopoulos, D.A. (1986) *Handbook of the Band Structure of Elemental Solids* (Plenum) [6.3.4].

Parker, E.H.C. (Ed.) (1985) *The Technology and Physics of Molecular Beam Epitaxy* (Plenum) [2.5.3].

Parkin, S.S.P. (1994) in *Ultrathin Magnetic Structures* (Eds. B. Heinrich and J.A.C. Bland, Springer) **2** chapter 2.4, pp. 148–186 [8.3.3].

Parkin, S.S.P., R. Bhadra & K.P. Roche (1991) *Phys. Rev. Lett.* **66** 2152–2155 [8.3.3].

Pashley, M.D., K.W. Haberen, W. Friday, J.M. Woodall & P.D. Kirchner (1988) *Phys. Rev. Lett.* **60** 2176–2179 [7.3.4].

Pavlovska, A., K. Faulian & E. Bauer (1989) *Surface Sci.* **221** 233–243 [1.2.3, 6.1.4].

Pavlovska, A., D. Dobrev & E. Bauer (1994) *Surface Sci.* **314** 331–352 [6.1.4].

Payne, M.C. (1987) *J. Phys.* C**20** L983-L987 [7.2.3].

Payne, M.C., N. Roberts, R.J. Needs, M. Needels & J.D. Joannopoulos (1989) *Surface Sci.* **211** 1–20 [7.2.3].

Payne, M.C., M.P. Teter, D.C. Allan, T.A. Arias & J.D. Joannopoulos (1992) *Rev. Mod. Phys.* **64** 1045–1097 [7.1.3].

Pendry, J.B. (1974) *Low Energy Electron Diffraction* (Academic) [3]; (1994) *Surface Sci.* **299/300** 375–390 [3.2.1]; (1997) *Surface Rev. Lett.* **4** 901–905 [3.2.1, 3.2.3].

Peng, L.M., S.L. Dudarev & M.J. Whelan (1996) *Acta Cryst.* A**52** 909–922 [3.2.3].

Peng, X., M.C. Schlamp, A.V. Kadavanich & A.P. Alivisatos (1997) *J. Amer. Chem. Soc.* **119** 7019–7029 [8.4.2].

People, R. & J.C. Bean (1985) *Appl. Phys. Lett.* **47** 322–324; (1986) *Appl. Phys. Lett.* **49** 229 [7.3.2].

Perdew, J.P. (1995) *Prog. Surface Sci.* **48** 245–259 [6.1.3–4].

Perdew, J.P., H.Q. Tran & E.D. Smith (1990) *Phys. Rev.* B**42** 11627–11636 [6.1.1, 6.1.3–4].

Peressi, M., N. Binggeli & A. Baldereschi (1998) *J. Phys. D: Appl. Phys.* **31** 1273–1299 [8.2.2–3].

Persaud, R., H. Noro, M. Azim, R.H. Milne & J.A. Venables (1994) *Scanning Mic.* **8** 803–812 [6.2.4].

Persaud, R., H. Noro & J.A. Venables (1998) *Surface Sci.* **401** 12–21 [3.4.2, 5.4.1, 5.5.3].

Perutz, M.F. (1964) *Scientific American* November issue pp. 64–76 [4.5.4].

Petersen, L., P. Laitenberger, E. Laegsgaard & F. Besenbacher (1998) *Phys. Rev.* B**58** 7361–7366 [problem 6.1].

Petroff, P. (1994) in *Epitaxial Microstructures* (Ed. A.C. Gossard) *Semiconductors and Semimetals* **40** 219–258 [8.2.4].

Pettifor, D.G. (1995) *Bonding and Structure of Molecules and Solids* (Oxford University Press) [6.1, 6.3, 7.1, 9.2, Appendix K].

Phillips, J.M., L.W. Bruch & R.D. Murphy (1981) *J. Chem. Phys.* **75** 5097- 5109 [4.2.4].

Pierce, D.T. (1995) in *Experimental Methods in the Physical Sciences* (Eds. F.B. Dunning and R.G. Hulet, Academic) **29A** 1–38 [6.3.3].

Pierce, D.T., R.J. Celotta, G.C. Wang, W.N. Unertl, A. Galejs, C.E. Kuyatt & S.R. Mielczarek (1980) *Rev. Sci. Inst.* **51** 478–499 [6.3.3].

Pierce, D.T., J. Unguris & R.J. Cellotta (1994) in *Ultrathin Magnetic Structures* (Eds. B. Heinrich and J.A.C. Bland, Springer) **2** chapter 2.3 pp. 117–147 [6.3.4].

Pimpinelli, A., J. Villain, D.E. Wolf, J.J. Métois, J.C. Heyraud, I. Elkinani & G. Uimin (1993) *Surface Sci.* **295** 143–153 [7.3.1].

Poensgen, M., J.F. Wolf, J. Frohn, M. Giesen & H. Ibach (1992) *Surface Sci.* **274** 430–440 [5.5.3].

Ponce, F.A. & D.P. Bour (1997) *Nature* **386** 351–359 [7.3.4].

Poppa, H.R. (1983) *Ultramicroscopy* **11** 105–116; (1984) *Vacuum* **34** 1081–1095 [4.5.4].

Poppendieck, T.D., T.C. Ngoc & M.B. Webb (1978) *Surface Sci.* **78** 287–315 [1.4.4].

Pouthier, V., C. Ramseyer, C. Giradet, K. Kuhnke, V. Marisco, M. Blanc, R. Schuster & K. Kern (1997) *Phys. Rev.* **B56** 4211–4223 [4.4.4].

Powell, C.J., *Surface Sci.* (1994) **299/300** 34–48 [3.3.4, 3.4.2].

Powell, C.J., A. Jablonski, S. Tanuma & D.R. Penn (1994) *J. Elect. Spectrosc.* **68** 605–616 [3.3.4].

Powell, C.J. & M.P. Seah (1990) *J. Vac. Sci. Tech.* **A8** 735–763 [3.4.2].

Price, G.L. (1974) *Surface Sci.* **46** 697–702 [problem 4.2].

Price, G.L. & J.A. Venables (1976) *Surface Sci.* **59** 509–532 [4.4.1, Appendix E].

Prietsch, M. (1995) *Phys. Rep.* **253** 163–233 [8.1.3].

Prokes, S.M. & K.L. Wang, (Eds.) (1999) *Materials Research Bulletin*, August issue, pp. 13–49 [8.2.4].

Prutton, M. (1994) *Introduction to Surface Physics* (Oxford University Press) [1.4, 3.2–3, 3.5].

Prutton, M., L.A. Larson & H. Poppa (1983) *J. Appl. Phys.* **54** 374–381 [3.5.1].

Prutton, M., I.R. Barkshire & M. Crone (1995) *Ultramicroscopy* **59** 47–62 [3.5.2].

Qian, G.X. & D.J. Chadi (1987) *J. Vac. Sci. Tech.* **A5** 906–909; (1987) *Phys. Rev.* **B35** 1288–1293 [both 7.2.3].

Rabo, J.A. (1993) *Proc. 10th Int. Cong. on Catalysis* (Budapest, Elsevier) pp. 1–31 [2.4.4].

Ramstad, A., G. Brocks & P.J.Kelly (1995) *Phys. Rev.* **B51** 14504–14523 [7.2.4].

Raynerd, G., M. Hardiman & J.A. Venables (1991) *Phys. Rev.* **B44** 13803–13806 [3.4.2].

Redhead, P.A. (1996) *J. Vac. Sci. Tech.* **A14** 2599–2609 [2.3.4].

Redhead, P.A., J.P. Hobson & E.V. Kornelsen (1968) *The Physical Basis of Ultra-high Vacuum* (Chapman and Hall), reprinted 1993 (American Inst. of Physics) [2.3.4–5].

Reinhard, H.P. (1983) *Proc. 9th Int. Vacuum Cong. (Madrid)* pp. 273–282 [2.1.3, 2.3.2].

Remler, D.K. & P.A. Madden (1990) *Molecular Physics* **70** 921–966 [7.1.3].

Renaud, G. (1998) *Surface Sci. Rep.* **32** 1–90 [1.4.8, 3.2.1].

Rhoderick, E.H. & R.H. Williams (1988) *Metal-Semiconductor Contacts* (Oxford University Press) [8.2.1].

Ribiero, F.H. & G.A. Somorjai (1995) in *Handbook of Surface Imaging and Visualization* (Ed. A.T. Hubbard, CRC Press) chapter 56 pp. 767–783 [2.4.4].

Rikvold, P.A., K. Kaski, J.D. Gunton & M.C. Yabalik (1984) *Phys. Rev.* **B29** 6285–6294 [4.5.1].

Rikvold, P.A. (1985) *Phys. Rev.* **B32** 4756–4759 [4.5.1].

Rivière, J.C. (1990) *Surface Analytical Techniques* (Oxford University Press) [3.1, 3.3].

Robertson, J. (1997) *Mater. Res. Soc. Symp.* **471** 217–229 [6.2.2].

Robertson, J. & W.I. Milne (1997) *Mater. Res. Soc. Symp.* **424** 381–386; (1998) *J. Non-Crystalline Solids* **227/230** 558–564 [6.2.2].

Robins, J.L. (1988) *Appl. Surface Sci.* **33/34** 379–394 [5.3.1–2].

Robinson, I.K & D.J. Tweet (1992) *Rep. Prog. Phys.* **55** 599–651 [3.2.1, 7.2.3].

Robinson, I.K, W.K. Waskiewicz, P.H. Fuoss & L.J. Norton (1988) *Phys. Rev. B***37** 4325–4328 [7.2.3].

Roelofs, L.D. (1982) in *Chemistry and Physics of Solid Surfaces* (Eds. R. Vanselow and R.F. Howe, Springer) **4** 218–249 [4.5.1]; (1996) in *Handbook of Surface Science* (Ed. W.N. Unertl, Elsevier) **2** chapter 13, pp. 713–807 [4.5.1, 6.3.2].

Roland, G.H. & G.H. Gilmer (1992) *Phys. Rev. B***46** 13428–13451 [7.3.2].

Rose, J.H., J.R. Smith & J. Ferrante (1983) *Phys. Rev. B***28** 1835–1845 [6.1.4].

Ross, F.M., J.M. Gibson & R.D. Twesten (1994) *Surface Sci.* **310** 243–266 [8.1.3].

Ross, F.M., J. Tersoff & R.M. Tromp (1998) *Phys. Rev. Lett.* **80** 984–987 [7.3.3].

Rossiter, P.L. (1987) *The Electrical Resistivity of Metals and Alloys* (Cambridge University Press) [8.3].

Rotermund, H.H. (1997) *Surface Sci. Rep.* **29** 265–364 [4.5.4].

Rotermund, H.H., W. Engel, M. Kordesch & G. Ertl (1990) *Nature* **343** 355–357 [4.5.4].

Rotermund, H.H., G. Haas, R.U. Franz, R.M. Tromp & G. Ertl (1995) *Science* **270** 608–610 [4.5.4].

Roth, A. (1990) *Vacuum Technology* (3rd Edn, North-Holland) [2.1–3, Appendix G].

Rottman, C. & M. Wortis (1984) *Phys. Rep.* **103** 59–79 [1.2.3].

Roy, D. & J.D. Carette (1977) in *Electron Spectroscopy for Surface Analysis* (Ed. H. Ibach, Springer) pp. 13–58 [3.3.1].

Rugar, D., H.J. Mamin, P. Guethner, S.E. Lambert, J.E. Stern, I. McFayden & T. Yogi (1990) *J. Appl. Phys.* **68** 1169–1183 [6.3.3].

Ruggerone, P., C. Ratsch & M. Scheffler (1997) in *The Chemical Physics of Solid Surfaces and Heterogeneous Catalysis* (Eds. D.A. King and D.P. Woodruff, Elsevier) **8** chapter 13 pp. 490–544 [4.5.2, project 4.4, 5.4.3, 5.5.2, 6.1.4].

Sakamoto, T. (1988) in *Physics, Fabrication, and Applications of Multilayered Structures* (Eds. P. Dhez and C. Weisbuch, Plenum) NATO ASI **B182** 93–110 [3.2.2].

Saldin, D.K. (1997) *Surface Rev. Lett.* **4** 441–457 [3.2.3].

Sambles, J.R. (1983) *Thin Solid Films* **106** 321–33 [8.3.2].

Sambles, J.R. & T.W. Preist (1982) *J. Phys. F: Metal Physics* **12** 1971–1987 [8.3.2].

Sambles, J.R., K.C. Elsom & D.J. Jarvis (1982) *Phil. Trans. R. Soc. Lond. A***304** 365–396 [8.3.2].

Sambles, J.R., L.M. Skinner & N.D. Lisgarten (1970) *Proc. R. Soc. Lond. A***318** 507–522 [6.1.4].

Sankey, O. & D.J. Niklewski (1989) *Phys. Rev. B***40** 3979–3995 [7.1.3].

Schabes-Retchkiman, P.S. & J.A. Venables (1981) *Surface Sci.* **105** 536–564 [4.4.2–4].

Scheinfein, M.R., J. Unguris, M.H. Kelly, D.T. Pierce & R.J. Celotta (1990) *Rev. Sci. Inst.* **61** 2501–2526 [6.3.3].

Scheithauer, U., G. Meyer & M. Henzler (1986) *Surface Sci.* **178** 441–451 [3.2.1].

Schick, M. (1981) *Prog. Surface Sci.* **11** 245–292 [4.5.1, 6.3.2].

Schlamp, M.C., X. Peng & A.P. Alivisatos (1997) *J. Appl. Phys.* **82** 5837–5842 [8.4.4].

Schmidt, A., V. Schunemann & R. Anton (1990) *Phys. Rev. B***41** 11875–11880 [5.3.2].

Schneider, H.G. & V. Ruth (Eds.) (1971) *Advances in Epitaxy and Endotaxy* (VEB, Leipzig) [5.1.1].

Schneider, T.N., S. Katsimichas, C.R.E. de Oliveira & A.J.H. Goddard (1998) *J. Vac. Sci. Tech.* A**16** 175–180 [2.3.2].

Schöbinger, M. & F.F. Abraham (1985) *Phys. Rev.* B**31** 4590–4596 [4.4.2].

Schlösser, D.C., L.K. Verheij, G. Rosenfeld & G. Comsa (1999) *Phys. Rev. Lett.* **82** 3843–3846 [6.1.4].

Schottky, W. (1938) *Naturwissenschaften* **26** 843 (in German) [8.2.1].

Schroder, D.K. (1998) *Semiconductor Material and Device Characterization* (John Wiley) [8.1].

Schubert, E.F. (1994) in *Epitaxial Microstructures* (Ed. A.C. Gossard) *Semiconductors and Semimetals* **40** 2–152 [8.1.3, 8.2.4].

Schwarzchild, B. (1998) *Physics Today*, December issue pp. 17–19 [8.3.2].

Schwoebel, R.L. & E.J. Shipsey (1966) *J. Appl. Phys.* **37** 3682–3686 [5.5.1].

Seah, M.P. (1996) *Phil. Trans. R. Soc. Lond.* A**354** 2765–2780 [3.4.2].

Seah, M.P. & W.A. Dench (1979) *Surf. Interface Anal.* **1** 2–11 [3.3.4].

Seah, M.P. & I.S. Gilmore (1996) *J. Vac. Sci. Tech.* A**18** 1401–1407 [3.3.3, 3.4.1–2].

Shah, S.I. (1995) in *Handbook of Thin Film Process Technology* (Eds. D.A. Glocker and S.I. Shah, Institute of Physics) section A3 [2.5.4].

Shaw, C.G., S.C. Fain, Jr & M.D. Chinn (1978) *Phys. Rev. Lett.* **41** 955–957 [4.4.2].

Shchukin, V.A., N.N. Ledentsov, P.S. Kop'ev & D. Bimberg (1995) *Phys. Rev. Lett.* **75** 2968–2971 [7.3.3].

Shiba, H. (1979) *J. Phys. Soc. Japan* **46** 1852–1860; (1980) *J. Phys. Soc. Japan* **48** 211–218 [both 4.4.2].

Shih, W.C. & W.M. Stobbs (1991) *Ultramicroscopy* **35** 197–215 [8.1.3].

Shimizu, R. & Z.-J. Ding (1992) *Rep. Prog. Phys.* **55** 487–531 [3.5.1].

Shinjo, T. & T. Ono (1996) *J. Mag. Mag. Mat.* **156** 11–14 [8.3.3].

Shirley, D.A. (1973) *Phys. Rev.* A**7** 1520–1528 [3.3.3].

Shitara, T., D.D. Vvedensky, M.R. Wilby, J. Zhang, J.H. Neave & B.A. Joyce (1992) *Phys. Rev.* B**46** 6815–6833 [7.3.4].

Shkrebtii, A.I., R. Di Felice, C.M. Bertoni & R. Del Sole (1995) *Phys. Rev.* B**51** 11201–11204 [7.2.4, project 7.4].

Shrimpton, N. & B. Joós (1989) *Phys. Rev.* B**40** 10564–10576 [4.4.2].

Siegbahn, K., C. Nordling, A. Fahlman, R. Nordberg, K. Hamrin, J. Hedman, G. Johansson, T. Bergmark, S-E Karlsson, I. Lindgren & B. Lindberg (1967) *ESCA: Atomic, Molecular and Solid State Structure Studied by Means of Electron Spectroscopy* (Uppsala, Almqvist and Wiksells) [3.3.1–2].

Siegert, M. & M. Plischke (1996) *Phys. Rev.* B**53** 307–318 [7.3.4].

Skriver, H.L. & N.M. Rosengaard (1992) *Phys. Rev.* B**46** 7157–7168 [6.1.3–4, 6.3.4].

Slusarczuk, M.M.G. (1997) *Mater. Res. Soc. Symp.* **424** 363–369 [6.2.2, 8.4.3].

Smilauer, P. & D.D. Vvedensky (1993) *Phys. Rev.* B**48** 17603–17606 [7.3.4].

Smith, D.J. (1997) *Rep. Prog. Phys.* **60** 1513–1580 [3.1.3].

Smith, D.L. (1995) *Thin-Film Deposition: Principles and Practice* (McGraw-Hill) [2.5].

Smith, G.C. (1994) *Surface Analysis by Electron Spectroscopy* (Plenum) [3.3–5].

Soffer, S.B. (1967) *J. Appl. Phys.* **38** 1710–1715 [8.3.2].

Sondheimer, E.H. (1952) *Adv. Phys.* **1** 1–42 [8.3.2].

Spence, D.J. & S.P. Tear (1998) *Surface Sci.* **398** 91–104 [3.4.2].

Spence, J.C.H. (1999) *Materials Sci. & Eng.* **R26** 1–49 [3.1.3].

Spiller, G.D.T. (1982) *Phil. Mag.* **A46** 535–549 [4.5.4].

Spiller, G.D.T., P. Akhter & J.A. Venables (1983) *Surface Sci.* **131** 517–533 [5.4.1].

Springholz, G., Z. Shi & H. Zogg (1999) in *Thin Films: Heteroepitaxial Systems* (Eds. W.K. Liu & M.B. Santos, World Scientific) chapter 13 pp. 612–688 [7.3.4].

Srivastava, G.P. (1997) *Rep. Prog. Phys.* **60** 561–613 [7.2].

Stanley, H.E. (1971) *Introduction to Phase Transitions and Critical Phenomena* (Oxford University Press) [4.5, 6.3].

Stich, I., R. Car & M. Parinello (1991) *Phys. Rev.* **B44** 4262–4274 [7.1.1].

Stich, I., M.C. Payne, R.D. King-Smith, J.-S. Lin & L.J. Clarke (1992) *Phys. Rev. Lett.* **68** 1351–1354; see also *Phys. Rev. Lett.* **71** (1993) 3612–3613 for ensuing correspondence [7.2.3].

Stiles, M.D. (1993) *Phys. Rev.* **B48** 7238–7258; (1996) *J. Appl. Phys.* **79** 5805–5810 [both 6.3.4].

Stoltze, P. (1994) *J. Phys. Condens. Matter* **6** 9495–9517 [6.1.2]; (1997) *Simulations in Atomic Scale Materials Physics* (Polyteknisk Verlag, Copenhagen) [5.2.1, 6.1.2].

Stoneham, A.M. & S.C. Jain (1995) *GeSi Strained Layers and their Applications* (Institute of Physics Publishing) [7.3.3, 8.2.4].

Stowell, M.J. (1972) *Phil. Mag.* **26** 361–374 [5.3.1]; (1974) *Thin Solid Films* **21** 91–105 [5.3.1–2].

Stoyanov, S. & D. Kaschiev (1981) in *Current Topics in Materials Sci.* (Ed. E. Kaldis, North-Holland) **7** 69–141 [5.2.2].

Strandburg, K.J. (Ed.) (1992) *Bond-orientational Order in Condensed Matter Systems* (Springer) [4.4.3].

Stroscio, J. & E. Kaiser (Eds.) (1993) *Scanning Tunneling Microscopy* (Methods of Experimental Physics, Academic) vol. 27 [3.1].

Stroscio, J. & D.T. Pierce (1994) *Phys. Rev.* **B49** 8522–8525 [5.4.3].

Stroscio, J., D.T. Pierce & R.A. Dragoset (1993) *Phys. Rev. Lett.* **70** 3615–3618 [5.4.3].

Stroscio, J., D.T. Pierce, R.A. Dragoset & P.N. First (1992) *J. Vac. Sci. Tech.* **A10** 1981 [5.5.3].

Stryer, L. (1995) *Biochemistry* (4th Edn, Freeman) chapter 7 pp. 147–180 [4.5.4].

Stuckless, J.T., N.A. Frei & C.T. Campbell (1998) *Rev. Sci. Inst.* **69** 2427–2438 [4.4.4].

Sudijono, J., M.D. Johnson, M.B. Elowitz, C.W. Snyder & B.G. Orr (1993) *Surface Sci.* **280** 247–257 [7.3.4].

Sugawara, A. & M.R. Scheinfein (1997) *Phys. Rev.* **B56** R8499–R8502 [8.3.3].

Sugawara, A., S.T. Coyle, G.G. Hembree & M.R. Scheinfein (1997) *Appl. Phys. Lett.* **70** 1043–1045; (1997) *J. Appl. Phys.* **82** 5662–5669 [both 8.3.3].

Sundquist, B.E. (1964) *Acta Metall.* **12** 67–86; see *Acta Metall.* **12** 585–592, on some effects of impurities [1.2.3].

Sutton, A.P. (1994) *Electronic Structure of Materials* (Oxford University Press) [6.1, 6.3, 7.1].

Sutton, A.P. & R.W. Balluffi (1995) *Interfaces in Crystalline Materials* (Oxford University Press) [1.2, 6.1, 7.1, 8.2].

Suzanne, J., J.P. Coulomb and M. Bienfait (1973) *Surface Sci.* **40** 414–417; (1974) *Surface Sci.* **44** 141–156; (1975) *Surface Sci.* **47** 204–205 [all 4.4.1].

Suzanne, J. & J.M. Gay (1996) in *Handbook of Surface Science* (Ed. W.N. Unertl, Elsevier) **1** pp. 503–575 [4.3.1, 4.4.4].

Suzuki, T., J.J. Métois & K. Yagi (1995) *Surface Sci.* **339** 105–113 [7.3.1].

Swan, A.K., Z-P Shi, J.F. Wendelken & Z. Zhang (1997) *Surface Sci.* **391** L1205–L1211 [5.4.3].

Swanson, L.W. (1984) in *Electron Optical Systems for Microscopy, Microanalysis and Microlithography* (SEM Pfefferkorn Conference Series) **3** 137–147 [6.2.3].

Swanson, L.W., M.A. Gesley & P.R. Davis (1981) *Surface Sci.* **107** 263–289 [6.2.1].

Swanson, L.W. & P.R. Davis (1985) in *Solid State Physics: Surfaces* (Eds. R.L. Park and M.G. Lagally, Methods of Experimental Physics, Academic) **22** 1–22 [6.1.3].

Swartzentruber, B.S. (1996) *Phys. Rev. Lett.* **76** 459–462; (1997) *Phys. Rev. B***55** 1322–1325; (1997) *Surface Sci.* **374** 277–282 [all 7.3.2].

Swartzentruber, B.S., Y-W. Mo, M.B. Webb & M.G. Lagally (1989) *J. Vac. Sci. Tech.* *A***7** 2901–2905 [7.3.1].

Swartzentruber, B.S., Y-W. Mo, R. Kariotis, M.G. Lagally & M.B. Webb (1990) *Phys. Rev. Lett.* **65** 1913–1916 [7.3.2].

Swartzentruber, B.S., A.P. Smith & H. Jónsson (1996) *Phys. Rev. Lett.* **77** 2518–2521 [7.3.2].

Sze, S.M. (1981) *Physics of Semiconductor Devices* (2nd Edn, John Wiley) [8.1].

Tagaki, T. (1986) *Vacuum* **36** 27–31 [2.5.4].

Tagaki, T. & I. Yamada (1989) in *Ion Beam Assisted Film Growth* (Ed. T. Itoh, Elsevier) pp. 253–288 [2.5.4].

Takashita, H., Y. Suzuki, H. Akinaga, W. Mitzutani, K. Tanaka, T. Katayama & A. Itoh (1996) *Appl. Phys. Lett.* **68** 3040–3042 [8.3.3].

Takayanagi, K., Y. Tanashiro, M. Takahashi & S. Takahashi (1985) *J. Vac. Sci. Tech.* *A***3** 1502–1506 [1.4.5, 7.2.3]; (1985) *Surface Sci.* **164** 367–392 [7.2.3].

Takayanagi, K. & Y. Tanashiro (1986) *Phys. Rev. B***34** 1034–1040 [7.2.3].

Tanaka, S., N.C. Bartelt, C.C. Umbach, R.M. Tromp & J.M. Blakely (1997) *Phys. Rev. Lett.* **78** 3342–3345 [7.3.2].

Tang, S & A.J. Freeman (1994) *Phys. Rev. B***50** 10941–10946 [7.2.4].

Tans, S.J., A.R.M Verschueren & C. Dekker (1998a) *Nature* **393** 49–51 [8.4.2].

Tans, S.J., M.H. Devoret, R.J.A. Groeneveld & C. Dekker (1998b) *Nature* **394** 761–764 [8.4.2].

Tanuma, S., C.J. Powell & D.R. Penn (1991) *Surf. Interface Anal.* **17** 911–939; *Surf. Interface Anal.* (1993) **21** 165–176 [all 3.3.4].

Temperly, H.N.V. (1972) in *Phase Transitions and Critical Phenomena* (Ed. C. Domb, Academic) **1** 227–267 [4.5.1, 6.3.1].

Terrones, M., W.K. Hsu, A. Schilder, H. Terrones, N. Grober, J.P. Hare, Y.Q. Zhu, M. Schwoerer, K. Prassides, H.W. Kroto & D.R.M. Walton (1998) *Appl. Phys. A***66** 307–317 [8.4.3].

Terrones, M., W.K. Hsu, H.W. Kroto & D.R.M. Walton (1999) *Topics in Current Chemistry* **199** 189–234 [8.4.3].

Tersoff, J. (1984) *Phys. Rev. Lett.* **52** 465–468; *Phys. Rev. B***30** 4874–4877; (1985) *Phys. Rev. B***32** 6968–6971; (1986) *Surface Sci.* **168** 275–283 [all 8.2.1]; (1987) in *Heterostructure Band Discontinuities: Physics and Device Applications* (Eds. F. Capasso and G. Margaritondo, North Holland), chapter 1, pp. 3–57 [8.2.1–3].

Tersoff, J. & F.K. LeGoues (1994) *Phys. Rev. Lett.* **72** 3570–3573 [7.3.3].

Tersoff, J., M.D. Johnson & B.G. Orr (1997) *Phys. Rev. Lett.* **78** 282–285 [7.3.4].

Tersoff, J., C. Tiechert & M.G. Lagally (1996) *Phys. Rev. Lett.* **76** 1675–1678 [7.3.3].

Theis, W. & R.M. Tromp (1996) *Phys. Rev. Lett.* **76** 2770–2773 [7.3.2].

Thomy, A. & X. Duval (1994) *Surface Sci.* **299/300** 415–425 [4.3.1].

Thomy, A., X. Duval & J. Regnier (1981) *Surface Sci. Rep.* **1** 1–38 [4.3.1].

Tinkham, M. (1996) *Introduction to Superconductivity* (2nd Edn, McGraw-Hill) [8.3].

Titmus, S., A. Wander & D.A. King (1996) *Chemical Reviews* **96** 1291–1305 [6.1.2].

Todd, C.J. & T.N. Rhodin (1974) *Surface Sci.* **42** 109–138 [6.2.3].

Tölkes, C., P. Zeppenfeld, M.A. Kryzowski, R. David & G. Comsa (1997) *Surface Sci.* **394** 170–184; (1997) *Phys. Rev. B***55** 13932–13937 [both 5.5.3].

Tong, S.Y., H. Huang, C.M. Wei, W.E. Packard, F.K. Men, G. Glander & M.B. Webb (1988) *J. Vac. Sci. Tech. A***6** 615–624 [7.2.3].

Tong, X. & P.A. Bennett (1991) *Phys. Rev. Lett.* **67** 101–104 [7.3.2].

Tringides, M.C. (Ed.) (1997) Surface Diffusion: Atomistic and Collective Processes (Plenum NATO ASI) **B360** [5.1].

Tringides, M. & R. Gomer (1985) *Surface Sci.* **155** 254–278 [6.2.3].

Tromp, R.M & M.C. Reuter (1992) *Phys. Rev. Lett.* **68** 820–822; (1993) *Phys. Rev. B***47** 7598–7601 [both 7.3.1].

Tromp, R.M. & M. Mankos (1998) *Phys. Rev. Lett.* **81** 1050–1053 [7.3.2].

Tsao, J.Y. (1993) *Materials Fundamentals of Molecular Beam Epitaxy* (Academic) [2.5, 7.3].

Tsao, J.Y., E. Chason, U. Koehler & R. Hamers (1989) *Phys. Rev. B***40** 11951–11954 [7.3.2].

Tsong, T.T. & C. Chen (1997) in *The Chemical Physics of Solid Surfaces and Heterogeneous Catalysis* (Eds. D.A. King and D.P. Woodruff, Elsevier) **8** chapter 4 pp. 102–148 [3.1.3, 5.4.2].

Tuck, R.A. (1983) *Vacuum* **33** 715–721 [6.2.1].

Turchi, P.E.A., A. Gonis & L. Colombo (Eds.) (1998) *Mater. Res. Soc. Symp.* **491** 1–542 [7.1.3, Appendix K].

Tyson, W.R. & W.A. Miller (1977) *Surface Sci.* **62** 267–276 [6.1.4].

Unguris, J., R.J. Celotta & D.T. Pierce (1991) *Phys. Rev. Lett.* **67** 140–143; (1994) *J. Appl. Phys.* **75** 6437–6439 [both 6.3.4].

Usher, B.F. & J.L. Robins (1987) *Thin Solid Films* **149** 351–383; *Thin Solid Films* **155** 267–283 [both 5.3.2–3].

Van de Walle, C.G. (1989) *Phys. Rev.* **B39** 1871–1883 [8.2.2].
Van de Walle, C.G. & R.M. Martin (1986) *J. Vac. Sci. Tech.* **B4** 1055–1059 [8.2.3]; (1987) *Phys. Rev.* **B35** 8154–8165 [8.2.2–3].
Vanderbilt, D. (1987) *Phys. Rev.* **B36** 6209–6212 [7.2.3].
Van Hove, M.A. & G.A. Somorjai (1994) *Surface Sci.* **299/300** 487–501 [1.4.1].
Van Hove, M.A., R.J. Koestner, P.C. Stair, J.P Biberian, L.L. Kesmodel, I. Bartos & G.A. Somorjai (1981) *Surface Sci.* **103** 189–238 [6.1.2].
Velfe, H.D., H. Stenzel & M. Krohn (1982) *Thin Solid Films* **98** 115–138 [5.3.2].
Venables, J.A. (1973) *Phil. Mag.* **27** 697–738 [5.2.1–2, 5.3.1–2]; (1987) *Phys. Rev.* **B36** 4153–4162 [5.1.4, 5.2.1–5, 5.4.3]; (1994) *Surface Sci.* **299/300** 798–817 [5.1.4, 5.3.2]; (1997) *Physica A***239** 35–46; (1997) *Mater. Res. Soc. Symp. Proc.* **440** 129–140; (1997) in *The Chemical Physics of Solid Surfaces and Heterogeneous Catalysis* (Eds. D.A. King and D.P. Woodruff, Elsevier) **8** chapter 1, pp. 1–45 [all 5.3.2–3]; (1998) *Phys. Ed.* **33** 157–163; *J. Mater. Ed.* **20** 57–66 [both Appendix D]; (1999) in *Thin Films:Heteroepitaxial Systems* (Eds. W.K. Liu and M.B. Santos, World Scientific) chapter 1, p. 1–63 [5.3.2–3].
Venables, J.A. and J.H. Harding (2000) *J. Cryst. Growth* **211** 27–33 [5.3.3.].
Venables, J.A. & R. Persaud (1997) *J. Phys. D: Appl. Phys.* **30** 3163–3165 [3.4.2, 5.5.3].
Venables, J.A. & G.L. Price (1975) in *Epitaxial Growth* (Ed. J.W. Matthews, Academic) part B, chapter 4, pp. 381–436 [5.2.2, 5.3.1].
Venables, J.A. & P.S. Schabes-Retchkiman (1978) *Surface Sci.* **71** 27–41 [1.4.2, 4.4.2]; (1978) *J. Phys* C**11** L913–L918 [4.4.2, 4.4.3].
Venables, J.A., A.P. Janssen, P. Akhter, J. Derrien & C.J. Harland (1980) *J. Microscopy* **118** 351–365 [3.3.1].
Venables, J.A., G.D.T. Spiller & M. Hanbücken (1984) *Rep. Prog. Phys.* **47** 399–459 [5.1.3–4, 5.2.2, 5.2.4, 5.3.1, 5.4.1, 5.5.1].
Venables, J.A., D.R. Batchelor, M. Hanbücken, C.J. Harland & G.W. Jones (1986) *Phil. Trans. R. Soc. Lond.* **A318** 243–257 [3.5.1].
Venables, J.A., D.J. Smith & J.M. Cowley (1987) *Surface Sci.* **181** 235–249 [3.1.3].
Venables, J.A., Y. Li, G.G. Hembree, H. Noro & R. Persaud (1996) *J. Phys. D: Appl. Phys.* **29** 240–245 [3.4.2, 5.5.3].
Venkatesan, T., X.D. Wu, A. Iman & J.B. Wachtman (1988) *Appl. Phys. Lett.* **52** 1193–1195 [2.5.2].
Ventrice, C.A. Jr., V.P. LaBella, G. Ramaswamy, H.P. Yu & L.J. Schowalter (1996) *Phys. Rev.* **B53** 3952–3959 [8.1.3].
Verdozzi, C. & M. Cini (1995) *Phys. Rev.* **B51** 7412–7420 [3.3.4].
Verdozzi, C., D.R. Jennison, P.A. Schultz & M.P. Sears (1999) *Phys. Rev. Lett.* **82** 799–802 [1.4.8].
Vescan, L. (1995) in *Handbook of Thin Film Process Technology* (Eds. D.A. Glocker and S.I. Shah, Institute of Physics) section B1 [2.5.4].
Vincent, R.A. (1971) *Proc. R. Soc. Lond.* **A321** 53–68 [5.2.4].

Vitos, L., A.V. Ruban, H.L. Skriver & J. Kollár (1998) *Surface Sci.* **411** 186–202 [6.1.4].

Vitos, L., H.L. Skriver & J. Kollár (1999) *Surface Sci.* **425** 212–223 [6.1.4].

Voigtländer, B. (1999) *Micron* **30** 33–39 [3.1.3].

Voigtländer, B. & Zinner, A. (1993) *Appl. Phys. Lett.* **63** 3055–3057 [3.1.3].

Voigtländer, B., G. Meyer & N.M. Amer (1991) *Phys. Rev.* **B44** 10354–10357 [5.5.3, 8.3.3].

Vossen, J.L. & W. Kern (Eds.) (1991) *Thin Film Processes II* (Academic) [2.5.4].

von Känel, H., E.Y. Lee, H. Sirringhaus & U. Kafader (1995) *Thin Solid Films* **267** 89–94 [8.1.3].

von Känel, H., T. Meyer, H. Sirringhaus & E.Y. Lee (1997) *Surface Rev. Lett.* **4** 307–318 [8.1.3].

Voorhees, P.W. (1985) *J. Stat. Phys.* **38** 231–252 [5.5.2].

Vossmeyer, T., S. Jia, E. DeIonno, M.R. Diehl, S.H. Kim, X. Peng, A.P. Alivisatos & J.R. Heath (1998) *J. Appl. Phys.* **84** 3664–3670 [8.4.4].

Wagner, H. (1979) in *Solid Surface Physics* (G. Höhler, Springer Tracts in Modern Physics **85**) 151–221 [6.1.3].

Wahnström, G., A.B. Lee & J Strömqvist (1996) *J. Chem. Phys.* **105** 326–336 [4.5.3].

Walker, C.G.H., D.C. Peacock, M. Prutton & M.M. El-Gomati (1988) *Surf. Interface Anal.* **11** 266–278 [3.5.2].

Walls, J.M. (Ed.) (1990) *Methods of Surface Analysis* (Cambridge University Press) [3.1].

Walton, D. (1962) *J. Chem. Phys.* **37** 2182–2188 [5.2.2].

Wang, C.Z. & K.M. Ho (1996) *Adv. Chem. Phys.* **93** 651–702 [7.1.3].

Wang, S.C. & G. Ehrlich (1988) *Surface Sci.* **206** 451–474 [5.4.2].

Watanabe, F. & G. Ehrlich (1992) *J. Chem. Phys.* **96** 3191–3199 [5.4.2].

Watson, J. & K. Ihokura (Eds.) (1999) *Materials Research Bulletin*, June issue pp. 14–59 [9.3].

Watson, P.R., M.A. Van Hove & K. Hermann (1996) *Atlas of Surface Structures, vols. 1A and 1B*, (*J. Phys. Chem. Ref. Data*, Monograph 5, ACS publications) [1.4.1–2, 1.4.4, 1.4.8, 3.2.1].

Webb, M.B. (1994) *Surface Sci.* **299/300** 454–468 [7.3.1–2].

Webb, M.B. & M.G. Lagally (1973) *Solid State Phys.* **28** 301–405 [3.2.1].

Webb, M.B., F.K. Men, B.S. Swartzentruber, R. Kariotis & M.G. Lagally (1991) *Surface Sci.* **242** 23–31 [7.3.1].

Weeks, J.D. & G.H. Gilmer (1979) *Adv. Chem. Phys.* **40** 157–227 [1.3.1–2, 5.1.4].

Weierstall, U., J.M. Zuo, T. Kjorsvik & J.C.H. Spence (1999) *Surface Sci.* **442** 239–250 [3.2.2].

Weightman, P. (1982) *Rep. Prog. Phys.* **45** 753–814 [3.3.3–4]; (1995) *Microsc. Microanal. Microstruct.* **6** 263–288 [3.3.4].

Weinert, M., E. Wimmer & A.J. Freeman (1982) *Phys. Rev.* **B26** 4571–4578 [6.3.4].

Welch, K.M. (1994) *J. Vac. Sci. Tech.* **A12** 915–920 [2.3.2].

Werner, H.W. & R.P.H. Garten (1984) *Rep. Prog. Phys.* **47** 221–344 [3].

Whall, T.E. & E.H.C. Parker (1998) *J. Phys. D: Appl. Phys.* **31** 1397–1416 [7.3.3].

Wiesendanger, R. (1994) *Scanning Probe Microscopy and Spectroscopy* (Cambridge University Press) [3.1, 7.2].

Williams, A.A., J.M.C. Thornton, J.E. McDonald, R.G. van Silfhout, J.F. van der Veen, M.S. Finney, A.D. Johnson & C. Norris (1991) *Phys. Rev.* B43 5001–5011 [7.3.3].

Williams, E.D. (1994) *Surface Sci.* 299/300 502–524 [7.3.1].

Williams, E.D., R.J. Phaneuf, J. Wei, N.C. Bartelt & T.L. Einstein (1993) *Surface Sci.* 294 219–242 [7.3.1–2].

Williams, R.S. (1999) *Physics World* December issue 49–51 [9.1].

Wintterlin, J., R. Schuster & G. Ertl (1996) *Phys. Rev. Lett.* 77 123–126 [4.5.4].

Wollschläger, J. (1995) *Surface Sci.* 328 325–336 [3.2.1, 3.2.3].

Wood, E.A. (1964) *J. Appl. Phys.* 35 1306–1311 [1.4.2].

Woodruff, D.P. (1973) *The Solid–Liquid Interface* (Cambridge University Press) chapters 1–3 and 8 [1.3.2].

Woodruff, D.P. & T.A. Delchar (1986, 1994) *Modern Techniques of Surface Science* (Cambridge University Press) [3.2, 6.1].

Wrigley, J.D. & G. Ehrlich (1980) *Phys. Rev. Lett.* 44 661–663 [5.4.2].

Wu, F. & M.G. Lagally (1995) *Phys. Rev. Lett.* 75 2534–2537 [7.3.1].

Wu, F.Y. (1982) *Rev. Mod. Phys.* 54 235–268 [4.5.1].

Wu, R., D.S. Wang & A.J. Freeman (1995) in *Handbook of Surface Imaging* (Ed. A.T. Hubbard, CRC Press) chapter 27, pp. 385–400 [6.3.4].

Wulfhekel, W., N.N. Lipkin, J. Kliewer, G. Rosenfeld, L. C. Jorritsma, B. Poelsema & G. Comsa (1996) *Surface Sci.* 348 227–242 [5.4.3].

Wulfhekel, W., I. Beckmann, G. Rosenfeld, B. Poelsema & G. Comsa (1998) *Surface Sci.* 395 161–181 [5.4.3].

Xia, Y. & G.M. Whitesides (1998) *Ann. Rev. Mater. Sci.* 28 153–184 [8.4.4].

Yagi, K. (1988) in *High Resolution Transmission Electron Microscopy and Associated Techniques* (Eds. P. Buseck, J.M. Cowley and L. Eyring, Oxford University Press) chapter 13 pp. 568–606; (1989) *Adv. Optical & Electron Microscopy* 11 57–100; (1993) *Surface Sci. Rep.* 17 305–362 [all 3.1.3, 3.2.3, 5.4.3, 8.1.3].

Yang, Y.N. & E.D. Williams (1994) *Phys. Rev. Lett.* 72 1862–1865; (1994) *Scanning Microscopy* 8 781–794 [7.2.3].

Yang, Y.N., E.D. Williams, R.L. Park, N.C. Bartelt & T.L. Einstein (1990) *Phys. Rev. Lett.* 64 2410–2413 [4.5.1].

Yasunaga, H. & A. Natori (1992) *Surface Sci. Rep.* 15 205–280 [9.1].

Yates, J.T. Jr. (1997) *Experimental Innovations in Surface Science* (Springer-AIP) [2.1.3, 2.3.1, Appendix G].

Yin, M.T. & M.L. Cohen (1982) *Phys. Rev.* B26 5668–5687 [7.1.1, 7.1.3].

Young, A.P. (1979) *Phys. Rev.* B19 1855–1866 [4.4.3].

Yu, E.T., J.O. McCaldin & T.C. McGill (1992) *Solid State Physics* (Eds. H. Ehrenreich and D. Turnbull, Academic) 46 1–146 [8.2.3].

Yu, P.Y. & M. Cardona (1996) *Fundamentals of Semiconductors: Physics and Materials Properties* (Springer) [7.1].

Zambelli, T., J.V. Barth, J. Wintterlin & G. Ertl (1997) *Nature* **390** 495–497 [4.5.4].

Zandvliet, H.J.W., B. Poelsma & H.B. Elswijk (1995) *Phys. Rev.* **B51** 5465–5468 [7.3.2].

Zangwill, A. (1988) *Physics at Surfaces* (Cambridge University Press) [preface, 4.4–5, 6.1, 7.1–2].

Zangwill, A. & E. Kaxiras (1995) *Surface Sci.* **326** L483–L488 [5.2.5, 5.4.3, 5.5.3].

Zerrouk, T.E.A., M. Hamichi, J.D.H. Pilkington & J.A. Venables (1994) *Phys. Rev.* **B50** 8946–8949 [4.4.3].

Zeysing, J.H. & R.L. Johnson (1999) unpublished mass spectrum from laboratory notebook [2.3.5].

Zhang, Q.M., C. Roland, P. Bogulawski & J. Bernholc (1995) *Phys. Rev. Lett.* **75** 101–104 [7.3.2].

Zhang, S.B. & A. Zunger (1996) *Phys. Rev.* **B53** 1343–1356 [7.1.3].

Zimmermann, C.G., M. Yeadon, K. Nordlund, J.M. Gibson, R.S. Averback, U. Herr & K. Samwer (1999) *Phys. Rev. Lett.* **83** 1163–1166 [5.5.3].

Zinke-Allmang, M. (1999) *Thin Solid Films* **346** 1–68 [5.5.2].

Zinke-Allmang, M., L.C. Feldman & M.H. Grabow (1992) *Surface Sci. Rep.* **16** 377–463 [5.5.2].

Index

Note: Page numbers for text references are in upright type, whereas page numbers for figures are in sloping type. Problems and projects are indicated (p), appendices are indicated (a), with Appendix D indicated (w) for web. Not all techniques and acronyms are indexed; see Appendix B for a guide.

Printed in the United States
By Bookmasters